# DISRUPTIVE TECHNOLOGIES FOR SOCIETY 5.0

# DISRUPTIVE TECHNOLOGIES FOR SOCIETY 5.0

Exploration of New Ideas, Techniques, and Tools

Edited by
Vikram Bali, Vishal Bhatnagar, Sapna Sinha, and Prashant Johri

CRC Press is an imprint of the
Taylor & Francis Group, an **informa** business

First edition published 2022
by CRC Press
6000 Broken Sound Parkway NW, Suite 300, Boca Raton, FL 33487-2742

and by CRC Press
2 Park Square, Milton Park, Abingdon, Oxon, OX14 4RN

© 2022 selection and editorial matter, Vikram Bali, Vishal Bhatnagar, Sapna Sinha, and Prashant Johri; individual chapters, the contributors

CRC Press is an imprint of Taylor & Francis Group, LLC

Reasonable efforts have been made to publish reliable data and information, but the author and publisher cannot assume responsibility for the validity of all materials or the consequences of their use. The authors and publishers have attempted to trace the copyright holders of all material reproduced in this publication and apologize to copyright holders if their permission to publish in this form has not been obtained. If any copyright material has not been acknowledged, please write and let us know so we may rectify in any future reprint.

Except as permitted under U.S. Copyright Law, no part of this book may be reprinted, reproduced, transmitted, or utilized in any form by any electronic, mechanical, or other means, now known or hereafter invented, including photocopying, microfilming, and recording, or in any information storage or retrieval system, without written permission from the publishers.

For permission to photocopy or use material electronically from this work, access www.copyright.com or contact the Copyright Clearance Center, Inc. (CCC), 222 Rosewood Drive, Danvers, MA 01923, 978-750-8400. For works that are not available on CCC, please contact mpkbookspermissions@tandf.co.uk

*Trademark notice*: Product or corporate names may be trademarks or registered trademarks and are used only for identification and explanation without intent to infringe.

*Library of Congress Cataloging-in-Publication Data*
Names: Bali, Vikram, editor. | Bhatnagar, Vishal, 1977- editor. |
Sinha, Sapna, editor. | Johri, Prashant, editor.
Title: Disruptive technologies for society 5.0 : exploration of new ideas, techniques, and tools / edited by Vikram Bali, Vishal Bhatnagar, Sapna Sinha, and Prashant Johri.
Description: Boca Raton, FL : CRC Press, 2022. | Includes bibliographical references and index.
Identifiers: LCCN 2021024821 (print) | LCCN 2021024822 (ebook) |
ISBN 9780367724078 (hbk) | ISBN 9780367724092 (pbk) | ISBN 9781003154686 (ebk)
Subjects: LCSH: Disruptive technologies. | Technology–Social aspects. | Industry 4.0.
Classification: LCC T14.5 .D59 2022 (print) | LCC T14.5 (ebook) | DDC 303.48/3–dc23
LC record available at https://lccn.loc.gov/2021024821
LC ebook record available at https://lccn.loc.gov/2021024822

ISBN: 978-0-367-72407-8 (hbk)
ISBN: 978-0-367-72409-2 (pbk)
ISBN: 978-1-003-15468-6 (ebk)

DOI: 10.1201/9781003154686

Typeset in Times
by MPS Limited, Dehradun

# Contents

Preface .................................................................................................................... ix
Editor Biographies ............................................................................................... xv

## Section A  Disruptive Technologies: Introduction and Innovation ........................................... 1

**Chapter 1**  Blockchain and Internet of Things: An Amalgamation of Trending Techniques ........................................................................ 3

*Shilpi Bisht, Neeraj Bisht, Pankaj Singh, and Shray Dasila*

**Chapter 2**  Software-Defined Networking: Evolution, Open Issues, and Challenges ..................................................................................... 31

*Sukhvinder Singh and Preeti Gupta*

**Chapter 3**  Performance Enhancement in Cloud Computing Using Efficient Resource Scheduling ......................................................... 49

*Prashant Lakkadwala and Priyesh Kanungo*

**Chapter 4**  Hyper-Personalized Recommendation Systems: A Systematic Literature Mapping ................................................................... 69

*Bijendra Tyagi and Dr. Vishal Bhatnagar*

**Chapter 5**  Evolutionary Computational Technique for Segmentation of Bilingual Roman & Gurmukhi Handwritten Script ........................ 87

*Gurpreet Singh and Manoj Sachan*

**Chapter 6**  A Metric to Determine the Change Proneness of Software Classes Using GMDH Networks ............................................... 109

*Ashu Jain, Dhyanendra Jain, and Dr. Prashant Singh*

v

## Section B  Application of Disruptive Technology in Various Sectors ................................. 121

**Chapter 7** Application of IoT Technology in the Design and Construction of an Android Based Smart Home System ................................... 123

Adeyemi Abel Ajibesin and Ahmed Tijjani Ishaq

**Chapter 8** Privacy-Preserved Access Control in E-Health Cloud-Based System .................................................................................................. 145

Suman Madan

**Chapter 9** Eye Gaze Mouse Empowers People with Disabilities ................... 163

Sonia Rathee, Amita Yadav, Harvinder Rathee, and Navdeep Bohra

**Chapter 10** Strategies for Resource Allocation in Cloud Computing Environment ............................................................................. 179

Nikky Ahuja, Priyesh Kanungo, and Sumant Katiyal

**Chapter 11** Optimization Mechanism for Energy Management in Wireless Sensor Networks (WSN) Assisted IoT ........................................... 201

Urmila Shrawankar and Kapil Hande

**Chapter 12** Intelligent Systems for IoT and Services Computing ................... 213

Dr. Preeti Arora, Dr. Laxman Singh, Saksham Gera, and Dr. Vinod M Kapse

**Chapter 13** Framework for the Adoption of Healthcare 4.0 – An ISM Approach ........................................................................................ 235

Vinaytosh Mishra and Sheikh Mohammed Shariful Islam

## Section C  Impact of Disruptive Technologies in Society 5.0 ............................................. 249

**Chapter 14** E-Commerce Security for Preventing E-Transaction Frauds ......... 251

Reshu Agarwal, Manisha Pant, and Shylaja Vinaykumar Karatangi

Contents

**Chapter 15** Botnet Forensic Analytics for Investigation of Disruptive Botherders .................................................................................. 265

*Kapil Kumar, Shyla, and Vishal Bhatnagar*

**Chapter 16** Design of a Toolbox to Give One Stop Solution for Multidimensional Data Analysis ...................................................... 277

*Dr. Prarthana A. Deshkar*

**Chapter 17** IoT Based Intelligent System for Home Automation ..................... 299

*Navjot Kaur Sekhon, Surinder Kaur, and Hatesh Shyan*

**Chapter 18** Digital Learning Acceptance during COVID-19: A Sustainable Development Perspective ................................................................ 317

*Praveen Srivastava, Shelly Srivastava, S.L. Gupta, and Niraj Mishra*

**Chapter 19** A Framework for Real-Time Accident Prevention using Deep Learning ......................................................................................... 327

*Pashmeen Kaur, Dr. K.C. Tripathi, and Prof. (Dr.) M.L. Sharma*

**Chapter 20** Multi-Modality Medical Image Fusion Using SWT & Speckle Noise Reduction with Bidirectional Exact Pattern Matching Algorithm ......................................................................................... 339

*Kapil Joshi, Minakshi Memoria, Laxman Singh, Parag Verma, and Archana Barthwal*

**Index** ............................................................................................................... 361

# Preface

Greetings!

The Industry 4.0 is the era of emergence of technologies changing the way existing technologies were used by the industries. The arrival of technologies has already created disruption in the sectors, some of these technologies are 3D printing, IoT, cloud, block chain, machine leaning, sensors, and artificial intelligence. From their adoption has increased the significant complexity in the non-IT sectors. Issues like adoption, implementation, security, and management have a need to be addressed. Industry 4.0 is a massive shift towards automation and digitization, utilizing the Internet of Things and cyber-physical systems, such as sensors, to aid in data-collection for manufacturing verticals. Industry 4.0 involves a hyper-connected system which advocates the smart use of robotics to effectively and efficiently move manufacturing to new heights. With the use of these technological systems, it is imperative to ensure that cybersecurity plays a role during the rise of this digital industrial revolution. Advances in Machine Learning is changing the traditional manufacturing era into smart manufacturing era of Industry 4.0.

This book addresses the topics related to IoT, Machine Learning, Sensors, Artificial Intelligence, and Cloud Computing for industry 4.0. The aim of this book is to bring together researchers, developers, practitioners, and users who are interested in these disruptive technologies to explore new ideas, techniques, and tools, and to exchange their experiences.

"Disruptive Technologies for Society 5.0: Exploration of New Ideas, Techniques, and Tools" is an important book for understanding the latest trends in the use of disruptive technologies in shaping our world. It provides a way of analyzing the historic changes that are taking place, so that we can collectively create an empowered and human-centered future for all. The reader will gain valuable insights for navigating the future using this interesting book.

In Chapter 1 the authors provide a brief literature review in this field which is the motivation to understand concepts of Blockchain and IoT. A detailed description of Blockchain Technology, its structure, and advantages and disadvantages of Blockchain technology have been provided. Internet of Things, its structure, and the integration of IoT with Blockchain are covered in detail. We have discussed the potential areas of applications of integration of IoT and Blockchain. The security and the privacy issues arising from this amalgamation of two technologies are of major concern. These aspects have been explored and discussed to resolve, which provides an open scope to the researchers to meet the challenges faced by integration of these techniques.

In Chapter 2 the authors describe the changing networking scenarios that lead to reaching the current state of SDN. The various open issues while establishing the role of SDN in areas like network security and cloud computing are also discussed. Apart from targeting SDN's role in large-scale network, the chapter also highlights its usefulness to pave a path for implementing Society 5.0. The future of SDN is driven by dynamic operational needs and innovation in software, which is discussed in the chapter while also delineating the challenges and what holds in the future for Software-Defined Networks.

In Chapter 3, The objective is to investigate and introduce an efficient resource scheduling technique that has a positive impact on computational cloud performance. The aim of the proposed scheduling algorithm is to achieve and demonstrate the two types of resource management techniques viz. priority-based algorithm and an algorithm based on dynamic workload. Among them, a priority-based resource scheduling algorithm assigns priority to each job in the process queue based on their execution time and submits these jobs to the data center for execution. The second algorithm is based on scheduling the dynamically appeared workload, which is an extension over the priority-based algorithm. Both the solutions have been implemented using the CloudSim tool with a random workload generation system. The experimental analysis shows that these approaches are useful for dealing with the cloud's increasing workloads and also demonstrates the improvement in the cloud's computational performance.

In Chapter 4 the authors, "Discuss hyper personalization used in the recommendation system which is different from the common strategy used in various recommendation systems. Traditional personalization approach uses basic customer data points, limiting itself to easy tactics like to use customer's first name in the subject line to address them. Whereas, hyper-personalization moves a step further by using real-time data points to deliver more appropriate communications/messages to consumers. It connotes having to re-evaluate interaction with customer on an individual basis, where each and every customer is considered unique and with a customized experience being designed for each one of them. The chapter discusses how hyper-personalization helps in creating a clear and simple shopping experience for the consumers by reducing their effort required to get what they are searching for.

In Chapter 5 the authors discuss how, Handwriting recognition under the bilingual script environment is considered to be one of the applications of evolutionary computation and intelligent systems. It is a complex task due to the existence of a huge number of languages for the purpose of communication all over the globe. In this work, the bilingual system for Roman and Gurmukhi script accepts online handwritten text. For the process to convert the handwritten data to its equivalent digital form, one of the important steps after pre-processing is segmentation. Different levels of segmentation have to be performed to get the digital data from handwritten samples. Script identification process directly depends on the results of segmentation phase at the level of strokes. After the identification of the script of the word under consideration, the recognition engine of the respective script helps in the extraction of refined information about the presence of different characters in the segmented word. The results obtained at different levels of segmentation are very much considerable.

In Chapter 6 the authors discuss how, Software class change proneness is the possibility that a class will undergo change in the future versions of the software. Class change proneness can be predicted by using the internal quality attributes of the software like cohesion, coupling, inheritance, and polymorphism. Authors have empirically explored the interrelation between dynamic software metrics and class change proneness. Machine Learning (ML) techniques have been used to construct the prediction models. Open source software (OSS) "SoundHelix" has been used for building the prediction model by using dynamic metrics as internal quality attributes of the software.

In Chapter 7 the authors discuss how, Most of the smart home systems were designed for entertainment and luxury, and having advanced settings that needed getting used to. Recently, it was realized that there is a need to develop a smart home that targeted special people, such as persons with disabilities (PWD). This work designs an android based smart home system that addresses the problems in conventional switching or control system. The conventional switches are not easy to operate in terms of mobility. In order to solve this problem, automated devices, such as sensors, actuators, and biomedical monitors are mounted in the houses, offices, and areas that needed to be covered for purposes of surveillance and reporting. Some devices can also be worn by an individual. As a result, a user-friendly remote controllability system that allows the end-user to control appliances such as lighting points is considered. Besides the comfort, the proposed design is cost and energy-efficient.

In Chapter 8 the author discusses how, The advancements in information and communication technology has certainly enhanced services in all segments in the world. Unambiguously, information technology has directed a very essential development in health sector called e-Health. So as to augment complete and outstanding benefits of this advancement, a cloud-based environmental implementation will be significant. Securing the e-health data is a difficult responsibility for the data owners and numerous associated organizations totally rely upon this e-health data. Before sending to the cloud, the data is normally encrypted and each time the client changes the accreditations, re-encryption will be done. This weakness is eliminated when the data owner does one degree of encryption and the cloud does the second degree of encryption accordingly decentralizing the control. In the chapter, the author has proposed a framework, so that for adept maintenance of e-health record in clinics, the group-key administration is used effectively, to deal with access control arrangements with privacy and security protection of clients preserved in cloud.

In Chapter 9 the authors outline the study and development of an Eye Mouse, which would help control the mouse cursor using Eye Gaze Tracking with the simple inbuilt webcam of a computer or laptop or by using a simple additional web camera. This will provide an easy interaction system for people with motor disabilities such as ALS, cerebral palsy, or physical disabilities to easily communicate and interact with the computer, making things easier for them. Usage of the popular webcams makes the system inexpensive. The eye gaze mouse enables these motor-impaired patients to use the mouse of a computer by means of moving their eye or head.

In Chapter 10, It is an attempt to present some resource allocation strategies for optimum utilization of cloud systems' resources. Proposed strategies with surprise element of loyalty point scheme for customers entering into multiple contracts with provider or those placing consecutive requests will help the provider earn customer loyalty. SLA is a vital element for the success of strategies that will help in reducing penalty cost for any work delayed due to violation of SLA terms. The focus is to reduce request rejection rate, minimize request response time, scalability, and flexibility of the system, and optimize cost and profit. The work hopes to provide a structure for strengthening the cloud computing model and will hopefully motivate future researchers to come up with smarter and secured optimal resource allocation algorithms and a framework to strengthen the cloud computing paradigm.

In Chapter 11 the authors demonstrated the method of combining clustering and traffic allocation protocol to achieve load balancing that increases the network lifetime. The idea is to evenly distribute the load on all discovered paths which will lead to balanced energy consumption at each node, in general. The protocol effectively smooths out the traffic evenly on all discovered routes. Traffic allocation strategy is also implemented to distribute the data evenly in all discovered paths. The performance of the energy saving method is simulated and results are encouraging. The network lifetime increases by 30% in case of traffic allocation protocol than the traditional protocols used in WSN make possible. The method also delivers the messages at double the rate than that for traditional protocols. The chapter focuses on the solutions for efficient energy management in WSN-assisted IoT.

In Chapter 12 the authors discuss about the solution for generating of a secured and automated system to create a data billing and goods management system. Shopping for Goods in different systems nowaday's billing system is currently in manual mode. Due to abundance of anomalies and drawbacks in the current billing system, there are frequent chances of frauds or data entry inaccuracies. It may lead to incorrect results to highly sensitive data. For maintaining the accuracy of data, we have to find a better way of automated billing system rather than bills generated in manual setup. It will provide a better way of analyzing results.

In Chapter 13 the authors discuss how, Healthcare is facing a challenge of quality and affordability in developing countries like India. The adoption of digital technology can help in solving this dual problem. Traditionally, healthcare has been digitalization deficient. The falling price of the smartphone and wearable devices, and storage devices is driving the Healthcare 4.0. Using the systematic review of the literature (SLR), this study identifies factors influencing the adoption of Healthcare 4.0. The research further used Interpretive Structural Modelling to establish the association between these identified factors. The result of the study suggests that the Government's Effort is the most important factor, followed by Empowered Customer and Financial Investment. The findings of this study can be effectively utilized by policymakers to devise policies to enforce the digital transformation of the healthcare sector in the country. The study also discusses the policy measures to overcome the barriers identified in the chapter.

In Chapter 14, the authors use and assess various knowledge strategies (e.g. Data mining and AI) to distinguish scams in electronic exchanges, all the more explicitly in MasterCard activities done through web. E-commerce implies a wide range of exchanges through internet and it is categorized formally as Business. Many security methods are used to fulfil the necessities of online trade. Every year, a lot of fraud cases for online transactions are reported. The current extortion location prototypes are utilized to check dealer respectability in the exchange technique of ecommerce. In India, countryside individuals don't accept net banking, but in using our method, a straightforward dependable and simple to deal with ecommerce framework for a wide range of information is made possible, for which a new ecommerce model was proposed that can be held worthy for use by the clients.

In Chapter 15 the authors discuss how, The Bot-IoT dataset is used, which is designed over a network environment enclosing network traffic information, different types of attacks, and botnet traffic. The consistent adoption of emerging technologies determines the requirement of botnet forensics for malware recognition. The swift progress of interconnected technologies introduces the Internet of Things (IoT) in the form of smart cities, shrewd households, elegant healthcare, and smart infrastructure. The smart-physical system includes the interconnection of a group of devices, microcontrollers, computers, and software which makes the system most vulnerable to security threats, risks, malware, cyber-attacks, and Botherders.

In Chapter 16 the author discusses how, Recent developments in technology lead to the development in real-time applications, sensor technology, and various online services which in turn is responsible for generating large quantities of data which can be used for analysis. To accommodate and use this data in decision making systems is the big challenge. Decision making system demands the ad hoc querying and ad hoc reporting capabilities. This chapter focuses on the architecture of the analysis toolbox which uses the on-the-fly query generation technique. This chapter gives details of the numerous operations and customizations provided by the system to have efficient, accurate, and quick decision making. The functionality is demonstrated with the help of case studies using sensor data and business data.

In Chapter 17 the authors discuss how, Internet of Things (IoT) is a very trendy word these days. Basically it is originated from the field of Internet i.e. Network of networks, the research field of sensor networks, and the field of cloud computing. IoT is the combination of all these three fields. In our research, some physical objects have been considered, like Internet of things consists of heterogeneous objects somehow connected with each other and capable of passing on some valuable information to one another and in handling requests for actions without human intervention. This chapter deals with the highlights and the importance of the field of IoT, where a framework of IoT network is also presented which is a complex task to construct by joining various types of devices on a single network, and then highlights some of the related applications of IoT.

In Chapter 18 the authors discuss how, Digital Learning or E-learning has been identified as learning by means of computer, internet, animation, video etc. This form of learning was making its space in the Indian education system by the introduction of several online platforms like NPTEL, Swayam, Coursera, EdX, Udemy etc. However, with the unfortunate pandemic hitting the world, the learning and teaching process came to a standstill and with lockdown in place, online learning was the only option which could have ensured the learning-teaching activities took place during the lockdown phase. Hence, every educational institution switched to this mode in a hurry by using various virtual interaction platforms like Zoom, Google Meet, Microsoft Team, Cisco Webex, go to meeting etc. With all these efforts being taken from management side, one very important dimension of learning remained ignored. This was the challenge faced by students who while accepting this mode of learning had dobts whether significant learning output was there. With this backdrop, the current chapter is an attempt to find the acceptance of e-learning by undergraduate students of various programs during Covid-19 for sustainable development in higher education.

Data was collected from the undergraduate students and was analyzed to present the finding.

In Chapter 19 the authors introduce a real-time accident prevention system made by implementing modern technologies. In this chapter, we firstly discuss the importance of a road-accident prevention system, go through some statistics and facts, and look upon some needs for such a system for today's purpose. The chapter concisely describes how drowsiness is a large promoter of such accidents and how the detection of it at an early stage can save the lives of people. It then discusses the algorithm of the proposed system made by implementing Computer Vision and Deep Learning Neural Networks and walks the reader through the steps of creating such a system, mentioning all the requirements. Finally, the system's testing result metrics are presented in the figures. Thus, this chapter enables readers to understand the importance of such a system, helps them to create such a system, and to analyze it critically.

In Chapter 20 the authors discuss how, The multi-modality image fusion technique in association with speckle noise reduction (SNR) technique is applied to carry out image fusion. The SNR technique is used to eliminate the noise, thereby improving the quality of medical images. Further, different aspects of combining medical images into an informative image through image fusion are discussed along with their advantages and failures. In addition, in this study, a more powerful technique named stationary wavelet transformation (SWT) scheme is also explored for the same purpose of improved images. The results are evaluated in terms of performance metrics such as standard deviation, entropy, and average gradient, to name a few. The schemes presented in this work are extremely useful for extracting the detailed information of a patient from a single blended image.

We wish all our readers and their family members good health and prosperity.

**Editor(s)**

# Editor Biographies

**Vikram Bali** is Professor and Head-Computer Science and Engineering Department at JSS Academy of Technical Education, Noida, India. He has graduated from REC, Kurukshetra – B.Tech (CSE), Post Graduation from NITTTR, Chandigarh – M.E (CSE), and Doctorate (Ph.D) from Banasthali Vidyapith, Rajasthan. With more than 20 years of rich academic and administrative experience, he has published more than 50 research papers in International Journals/Conferences and edited Books, authored five text books, and published 06 Patents. On the editorial board and the review panel of many International Journals, he is SERIES EDITOR for three Book Series of CRC Press, Taylor & Francis Group, and a lifetime member of IEEE, ISTE, CSI, and IE. Awarded Green Thinker Z-Distinguished Educator Award 2018 for remarkable contribution in the field of Computer Science and Engineering at the 3rd International Convention on Interdisciplinary Research for Sustainable Development (IRSD) at Confederation of Indian Industry (CII), Chandigarh, he has also attended Faculty Enablement program organized by Infosys and NASSCOM. He is member of the board of studies of different universities in India and member of organizing committee for various National and International Seminars/Conferences. Presently working on four sponsored research projects funded by TEQIP-3 and Unnat Bharat Abhiyaan, his research interests include Software Engineering, Cyber Security, Automata Theory, CBSS, and ERP.

**Vishal Bhatnagar** holds B.E, MTech, and PhD in the Engineering field. With more than 21 years of teaching experience in various technical institutions, he is currently Professor in Computer Science & Engineering Department at Netaji Subhash University of Technology East Campus (Formerly Ambedkar Institute of Advanced Communication Technologies & Research), Delhi, India. His research interests include Data-Mining, Social Network Analysis, Data Science, Blockchain, and Big Data Analytics, and he has to his credit more than 130 research papers in various international/national journals, conferences, and Book Chapters, and is Associate Editor of a few Journals of IGI global and Inderscience. With the experience of handling special issues of many Scopus, ESCI, and SCIE Journals, and SERIES EDITOR for three Book Series of CRC Press, Taylor & Francis Group, he has edited many books of Springer, Elsevier, IGI Global, and CRC Press. He is lifetime member of the Indian Society for Technical Education (ISTE).

**Sapna Sinha** is an Associate Professor in the Amity Institute of Information Technology, Amity University Uttar Pradesh, Noida. Having 20+ years of teaching experience in teaching UG and PG computer science courses, she is a Ph.D. in Computer Science and Engineering from Amity University. With several book chapters and research papers authored in journals of repute to her credit, her areas of interest are Machine Learning, Big Data Analytics, Artificial Intelligence, Networking and Security. She is D-Link certified Switching and Wireless professional, and is also Microsoft Technology Associate in Database Management System, Software Engineering and Networking. Apart from being EMC Academic Associate in Cloud Infrastructure Services, she is lifetime member of The Institution of Electronics and Telecommunication Engineers (IETE). She has also edited several books.

**Prashant Johri** is Professor in School of Computing Science & Engineering, Galgotias University, Greater Noida, India. He completed his M.C.A. from Aligarh Muslim University and a Ph.D. in Computer Science from Jiwaji University, Gwalior, India. Previously, he has held the post of Professor and Director (M.C.A.), Galgotias Institute of Management and Technology, (G.I.M.T.), and the post of Professor and Director (M.C.A.), Noida Institute of Engineering and Technology, (N.I.E.T.) Gr.Noida. As Chair in many conferences and affiliated as member of the program committee of many conferences in India and abroad, he has published a number of papers in national and International Journals and Conferences, and has organized several Conferences/Workshops/Seminars at the national and international levels. Having published edited books in Elsevier and Springer, his research interests include data retrieval and predictive analytics, information security, privacy protection, big data open platforms, Software Reliability cloud computing, Mobile cloud, Machine learning, AR & VR, Soft computing, Fuzzy systems, Healthcare, Agriculture, Pattern recognition, Bio-inspired phenomena, and advanced optimization model & computation. https://orcid.org/0000-0001-8771-5700.

# Section A

*Disruptive Technologies:
Introduction and Innovation*

# 1 Blockchain and Internet of Things: An Amalgamation of Trending Techniques

*Shilpi Bisht, Neeraj Bisht, Pankaj Singh, and Shray Dasila*

## CONTENTS

1.1 Introduction .................................................................................................. 4
1.2 Literature Review ......................................................................................... 5
1.3 Blockchain .................................................................................................... 8
    1.3.1 Types of Blockchain ........................................................................ 12
        1.3.1.1 Permission Seeking Blockchain ...................................... 12
        1.3.1.2 Permission Less Blockchain ............................................ 13
    1.3.2 The Structure of Blockchain Technology ....................................... 13
    1.3.3 Advantages of Blockchain Technology ........................................... 15
    1.3.4 Disadvantages of Blockchain Technology ...................................... 16
1.4 Internet of Things (IoT) .............................................................................. 17
    1.4.1 Structure of IoT ................................................................................ 18
1.5 Amalgamation of Blockchain Technology and IoT .................................... 18
1.6 Application Area of Blockchains in IoT ..................................................... 19
    1.6.1 Agriculture Sector ............................................................................ 21
    1.6.2 Education Sector .............................................................................. 21
    1.6.3 Smart Homes and Smart Cities ....................................................... 22
    1.6.4 Healthcare Sector ............................................................................. 22
    1.6.5 Hydrocarbon Industry ...................................................................... 22
    1.6.6 E-Business ........................................................................................ 23
    1.6.7 Data Collection and Tracking ......................................................... 23
    1.6.8 Finance ............................................................................................. 23
    1.6.9 Tourism and Hospitality .................................................................. 24
1.7 Privacy and Security Related Issues ........................................................... 24
    1.7.1 Threats .............................................................................................. 24
    1.7.2 Attacks .............................................................................................. 25
    1.7.3 Private Key Security ........................................................................ 25
    1.7.4 Fifty One% Vulnerability Attacks ................................................... 25
    1.7.5 Updation Issues ................................................................................ 26

1.8 Limitations ................................................................................................26
1.9 Implications .............................................................................................26
1.10 Conclusion and Future Scope ..................................................................26
References ...........................................................................................................27

## 1.1 INTRODUCTION

The emergence of Artificial intelligent information systems has brought a phase of metamorphism in almost every aspect of our world, be it manufacturing, automobile, defence, agriculture, healthcare, finance, tourism and hospitality, or education. With the advent of new technologies everything around us is getting smarter day by day, and the two trending and most fascinating techniques that are elevating the service quality of our life, making our life easy and comfortable, along with taking care of our privacy and security are Blockchain and Internet of Things (IoT).

Satoshi Nakamoto who discovered the first Blockchain database created the first Blockchain transaction in 2009. As the years passed by and new applications of Blockchain came into existence, and despite the fact that there are debates about Nakamoto's actual character, still undoubtedly, he has given something progressive to the planet, and it depends upon the developer what they want to do with it (Vujičić et al., 2018). Use of Blockchain technology has become quite popular because of its user-friendly features such as transparency, reliability, proper management, decentralised nature, cost efficiency, and delectableness. These properties are remarkable and notable for managing any network such as 6G. "In particular, the integration of the Blockchain in 6G will enable the network to monitor and manage resource utilization and sharing efficiently" (Xu et al., 2020).

If Blockchain provides a more secure and safer environment to IoT, it's because of the fact that when Blockchain integrates with IoT it resolves problems like data leaks or privacy disturbances, mainly, with the presence and reachability of every single user or group being managed with ease, and by storing the data in Blockchain itself, which eliminates the risk of any privacy or security threats.

Blockchain innovation has demonstrated its impressive versatility recently when an assortment of market divisions searched for methods of fusing its capabilities into their tasks. In a report (The Global Competitiveness Report, 2015) by the World Economic Forum, 58% of all review respondents said that: "by 2025 10% of worldwide total national output are going to be put away utilizing Blockchain innovation". It is subsequently to be expected that Blockchain is drawing in crowds of speculators who are broadly putting resources into Blockchain-based new companies (Mettler, 2016) and the reason is not far behind. The advantage of Blockchain is that an attacker must bargain at least 51% of the frameworks to outperform the hashing intensity of the target organization, hence, it's computationally infeasible to dispatch an assault against the Blockchain network (Biswas & Muthukkumarasamy, 2016).

This chapter provides a brief literature review in section 1.2, which is the motivation to understand concepts of Blockchain and IoT. Blockchain Technology, its structure, advantages, disadvantages are discussed in detail in section 1.3. IoT, its structure and integration with Blockchain is provided in section 1.4. Area of deployment of integration of IoT and Blockchain, and security & privacy issues are covered in section 1.5. Limitations and implications are given in sections 1.6 and 1.7, respectively. Section 1.8 presents the conclusion and future scope in this area.

## 1.2 LITERATURE REVIEW

The "Internet of Things" (IoT) has brought a digital revolution in almost every aspect of our lives like meeting our daily needs, performing in academics, manufacture, finance, trading, and many more things, and specifically, it provides several comforts and conveniences to our daily lives, if we do not account for the privacy and security-related issues of IoT as major challenges to its spreading wings even further. Blockchain technology, which is a cryptographic technique-based decentralized database, has emerged as a promising solution for privacy- and security-related challenges in IoT. Among those researchers who have discussed on the applications of the Blockchains in the vast area of Internet of things (IoT), mention must be made of Huckle et al. discussed the benefits of IoT and Blockchain technology in shared economy applications, where the main emphasis of their work is to create decentralised, secure, and transparent shared economy distributed applications based on Blockchain technology, given such economic applications allow clients to securely monetise their risk to accelerate their wealth. They also explored the relationship of IoT with the research of Sussex's Shared Thing, a project about monetising shared economic applications (Huckle et al., 2016).

Dorri et al. (2017) pointed out the major pitfalls of Blockchains like high computational cost, high bandwidth delays, and overheads, that aren't compatible with most of the devices, and they proposed a lightweight Blockchain-based design for IoT which retains security and privacy features of Blockchain and virtually eliminates the overheads of the classical Blockchain. Since IoT devices enjoy a personal immutable ledger that acts virtually like a Blockchain but unlike Blockchain it is managed centrally, to optimize energy consumption (Dorri et al., 2017), and because an overlay network is generated by high resource devices to implement a distributed Blockchain which is publicly accessible and ensures peer-to-peer security and privacy, they proposed an architecture that reduces the block validation processing time through distributed trust. Their study on smart home setting represented wider IoT applications, whereas in this research they claimed that their method reduces packet and the processing overhead remarkably in comparison to the Blockchain implementation used earlier in Bitcoin. A number of researchers explored the security and privacy issues arising in Blockchain in their studies, and one survey article that devoted attention on security aspects of the Blockchain systems led its researchers to focus on the security-related threats to Blockchain in a systematic study and who examined popular Blockchain systems to identify corresponding real attacks (Li et al., 2020). They also explored the possibilities to enhance security for Blockchain, which might be used for the development of several Blockchain systems. Other researchers shed light on issues like privacy and confidentiality of the data or devices in Blockchain technology. Since it is evident that the content of the data could be accessed by anybody in the network for various unavoidable purposes like verification and mining (Rahulamathavan et al., 2017), they proposed a new Blockchain architecture which is privacy-preserving for IoT applications keeping in view these privacy issues, based on attribute-based encryption (ABE) technique. In support of their model, they also presented security, privacy, and numerical analyses of the same. Even as emerging idea of smart cities, smart homes, and smart gadgets is a stimulating factor for developing more and more IoT devices, still it cannot be ignored that these devices have their own limitations in terms of computational power consumption, storage, and network capacity, which results in devices' enhanced vulnerability to attacks in comparison to other endpoint devices like smartphones, tablets, or computers. Khan and Salah also presented and surveyed major security issues, attacks, threats, and state of art solutions for IoT, and reviewed and categorized common security issues with respect to the IoT-layered architecture, where they discussed the protocols used in networking, communication, and management. It is claimed that Blockchain can be a great facilitator for solving several security problems in IoT (Khan & Salah, 2018). A few researchers have provided an overview of common security issues of Software Defined Networking (SDN) with IoT clouds, and elaborated the design principles of the Blockchain technology and recommended Blockchain as a solution of security issues faced with SDN and IoT (Tselios et al., 2017). Applications of Blockchain technology and its major security issues in the field of cyber security have been studied in detail by Dai and others who discussed the advantages of Blockchain vis-a-vis cyber security, and in addressing the four important security issues of Blockchain and while performing deep analysis of every problem, their technique comprised of an encryption method based on attributes for an enhanced access control strategy (Dai et al., 2017).

By the year 2018, the concept of Internet of Things had stepped out of its infancy and gradually established itself as an integral component of the future Internet technology. Given the technologies for access management in IoT depend on centralized model, which introduces technical challenges in managing them at global platforms, Novo proposed a Blockchain technology-based control system with distributed access for different roles and several permissions in IoT. He implemented and evaluated this architecture on realistic IoT models, and for specific scalable IoT scenarios, results claimed that Blockchain technology can be a potential option for access management technology (Novo, 2018). A detailed study of Blockchain-based security in "Electric Vehicles Cloud and Edge Computing (EVCE)" is done by Liu et al. Addressing important security issues for such hybrid cloud and edge computing where EVs act as budding resource infrastructures for both energy and information interactions, they proposed Blockchain-based data and energy coins which apply frequency of data contribution and amount of energy contribution to attain the goal. For securing vehicular interactions in EVCE computing, they have proposed security solutions (Liu et al., 2018). Minoli and Occhiogrosso highlighted a few Blockchain Mechanisms (BCMs)-based IoT environments. For security of IoT-based applications, the Blockchain mechanism plays an important role in Castle approach or defences-in-depth in security. While the decentralization property of the Blockchain has gained attention of several organizations and corporations for the facility it provides of interaction to happen in a verified way between non-trusting members without any trusted intermediary member in a distributed peer to peer network (Minoli & Occhiogrosso, 2018), for sharing the sensors' data of the Internet of Things (IoT) on the other hand, Papadodimas et al. proposed a decentralized application "DApp" based on Blockchain technique, and illustrated various challenges faced during the development phase. This application is a combination of Blockchain technology with IoT in which smart contracts are executed on Ethereum ("a global, open-source platform for decentralized applications") Blockchain, which acts as the marketplace for IoT weather sensor data. The application deploys the "Sensing-as-a-Service (S2aaS)" business model in combination with Blockchain (Papadodimas et al., 2018). Atlam and Wills dedicated a chapter titled "Technical aspects of Blockchain and IoT" in Advances in Computers, in which they provide an exhaustive overview of the technical aspects of Blockchain and Internet of Things, while starting off by reviewing the Blockchain technology, its implications, and possible challenges, with following chapters including basic architecture, important characteristics, its applications, and also the challenges of the IoT system (Atlam & Wills, 2019). A few authors have explored privacy and the security concerns in IoT and designed a framework which integrates Blockchain with IoT, and claimed that it provides better security for IoT data and various operations, and proposed to attain desirable scalability including decentralized payment, authentication, and suggested few possible solutions to several privacy and security challenges faced in IoT based on Blockchain and Ethereum (Yu et al., 2018). Still many authors have discussed applications of Blockchain in IoT Technology, focusing on security and privacy issues, efficiency, and major challenges faced in their survey and review articles (Cho & Lee, 2019; Dai et al., 2019; Huynh et al., 2019; Jonathan & Sari, 2019; Karthikeyyan & Velliangiri, 2019; Shah & Sridaran, 2019). Pavithra et al. studied the applications of Blockchain in cloud computing, where they did analysis and compared several issues in cloud computing with the help of Blockchain (Pavithra et al., 2019). Given that an important property of Blockchain, its decentralized nature, is a key to many challenges like maintenance, authentication, and security issues of IoT systems, however, Blockchain-based IoT networks, because of their public nature, have several privacy issues.

Researchers have discussed these privacy issues and implementation of privacy preservation strategies (anonymization, private contract, mixing, encryption, and differential privacy) in Blockchain-based IoT systems, and have claimed to pave a way for future strategies of privacy preservation to address privacy issues with IoT systems operating over Blockchain (Hassan et al., 2019). For constituting the trust of process executions, researchers have conferred a smart contract that match the IoT environment, and introduced a similar approach extended from "Practical Byzantine Fault Tolerance (PBFT)" with selected validators, to mark time and meet challenges (Da & Viriyasitavat, 2019). Other researchers have conducted performance analysis for IoT applications which are subject to maximum delay constraints and proposed a mathematical model which calculates the peer-to-peer delay with different network configurations claiming that the distinctive characteristics of IoT traffic have an undeniable impact on the peer-to-peer delay requirement

(Alaslani et al., 2019). Several researchers have discussed applications of Blockchain in IoT and Industrial IoT (Puri et al. 2020; Sengupta et al., 2020; Wang, et. al., 2020). Banerjee (2019) has discussed the applications of Blockchain with IOT along with ERP in various business models particularly supply chain, where he explored the economic aspects and cost cutting in supply chain management through combination of these techniques. An intelligent IoT design with the help of Blockchain along with artificial intelligence was proposed by researchers (Singh et al., 2020), who performed qualitative analysis and quantitative analysis of proposed architecture and compared the results with already existing models to claim superiority of their architecture. Wu et al. elaborated on the dynamic nature of IoT applications and consensus in Blockchains, and to address the dynamic allocation of an IoT node in different Blockchains, they designed a software-defined Blockchain model based on a consensus function, to realize the dynamic configurations for Blockchains, claiming that their model enhances the flexibility and extensibility of Blockchains (Wu et al., 2020). In a data flow design IoB Health, for IoT based e-healthcare in combination with Blockchain, the proposed architecture can be used for overall management of e-healthcare data (Ray et al., 2020). Mhaisen et al. explored the significance of Smart Contacts for IoT monitoring applications to reduce monetary cost along with maximum advantage of security features of public Blockchain (Mhaisen et al., 2020). With the requirement for enhancing the computation efficiency of IoT devices, where researchers deployed the mobile edge computing at the "Small-cell Base Station (SBS)", it was proposed a DRL based solution for achieving efficient intelligent strategy, called "policy gradient based computing tasks scheduling (PG- CTS)" algorithm, in which they claimed better performance of PG-CTS over other strategies with the help of theory and experiments (Gao et al., 2020). The Blockchain- and IoT-dependent model for the intellectual property protection system that covers patents, trademarks, copyrights, industrial design, craft works, trade dress, trade secrets, geographical indications, and plant variety rights could revolutionize trade world (Lin et al., 2020). Security and privacy issues in Bitcoin and Blockchain and in Blockchain with IoT, respectively, must be dealt properly to ensure safer transactions (Du et al., 2020; Zaghloul, et al., 2020). A Blockchains-based, distributed protocol that relies on peer-to-peer network communication design for enabling the on-demand applications of IoT networks on top of the IoT devices would be more secure than the traditional communication protocols (Hamdaoui et al., 2020). The robustness against device failure and maliciousness is ensured by the self-recovery mechanism. Premkumar and Srimathi have reviewed the applications of Blockchain with IoT in the healthcare sector associated with the pharmaceutical industry, where, in a survey paper, the authors addressed various challenges faced with open issues in smart healthcare arising due to associated security measures (Tariq et al. 2020). Table 1.1 represents summary of Literature Review.

## 1.3 BLOCKCHAIN

Blockchain, as evident from its nomenclature, is a chain of blocks, or a collection of digital, verifiable, permanent, distributed records stored in blocks. Being, actually, the

## TABLE 1.1
### Summary of Literature Review

| S.No | Source | Title of Article | Major Findings |
| --- | --- | --- | --- |
| 1 | Huckle et al. (2016) | "Internet of things, blockchain and shared economy applications". | Benefits of IoT and Blockchain technology in shared economy applications. |
| 2 | Dorri et al. (2017) | "Towards an optimized blockchain for IoT". | Proposed a lightweight Blockchain based design for IoT which retains security and privacy features of Blockchain and virtually eliminates the overheads of classical Blockchain. |
| 3 | Li et al. (2020) | "A survey on the security of blockchain systems". | Security related threats to Blockchain in a systematic review study. |
| 4 | Rahulamathavan et al. (2017) | "Privacy-preserving blockchain based IoT ecosystem using attribute-based encryption". | Proposed a new Blockchain architecture which is privacy-preserving, for IoT applications. |
| 5 | Khan and Salah (2018) | "IoT security: Review, blockchain solutions, and open challenges". | Blockchain can be a great facilitator for solving several security related problems in IoT. |
| 6 | Tselios et al. (2017) | "Enhancing SDN security for IoT-related deployments through blockchain". | Blockchain as a solution of security issues faced with SDN and IoT. |
| 7 | Dai et al. (2017) | "From Bitcoin to cybersecurity: A comparative study of blockchain application and security issues". | Adopted an encryption method based on attributes for an enhanced access control strategy. |
| 8 | Novo (2018) | "Blockchain meets IoT: An architecture for scalable access management in IoT" | Proposed a Blockchain technology based control system with distributed access for different roles and several permissions in IoT. |
| 9 | Liu et al. (2018) | "Blockchain-enabled security in electric vehicles cloud and edge computing". | Detailed study of Blockchain-based security in "Electric Vehicles Cloud and Edge Computing (EVCE)". |
| 10 | Minoli and Occhiogrosso (2018) | "Blockchain mechanisms for IoT security" | Highlighted a few Blockchain Mechanisms (BCMs) based IoT environments. |
| 11 | Papadodimas et al. (2018) | "Implementation of smart contracts for blockchain based IoT applications". | Proposed a decentralized application "DApp" based on Blockchain, and illustrated various challenges faced during the development phase. |

*(Continued)*

## TABLE 1.1 (Continued)
## Summary of Literature Review

| S.No | Source | Title of Article | Major Findings |
|---|---|---|---|
| 12 | Atlam and Wills (2019) | "Technical aspects of blockchain and IoT". | Provided an elaborate overview of the technical aspects of Blockchain &IoT. |
| 13 | Yu et al. (2018) | "Blockchain-based solutions to security and privacy issues in the Internet of Things" | Explored privacy & security concerns in IoT and designed a framework which integrates Blockchain with IoT, also proposed to attain desirable scalability, privacy, and security challenges. |
| 14 | Cho and Lee (2019) and more | "Survey on the Application of BlockChain to IoT" | Discussed applications of Blockchain in IoT Technology, focusing on security and privacy issues, efficiency, and major challenges faced in their survey and review articles. |
| 15 | Pavithra et al. (2019) | "A survey on cloud security issues and blockchain" | Analysed and compared several issues in cloud computing with the help of Blockchain. |
| 16 | Hassan et al. (2019) | "Privacy preservation in blockchain based IoT systems: Integration issues, prospects, challenges, and future research directions". | Discussed privacy issues and implementation of privacy preservation strategies (anonymization, private contract, mixing, encryption, and differential privacy) in Blockchain-based IoT system. |
| 17 | Xu et al. (2020) | "Application of blockchain in collaborative Internet-of-Things services" | Conferred a smart contract that match the IoT environment. |
| 18 | Alaslani et al. (2019) | "Blockchain in IoT systems: End-to-end delay evaluation". | Proposed a mathematical model which calculates the peer-to-peer delay with different network configurations. |
| 19 | Puri et al. (2020) and more | "Blockchain meets IIoT: An architecture for privacy preservation and security in IIoT" | Discussed applications of Blockchain in IoT and Industrial IoT. |
| 20 | Banerjee (2019) | "Blockchain with IOT: Applications and use cases for a new paradigm of supply chain driving efficiency and cost". | He explored the economic aspects and cost cutting in supply chain management through combination of these techniques. |
| 21 | Singh et al. (2020) | "Blockiotintelligence: A blockchain-enabled | Proposed an intelligent IoT design with the help of Blockchain along |

## TABLE 1.1 (Continued)
## Summary of Literature Review

| S.No | Source | Title of Article | Major Findings |
|---|---|---|---|
| | | intelligent IoT architecture with artificial intelligence" | with artificial intelligence and compared the results with already existing models to claim superiority of their architecture. |
| 22 | Wu et al. (2020) | "Application-aware consensus management for software-defined intelligent blockchain in IoT". | Designed a software-defined Blockchain model based on a consensus function, to realize the dynamic configurations for Blockchains |
| 23 | Ray et al. (2020) | "Blockchain for IoT-Based Healthcare: Background, Consensus, Platforms, and Use Cases" | Proposed a data flow design IoB Health, for IoT based e-healthcare in combination with Blockchain which can be used for overall management of e-healthcare data |
| 24 | Mhaisen et al. (2020) | "To chain or not to chain: A reinforcement learning approach for blockchain-enabled IoT monitoring applications". | Explored the significance of Smart Contacts for IoT monitoring applications to reduce monetary cost along with maximum advantage of security features of public Blockchain |
| 25 | Gao et al. (2020) | "Deep Reinforcement Learning based Task Scheduling in Mobile Blockchain for IoT Applications". | Deployed the mobile edge computing at the" Small-cell Base Station (SBS)", proposed a "deep reinforcement learning (DRL)" based on "policy gradient based computing tasks scheduling (PG- CTS)" algorithm. |
| 26 | Lin et al. (2020) | "Blockchain and IoT-based architecture design for intellectual property protection". | Designed a Blockchain based IoT model for the intellectual property protection system that covers patents, trademarks, and copyrights. |
| 27 | Du et al. (2020) and more | "Spacechain: A Three-Dimensional Blockchain Architecture for IoT Security". | Addressed security and privacy issues in Bitcoin and Blockchain and in Block chain with IoT, respectively. |
| 28 | Hamdaoui et al. (2020) | "IoTShare: A Blockchain-Enabled IoT Resource Sharing On-Demand Protocol for Smart City Situation-Awareness Applications". | Proposed Blockchains-based, distributed protocol that relies on peer-to-peer network communication design for enabling the on-demand applications of IoT networks on top of the IoT devices |

*(Continued)*

**TABLE 1.1 (Continued)**
**Summary of Literature Review**

| S.No | Source | Title of Article | Major Findings |
| --- | --- | --- | --- |
| 29 | Premkumar and Srimathi (2020) | "Application of Blockchain and IoT towards Pharmaceutical Industry" | Reviewed the applications of Blockchain with IoT in the healthcare sector associated with the pharmaceutical industry |
| 30 | Tariq et al. (2020) | "Blockchain and Smart Healthcare Security: A Survey" | Addressed various challenges with open issues in smart healthcare arising due to associated security measures. |

blocks of digital information which have three parts: transaction details, participant details, and a unique hash code, these blocks are stored in a public database, each block referring to the previous block and hence forming a chain. A Blockchain implements a continuous control mechanism for data manipulation, quality, and errors.

Blockchain is an innovation fundamental to Bitcoin and different digital currencies, maintained by a decentralized PC organization. For example, use of a decentralized record in the Blockchain has framed an innovative model which can revolutionize the finance market (Nguyen, 2016). When it integrates with IoT, finance, healthcare, and other sectors, then Blockchain wipes out the impact of any third party or outsiders in these sectors which therefore face lesser threat of security and privacy (Yu et al., 2018). Also, it is hard to use Blockchain for false purposes, and which makes Bitcoin, and other sorts of cryptocurrency, that utilizes Blockchain-based open records easy to implement to form exchanges across distributed organizations (Tasatanattakool & Techapanupreeda, 2018).

### 1.3.1 Types of Blockchain

Blockchains can be categorized into two categories: Permission seeking or Permissionless, as defined below:

#### 1.3.1.1 Permission Seeking Blockchain

A permission-seeking Blockchain requires the granting of permission so as to operate it or have access to it. Known to prevent data loss and also maintain privacy, permission seeking Blockchains are further classified into: private and hybrid Blockchains.

    A. Private Blockchain:
        A private Blockchain is looking to form chains where minor hubs are restricted in number and IDs of every hub is known; so, along these lines, exchange preparation is restricted to known, allowed, or predefined users, and where transactions are visible only to the allowed participants. The advantages

of these Blockchains are adaptability and feasibility. Though these chains are more centralised in nature, cost efficiency is one of the darker sides. Hyperledger and R3 Corda are examples of a private Blockchain.

B. Hybrid or Consortium Blockchain:
Hybrid Blockchain can be considered a type of private blockchain being used by a group. It has features both of open and private Blockchain, and are known by names like Consortium Blockchain. These Blockchains aim at providing transparency, feasibility, and adaptability. Dragon chain is a hybrid Blockchain.

### 1.3.1.2 Permission Less Blockchain

If the single user or group does not require any permission to access the data, then it is a Permission less Blockchain, and it is also the reason for concerns of information leaks and security issues to surface as significant issues here. Public Blockchain is a kind of permission less Blockchain that licenses free and unlimited investment of all things considered, for example, it allows anyone to participate in a transaction in the role of user, developer, miner, or community member. While, in a public Blockchain, all the transactions are transparent and detectable, yet both feasibility and acceptance are poor. Two well-known public Blockchains are Bitcoin and Ethereum. Figure 1.1 represents the types of Blockchains.

### 1.3.2 THE STRUCTURE OF BLOCKCHAIN TECHNOLOGY

The Blockchain comprises a straight arrangement of blocks, which are added to the chain with the standard stretches. The data in the squares (blocks) relies upon the Blockchain organisation; however, the timestamp, exchange, and hash exist in all the Blockchain variations. Each square contains the cryptographic hash of the past square as shown in Figure 1.2. It includes the commitment of various machines (called hubs) to prevent or stop the information altering. That is, the entirety of the exchange records will be kept up as squares, with each square having

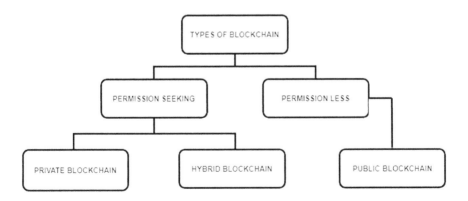

**FIGURE 1.1** Types of Blockchains.

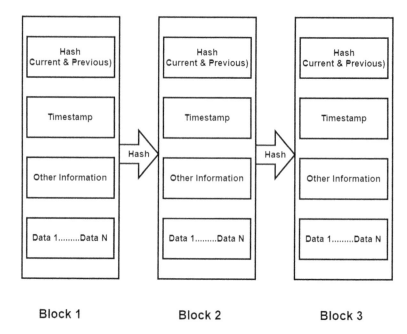

**FIGURE 1.2** Structure of Blockchain.

a signal element considerably marked by the square maker. Without the maker's private key, it is difficult to alter or adjust the square. Moreover, the chain extension gives a surety of an additional security to it. A Blockchain might be viewed as an open record and every one's dedicated exchanges are put away during a rundown of blocks. The chain develops as new blocks get attached without interruption. Bitcoin and Blockchain innovations have begun to shape and characterize new perspectives within software engineering and data innovation (Zheng et al., 2017).

The Blockchain has the following attributes:

Block Size: It refers to the size of a block in the Blockchain in bytes, and is the first field forming the structure of a block, is represented using 4-byte data, with capacity to enumerate the block's capacity which is a maximum of almost 4000 transactions in itself.

Block Header: In a Blockchain network, a block is identified with the help of a block header, which is made up of three fields: the first field is used for linking with the previous block in the Blockchain; the second field consists of the timestamp, and nonce; and the third field consists of a summary of all the transactions. Typical size of a block header is 80 bytes.

Exchange Size: Exchange provides an intermediary platform for transactions, and exchange size is the number of exchanges allowed in a Blockchain.

Miners in the network: A miner is a computational device which is responsible for adding new transactional data to the Blockchain's public ledger. It represents the total number of miners in a Blockchain network.

Mining power per hub: Representing the rate at which transactions are verified and added to a Blockchain, a high mining rate needs a high hash rate. It is measured in mega hashes per second.

Data Transfer capacity: In Blockchain the data transferring capability is quite high. This results in faster transfer of data from one system to another.

Crypto-puzzle: Crypto Puzzle is a puzzle in Blockchain which the miners need to encrypt before adding a block to a Blockchain.

### 1.3.3 Advantages of Blockchain Technology

Very popular and having its application in almost every field or industry, Blockchains have a lot of advantages which are discussed below:

  i. The primary advantage of Blockchain innovation is its decentralized framework which suggests that this framework works without a mediator and all participants of this Blockchain settle on the choice. It is an open, conveyed, collective, straightforward framework with no single purpose or weakness. Since, all members share admittance to the information hence, it can't be changed or ruined by a solitary player. Every member in the record holds indistinguishable duplicates of the information confirmed by the others, which guarantees the most trusted level of information being accessible to all the users.
  ii. Blockchain innovation has the biggest advantage to the medical services framework improving the nature of medical care by sharing information among all the members, upgrading specific security, and guaranteeing information well-being. A Blockchain development can be used to develop a balance and to allow patients to do designated tasks like performing standard enlistment and reporting their recuperating conditions. The never-ending nature of the record is similarly a charming advancement that could wind up being genuinely important.
  iii. Blockchain advances brilliant agreements, which expands the effectiveness of exchanges and instalments in the financial exchange. Since each block consists of a unique code known as hash value, which is unique to every single block in the chain, there is no chance of data leak. Suppose, a single user or group attacks the Blockchain, still they won't be successful in invading its security because each block has its own hash code and to decipher each block's code in an interminable chain of blocks is an impossible proposition making all the transactions in the Blockchain technology safe and secure. This is another reason why Blockchain technology is spreading like a forest fire.
  iv. Charges for unfamiliar trade exchanges, settlements, Visa exchanges, and different items can be decreased significantly which makes cost efficiency the major advantage of Blockchain technology. Since Blockchain acts as a direct link between the buyer and the seller, therefore, the role or the existence of others in between the seller and the buyer is no longer a reality.

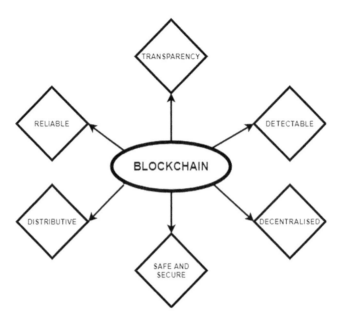

**FIGURE 1.3** Advantages of Blockchain.

As a result, it eliminates the need for commissions, and other factors, which give rise to increasing charges.

v. Recognizability of IoT information: In a Blockchain, the detectability of the chain is the ability to follow and confirm the temporary data of information blocks running through it. All the information that a block gives in a Blockchain has a connection with a noteworthy timestamp; henceforth, the chain guarantees the capacity to trace the information.

vi. Dependability of IoT information: given the need to protect the reliability of the IoT information, the Blockchain may ensure the trustworthiness of information by implementing cryptographic instruments including hash capacities, unbalanced encryption calculations, and advanced signatures, all intrinsic in Blockchains.

vii. Distributed nature of Blockchain: rather than the regular Peer-to-Peer network for sharing information documents between intrigued peers, Bitcoin uses the organization to quickly communicate information among all the associated hubs. This cycle is called flooding and proceeds until all hubs inside the organization get the broadcasted information.

Figure 1.3 illustrates the advantages of Blockchain technology.

### 1.3.4 Disadvantages of Blockchain Technology

Along with several important advantages of Blockchain, it has some demerits too. The major drawbacks of Blockchain technology are:

i. Incompleteness: with the need to comply with regulators, such as the legalities and guidelines for usage of Bitcoin and digital forms of money, there are limits to Blockchain innovation from being broadly applied.
ii. Secure and Synchronized Software Update: IoT gadgets are meant to work uninterrupted for significant stretches of time, with no updates in their product. This makes them an obvious target to any malware or suspicious activity focused on cyberattack.
iii. Lack of IoT driven Consensus Convention: The delays in agreement about Blockchain technology eventually closes or postpones encouraging exchanges, which cannot support IoT frameworks expansion because these need constant exchange encouragements.
iv. Cost issues: In Blockchain innovation, the information is put away in a lots of places. In short, there isn't one single spot where the information can be managed or stored. As a result, the cost issues have increased.
v. High vitality (energy) utilization: The Blockchain innovation network given its unending chain expends a great deal of vitality. The Bitcoin mining's yearly vitality utilization is 3.38 Terawatt hours.
vi. Protection and security issues: Popularity of different cryptographic forms of money has encouraged various clients, with private open keys to get to its digital money wallets, to join and explore Blockchain networks. Thus, it is important to deal with these keys safely.

Other security issues like forking, 51% weakness, spillage of protection, information breaks, and crimes are a significant reason for worry about the protection and security issues about the Blockchain technology.

## 1.4 INTERNET OF THINGS (IOT)

IoT is receiving loads of attention and recognition in the recent years, when it is actually the idea to collect the tiny bits of information, however reasonable, from a machine or device to be transferred from its source to its destination. The information assortment and sharing by the IoT devices are Blockchain-supported technologies and standards such as wireless communication standards and frequency identification (RFID). As multiple IoT devices cater information to an entranceway, leading the information to the Blockchain network that works on the distributed ledger technology, the Blockchain and IoT as discrete advances are however limited in their applications in many fields. Still, the Blockchain with IoT is a concept that is getting popular in almost every field, and it wouldn't be wrong to say that in the future, almost everything will use features of the integration of Blockchain with IoT. IoT, likewise known as the web of Everything or the economic web, is another innovation which could be viewed as a worldwide organization of machines and gadgets equipped to handle human activity as they perform in day to day life.

"A world where physical objects are seamlessly integrated into the information network, and where the physical objects can become active participants in business processes. Services are available to interact with these smart objects over the

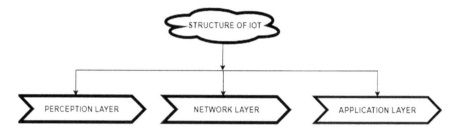

**FIGURE 1.4** Structure of IoT.

Internet, query their state and any information associated with them, taking into account security and privacy issues" (Haller et al., 2008).

It is estimated that the IoT can hit twenty six billion units by 2020, up from 0.9 billion in 2009. IoT has varied usage areas like transportation, agriculture, medical services, electricity generation, and distribution.

### 1.4.1 Structure of IoT

IoT structure is composed of the following layers:

  i. Perception Layer: Responsible for sensing environment and sending the collected data to network layer, there are various types of sensors used in this layer to collect data.
  ii. Network Layer: Meant to connect with other similar types of sensors, this layer is also responsible for collecting and transmitting sensor data.
  iii. Application Layer: Used to provide application-specific service to the end user, this layer has been used to develop user centric services like smart home, smart health, and smart city.

Structure of IoT is shown in Figure 1.4.

## 1.5 AMALGAMATION OF BLOCKCHAIN TECHNOLOGY AND IOT

With the amalgamation of Blockchain Technology and IoT, a lot of problems were solved as many benefits surfaced to tackle the problems. The integration of these two techniques is depicted in Figure 1.5. So, let's look at what all is actually an outcome of this integration.

  i. Transparency: integration of the two technologies provides transparency to the user. For example: the user can directly buy or sell his/her vehicle, furniture, and many other things to the real buyer/seller. This provides the user a transparent and a direct link.
  ii. Faster Financial Transfers: With the interlinking of IoT and Blockchain, the financial transfers are faster than usual. Nowadays, it takes only a few

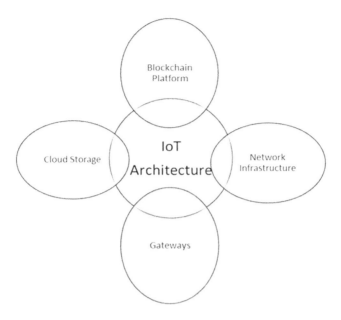

**FIGURE 1.5** Amalgamation of Blockchain and IoT.

seconds to transfer money from one account to the other account, with the financial sector seeing it like a boon.

iii. Lower Cost: Nowadays, since the user can evidently directly sell or buy anything of his/her choice, it eliminates the payment of commission and involvement of third parties between the buyer and the seller which reduces the final cost of the product or commodity.

iv. Reliability of the technology: The integration of these two technologies facilitates the manufacturer to track the location or status of his/her product, and thus creates a bond of confidence.

v. Security and Privacy: Everywhere in the world, with the blend of digitalisation with automation, the world is becoming prone to increasing number of threats, which the union of the two technologies reduce.

vi. Proper Management: If the manufacturer has a track and record of the products helping him modify the mechanism according to his will, it helps in proper management and secure delivery of the product. Figure 1.6 illustrates the benefits of amalgamation of Blockchain and IoT technology.

## 1.6 APPLICATION AREA OF BLOCKCHAINS IN IOT

IoT is extensively used in varied domains like health care, home applications, agriculture, transport, electrical instrumentation, natural disaster, town pollution, smart grid, water quality, logistics, and industrial transportation, and varied different Industries. Figure 1.7 shows the areas of deployment of Blockchain in IoT. Some of the major uses and applications of IoT with Blockchain are as follows:

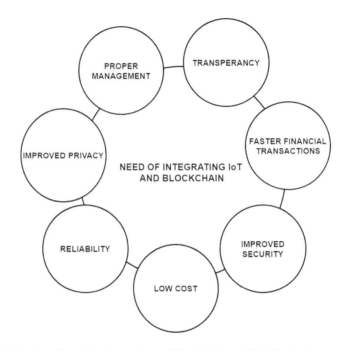

**FIGURE 1.6** Benefits of Amalgamation of Blockchain and IoT Technology.

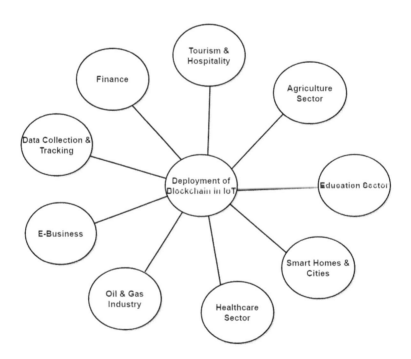

**FIGURE 1.7** Deployment of Blockchain in IoT.

### 1.6.1 Agriculture Sector

To deploy Blockchain with IoT, and as far as the GDP of our nation or financial improvement of our country is concerned, Agriculture sector features as the most promising since most of the population rely on this sector. As shown by the International Bank for Reconstruction and Development, about 65% of the population across the world relies upon farming, and efficiency of worldwide rural segments is anticipated to increase by 1.75% per annum, so that requests of people are met till 2050. With its being conceivable just when the water system framework is improved, and with evidence of ranchers seen to withstand enormous financial losses (which has its influence on the GDP) because of badly thought water system strategies and incorrect forecast of climate and rainfall by the meteorological division, it's an imperative to prevent the wastage of water, which requires immediate deployment of water system frameworks assisting ranchers with enhanced water assets. Exact and accurate water system methods are important with the goal that quantities of water should be used in manner judicious, ultimately moderating wastage of water. Essentially, it's a cycle of inappropriately providing crops with water, which becomes more acute in territories on the brink of no or unpredictable precipitation or little precipitation. The traditional PI framework comprises various stations, for instance, climate figure stations, base station, PI control framework, workstation, and infield stations, which improve the productivity of water use more than customary water system frameworks. It likewise kills overabundance of water and diminishes fermentation, even esterification of oceans, seas, and improves proficiency in crop well-being and farming potentiality. IoT plays an important role in modern farming: remote monitoring of crops can be done with help of drones; and, various sensors like humidity, light, and temperature can be placed in the fields for monitoring the soil quality and weather conditions. With the help of IoT, the cost of farming could be reduced significantly.

### 1.6.2 Education Sector

Recently, innovation in the customary school-centred study hall learning is gradually changing the format of learning methodologies. Correspondingly, deep-rooted learning, internet learning, portable learning, and circulated learning hooked on divinatory or sensible issues are popping up as increasingly normal modes of learning. Electronic Learning Contract (ELC) can represent a continuously working understanding among scholars and teachers to survey learner's comprehension outcomes, where in the chain of instructive squares, graduates under study have confirmation; however, received data has an established file size limit of graduates necessary during the training cycle. The course-learning results accomplishment system esteems that depending on the quantitative and subjective mixture of evaluations, cycle, and proof, the course name, learning result name (graduation prerequisite marker), and therefore the weight of the course are all recorded within the square. Blockchain would, henceforth, change the record-keeping of degrees, endorsements, and certificates, making qualifications up-to-date with technology and into the hands of the students, making work simpler and with fewer causes for unpredictability. While integration of IoT in the education sector brings many advantages to students and schools such as personalised learning, easy access to lesson plans, and other academic information, IoT on the other hand could also be used for improving the security of campus, student identification, and smart learning a reality for differently abled students.

### 1.6.3 Smart Homes and Smart Cities

With Smart Grid applications taking advantage of IoT to enhance the energy consumption of homes and buildings and to help power suppliers to regulate all the resources, the framework of a green city depends on the blend of the Internet of Things and distributed computing innovation supported by the smart city. IoT may be a key empowering influence of smart urban areas, and IoT detecting gadgets in urban areas are a big part of associating correspondence and systems administration gadgets. With the arrival of IoT improving green urban areas, everything is going to be composed legitimately and productively, and easily addressing the problems of road traffic of the region, environment pollution, land squandered, and water supply. Smart home automation, remote control and monitoring, better security, energy cost savings, and convenience are some of the key benefits of IoT for smart homes, as are smart city benefits from IoT in managing power consumption, environmental assessment, public transport, traffic, and city surveillance.

### 1.6.4 Healthcare Sector

In the healthcare domain, wearable IoT applications help in the real-time collection of data between various multiple devices and also in the accurate monitoring of the patients, and with the presentation of Blockchain, the medical services segment of the sector has seen a manifold improvement. For instance, while the patient's well-being records are often effortlessly put away and arranged easily, in clinical therapy measures, patients of chronic or acute ailments have pre-defined fields of utilization where Blockchain innovation can be very advantageous. Blockchain offers different application prospects within the field of maintaining and keeping records of patients, and wherein the patients as well as the doctors can record the data permanently like: pulse, circulatory strain, prescriptions taken, rest designs, and diet patterns. These details are often recovered from well-being applications, wearables, or doctor's visit and can be safely put away within the health bank Blockchain. Given these opportunities to ease patient lives, if he/she wants to consult other doctors, the integration of the two technologies with its fallout in medical services may assist the doctor who must contemplate the case with ease to do so, and to analyse better. Blockchain may remove the different issues ailing medical assistance services from before.

### 1.6.5 Hydrocarbon Industry

The hydrocarbon (Oil and Gas) sector is a fundamental need for the planet. As per the "BP Statistical Review, 2018" of World Energy, oil and petroleum gas represents 57% of the energy consumption of the world. As the oil and gas sector is moving toward information technology use and digitization, many huge oil and gas organizations are seeing long-term benefits with the Blockchain innovations put to use in the sector. The rationale behind this lies in the fact that Blockchain alongside IoT may be an efficient, straightforward, solid, and dependable source. IoT devices are used in the Oil and Gas industry for real-time monitoring, controlling physical assets, gathering data from sensors, fleet monitoring, floor management, and real time alerts. Remote monitoring of seismic activities, oil and gas tankers, oil exploration, and proactive maintenance could also benefit from integration of IoT with Blockchain helping in ensuring safety of personnel involved in these operations. By taking advantage of IoT, the Oil and Gas industry could optimize their operational efficiency and lower the overall cost of operations. As per a worldwide Blockchain study, 61% of respondents within this industry accept that "Blockchain is a money-related information base and a monetary help application and just 15% of respondents applied Blockchain to rehearse. There is a batch to be created, all things considered" (Pawczuk et al., 2018).

### 1.6.6   E-Business

E-business models use Blockchain-based IoT methods to enhance the income of the business, mainly to sell or buy a private possession or property like vehicle, bicycle, house, and electronic gadgets using Decentralized Autonomous Corporations (DAC). DAC gadgets using its innovation have been made to eliminate the human presence, and as a result, the problem of submitting to human tricks or commission is taken care of by the Blockchain. The gadgets utilizing the DAC innovation must be interconnected with the Blockchain and IoT helping in countering any security and protection issues afterward because Blockchain gives transparency in activities and acts as a direct link between the merchant and purchaser. Inventory management, supply chain management, customer experience, maintenance schedule, and product warranty are the areas where IoT could help in improving the e-business.

### 1.6.7   Data Collection and Tracking

IoT gadgets are related to freight trucks, delivery trucks, and warehouses helping the user to find and track the item's current area and other properties. Similarly, with the help sensors, one can take a record of the moisture, temperature, and different variables of item attributes as IoT gadgets connect individuals on the floor measuring information.

### 1.6.8   Finance

Due to the international money crisis in 2008, more and more attention is directed to doing away with the privacy and security issues in Blockchain, as the higher is the occurrence of IT helping clients demand for cash management using smartphones or mobiles the more is the use of Blockchain, which being an open supply technology, helping individuals to freely create monetary applications. In the current scenario, online banking and online transactions have become the need of the hour, and Blockchain technology can help to boost the speed and potency of execution, optimizing the time for transactions to be completed. However, Blockchain needs to be upgraded and improved in order to minimize security and privacy problems to a larger extent. In finance, self-checkout service, smart interaction, interaction between fintech devices, customized client service could benefit from IoT.

### 1.6.9 Tourism and Hospitality

In the time of Industry 4.0, the e-travel industry utilizes the main part of computerized instalments through applications upheld by heterogeneous instalment doors. Blockchain addresses the tourist's necessities as it offers trust, transparency, security, and attracts attention by including exchanges in a long-lasting record which cannot be altered. All exchanges made through digital forms of money are certain and detectable in the chain, which prompts secure checkpoints. Thus, tourist security and recovery from loss in case of unpleasant experiences are allowed from where they are maintained. To attract business, many travel agencies provide incentives and exciting offers to their customers. Baggage and Asset Tracking is another advantage of IoT and Blockchain, if used in the tourism sector. Overbooking which is a major issue in the tourism sector during the peak season, is resolved through IoT technology. Presently, the areas using IoT demand a system where the data and information can be protected from hackers or the other third parties. There are more possibilities that the hackers will initiate the attacks on information used by the applications. Thus, IoT security is of prime importance and a predominant demand for IoT developers.

## 1.7 PRIVACY AND SECURITY RELATED ISSUES

Security and privacy issues, the biggest disadvantage of using Blockchains and IoT together, are common for users. With continued efforts and research, these issues have been resolved to an extent.

### 1.7.1 Threats

This is most-common in Blockchain and IoT, and there are different types of threats in privacy and data leak issues (Dorri et al., 2017).

  i. Accessibility threats: a user finds difficulties under this threat in accessing the Blockchain or a single block. It is hence named as an accessibility threat.
  ii. Authentication threats: authentication and access management threats that are most common in a Blockchain allows the hacker to know the username, password, or access credentials of the user.
  iii. Confidentiality threats: here, the attacker gets the confidential information about a user with the help of various methods, with the result that the user's privacy is invaded and even the most sensitive information about the user becomes public.

iv. Integrity threats: the trust and reliability over a network or system is threatened under this threat and it becomes important to maintain the integrity of data as once the data is lost, it cannot be recovered.

### 1.7.2 Attacks

The different kinds of attacks commonly found are:

i. Denial of Service Attack (DOS): In this attack, the offender sends an oversized range of transactions to the target to choke its functioning and therefore interrupt its convenience to users.
ii. Modification Attack: During this attack, the attacker could ask to vary or delete stored information for a selected user.
iii. Dropping Attack: In order to begin this attack, the attacker ought to have management over a CH(Cluster Head) and then may drop all the received blocks and transactions.
iv. Appending Attack: To launch this attack, the opponent should manage multiple CH(Cluster Heads) that employ the hand and glove.

### 1.7.3 Private Key Security

While utilizing Blockchain, the character security qualification is the private key of a client, which is created and retained by the client instead of outside organizations or any third parties. Once the user's personal secret information become public or are leaked, it's nearly impossible to revert. If the personal information secrets fall in the hands of the criminals, the user's Blockchain account can face the chance of being tampered by others. The matter of personal key management isn't solved in the Blockchain, and existing Blockchain applications typically use personal keys to prove the identity of the user and complete a payment group action.

Since the user is anonymous (not known), it's very difficult to trace user behaviours, in addition to the need to maintain legal regulators. Unfortunately, the privacy protection measures in Blockchain aren't terribly well managed and built. Blockchain calculates the hash price of a user's public key to spot a singular user.

### 1.7.4 Fifty One% Vulnerability Attacks

During this kind of attack, the single-user or a group of users make the selected part of the Blockchain personal. The other members in the Blockchain can't have access to it as it's currently a personal Blockchain. A "51% attack" happens once the single or the mining group takes majority management of an indication of work-based Blockchain and double-spends a number of its coins (Saad et al., 2019).

The fifty-one% attacks are typically not in or include a long-lasting capability. Often this is caused by dominance of over fifty-one of the management or majority management of the only single-user or the group requiring tons of limited power and energy resources. It is terribly rare that one single labourer or the mining cluster takes the bulk of the management during a Blockchain.

### 1.7.5 Updation Issues

Fork downside is expounded to suburbanized node version agreement once the software system upgrades. There are two kinds of forks issues, generally: hard fork and soft fork. Hard fork problems arise once the system involves a brand new version or is updated so that the older nodes are not ready to alter with the newer nodes. This results in the forming of multiple Blockchains. A soft fork is simply the opposite of the hard fork, that is, during this issue, the new nodes aren't able to alter with the older ones.

## 1.8 LIMITATIONS

The above discussed privacy and security issues are still a challenge for Blockchain limiting the scope of the applications and restricts the extensive application of Blockchain to other needs. The capability of a block was set to one MB originally to resist attainable DDoS (Distributed Denial-of-Service) attacks. With the appearance of 5G technologies, IoT devices ought to be upgraded for compatibility with high-speed networks.

## 1.9 IMPLICATIONS

Blockchain and IoT have vast potential to increase the efficiency and productivity of the concerned IT systems, where the two techniques when applied together have positive implications in Healthcare sector, Finance, Education, Data Science, and Agriculture sectors. The integration of these two techniques may establish crypto currency at par with current fiduciary money by boosting the speed and potency of execution and optimizing the time for transactions to be completed. Patients' records will be more private and secure with quick accessibility. It will create smart cities, smart education system, and increase the proficiency and productivity of the agriculture sector and the Oil and Gas industry, and will enhance profits in E-business and tourism and hospitality sectors.

## 1.10 CONCLUSION AND FUTURE SCOPE

Blockchain and IoT are still an emerging technology. This chapter explains the concepts of Blockchain, IoT and their advantages and disadvantages, with the gradual development of the research in these two fields being addressed in detail, and presents the outlook on the need of integration of Blockchain and IoT in various sectors like Tourism and Hospitality, Oil and Gas Industry, Healthcare sector, Finance, Education, Data Science, and Agriculture Sectors. The amalgamation of these two technologies is aimed to provide more secure, reliable, transparent, and fast efficient systems at comparatively less cost in comparison to traditional systems in existence.

Blockchain's salient features like transparency and decentralisation have made it more important in managing networks like 6G. But issues like security updates, fork problems, and 51% vulnerability attacks are still a challenge for Blockchain.

For transformation from Internet of Things to Internet of Everything, we need to explore more on the above discussed technologies, and so, there is an open scope for the researchers to resolve the challenges as given above made open for implementation of Blockchain technology in a wider context and offering tamper-proof, more reliable, cost efficient, and secure systems.

## REFERENCES

Alaslani, M., Nawab, F., & Shihada, B. (2019). Blockchain in IoT systems: End-to-end delay evaluation. *IEEE Internet of Things Journal*, 6(5), 8332–8344.

Atlam, H. F., & Wills, G. B. (2019). Technical aspects of blockchain and IoT. *Advances in Computers*, 15, 1–39, Elsevier

Banerjee, A. (2019). Blockchain with IOT: Applications and use cases for a new paradigm of supply chain driving efficiency and cost. *Advances in Computers*, 115, 259–292, Elsevier

Biswas, K., & Muthukkumarasamy, V. (2016, December). *Securing smart cities using blockchain technology* [Conference session]. 2016 IEEE 18th international conference on high performance computing and communications; IEEE 14th international conference on smart city; IEEE 2nd international conference on data science and systems (HPCC/SmartCity/DSS), IEEE, pp. 1392–1393.

Cho, S., & Lee, S. (2019, January). *Survey on the Application of BlockChain to IoT* [Conference session]. Electronics, Information, and Communication (ICEIC), International Conference, IEEE, pp. 1–2.

Da Xu, L., & Viriyasitavat, W. (2019). Application of blockchain in collaborative Internet-of-Things services. *IEEE Transactions on Computational Social Systems*, 6(6), 1295–1305.

Dai, F., Shi, Y., Meng, N., Wei, L., & Ye, Z. (2017, November). *From Bitcoin to cybersecurity: A comparative study of blockchain application and security issues* [Conference session]. Systems and Informatics (ICSAI), 4th International Conference, IEEE, pp. 975–979.

Dai, H. N., Zheng, Z., & Zhang, Y. (2019). Blockchain for Internet of Things: A survey. *IEEE Internet of Things Journal*, 6(5), 8076–8094.

Dorri, A., Kanhere, S. S., & Jurdak, R. (2017, April). *Towards an optimized blockchain for IoT* [Conference session]. Internet-of-Things Design and Implementation (IoTDI), IEEE/ACM Second International Conference, IEEE, pp. 173–178.

Du, M., Wang, K., Liu, Y., Qian, K., Sun, Y., Xu, W., & Guo, S. (2020). Spacechain: A three-dimensional blockchain architecture for IoT security. *IEEE Wireless Communications*, 27(3), 38–45.

Gao, Y., Wu, W., Nan, H., Sun, Y., & Si, P. (2020, June). *Deep reinforcement learning based task scheduling in mobile blockchain for IoT applications* [Conference session]. Communications (ICC), ICC 2020-2020 IEEE International Conference, IEEE, pp. 1–7.

Haller, S., Karnouskos, S., & Schroth, C. (2008, September). *The internet of things in an enterprise context* [Symposium]. Future Internet, Springer, Berlin, Heidelberg, pp. 14–28.

Hamdaoui, B., Alkalbani, M., Rayes, A., & Zorba, N. (2020). IoTShare: A blockchain-enabled IoT resource sharing on-demand protocol for smart city situation-awareness applications. *IEEE Internet of Things Journal*, 7(10), 10548–10561.

Hassan, M. U., Rehmani, M. H., & Chen, J. (2019). Privacy preservation in blockchain based IoT systems: Integration issues, prospects, challenges, and future research directions. *Future Generation Computer Systems*, 97, 512–529.

Huckle, S., Bhattacharya, R., White, M., & Beloff, N. (2016). Internet of things, blockchain and shared economy applications. *Procedia Computer Science, 98*, 461–466.

Huynh, T. T., Nguyen, T. D., & Tan, H. (2019, July). *A survey on security and privacy issues of blockchain technology* [Conference session]. System Science and Engineering (ICSSE), 2019 International Conference, IEEE, pp. 362–367.

Jonathan, K., & Sari, A. K. (2019, December). *Security issues and vulnerabilities on a blockchain system: A review* [Seminar]. Research of Information Technology and Intelligent Systems (ISRITI), International Seminar, IEEE, pp. 228–232.

Karthikeyyan, P., & Velliangiri, S. (2019, July). *Review of blockchain based IoT application and its security issues* [Conference session]. Intelligent Computing, Instrumentation and Control Technologies, 2nd International Conference (ICICICT), IEEE, vol. 1, pp. 6–11.

Khan, M. A., & Salah, K. (2018). IoT security: Review, blockchain solutions, and open challenges. *Future Generation Computer Systems, 82*, 395–411.

Li, X., Jiang, P., Chen, T., Luo, X., & Wen, Q. (2020). A survey on the security of blockchain systems. *Future Generation Computer Systems, 107*, 841–853.

Lin, J., Long, W., Zhang, A., & Chai, Y. (2020). Blockchain and IoT-based architecture design for intellectual property protection. *International Journal of Crowd Science, 4*(3), 283–293.

Liu, H., Zhang, Y., & Yang, T. (2018). Blockchain-enabled security in electric vehicles cloud and edge computing. *IEEE Network, 32*(3), 78–83.

Mettler, M. (2016, September). *Blockchain technology in healthcare: The revolution starts here* [Conference session]. E-health Networking, Applications and Services (Healthcom), 18th International Conference, IEEE, pp. 1–3.

Mhaisen, N., Fetais, N., Erbad, A., Mohamed, A., & Guizani, M. (2020). To chain or not to chain: A reinforcement learning approach for blockchain-enabled IoT monitoring applications. *Future Generation Computer Systems, 111*, 39–51.

Minoli, D., & Occhiogrosso, B. (2018). Blockchain mechanisms for IoT security. *Internet of Things, 1*, 1–13.

Nguyen, Q. K. (2016, November). *Blockchain-a financial technology for future sustainable development* [Conference session]. Green Technology and Sustainable Development (GTSD), 3rd International Conference, IEEE, pp. 51–54.

Novo, O. (2018). Blockchain meets IoT: An architecture for scalable access management in IoT. *IEEE Internet of Things Journal, 5*(2), 1184–1195.

Papadodimas, G., Palaiokrasas, G., Litke, A., & Varvarigou, T. (2018, November). *Implementation of smart contracts for blockchain based IoT applications* [Conference session]. Network of the Future (NOF), 9th International Conference, IEEE, pp. 60–67.

Pavithra, S., Ramya, S., & Prathibha, S. (2019, February). *A survey on cloud security issues and blockchain* [Conference session]. Computing and Communications Technologies (ICCCT), 3rd International Conference, IEEE, pp. 136–140.

Pawczuk, L., Massey, R., & Schatsky, D. (2018). Deloitte's 2018 Global Blockchain Survey: Findings and Insights.

Premkumar, A., & Srimathi, C. (2020, March). *Application of blockchain and IoT towards pharmaceutical industry* [Conference session]. Advanced Computing and Communication Systems (ICACCS), 6th International Conference, IEEE, pp. 729–733.

Puri, V., Priyadarshini, I., Kumar, R., & Kim, L. C. (2020, March). *Blockchain meets IIoT: An architecture for privacy preservation and security in IIoT* [Conference session]. Computer Science, Engineering and Applications (ICCSEA), International Conference, IEEE, pp. 1–7.

Rahulamathavan, Y., Phan, R. C. W., Rajarajan, M., Misra, S., & Kondoz, A. (2017, December). *Privacy-preserving blockchain based IoT ecosystem using attribute-based encryption* [Conference session]. Advanced Networks and Telecommunications Systems (ANTS), IEEE International Conference, IEEE, pp. 1–6.

Ray, P. P., Dash, D., Salah, K., & Kumar, N. (2020). Blockchain for IoT-based healthcare: background, consensus, platforms, and use cases. *IEEE Systems Journal, 15*(1), 85–94.

Saad, M., Spaulding, J., Njilla, L., Kamhoua, C., Shetty, S., Nyang, D., & MohaisenA. (2019). Exploring the attack surface of blockchain: A systematic overview. arXiv preprint arXiv:1904.03487.

Sengupta, J., Ruj, S., & Bit, S. D. (2020). A comprehensive survey on attacks, security issues and blockchain solutions for IoT and IIoT. *Journal of Network and Computer Applications, 149*, 102481.

Shah, R., & Sridaran, R. (2019, March). *A study on security and privacy related issues in blockchain based applications* [Conference session]. Computing for Sustainable Global Development (INDIACom), 6th International Conference, IEEE, pp. 1240–1244.

Singh, S. K., Rathore, S., & Park, J. H. (2020). Blockiotintelligence: A blockchain-enabled intelligent IoT architecture with artificial intelligence. *Future Generation Computer Systems, 110*, 721–743.

Tariq, N., Qamar, A., Asim, M., & Khan, F. A. (2020). Blockchain and smart healthcare security: A survey. *Procedia Computer Science, 175*, 615–620.

Tasatanattakool, P., & Techapanupreeda, C. (2018, January). *Blockchain: Challenges and applications* [Conference session]. Information Networking (ICOIN), International Conference, IEEE, pp. 473–475.

The Global Competitiveness Report (2015). https://www.weforum.org/reports/global-competitiveness-report-2015

Tselios, C., Politis, I., & Kotsopoulos, S. (2017, November). *Enhancing SDN security for IoT-related deployments through blockchain* [Conference session]. Network Function Virtualization and Software Defined Networks (NFV-SDN), IEEE, pp. 303–308.

Vujičić, D., Jagodić, D., & Ranđić, S. (2018, March). *Blockchain technology, bitcoin, and Ethereum: A brief overview* [Symposium]. Infoteh-jahorina (infoteh), 17th International Symposium, IEEE, pp. 1–6.

Wang, Q., Zhu, X., Ni, Y., Gu, L., & Zhu, H. (2020). Blockchain for the IoT and industrial IoT: A review. *Internet of Things, 10*, 100081.

Wu, J., Dong, M., Ota, K., Li, J., & Yang, W. (2020). Application-aware consensus management for software-defined intelligent blockchain in IoT. *IEEE Network, 34*(1), 69–75.

Xu, H., Klaine, P. V., Onireti, O., Cao, B., Imran, M., & Zhang, L. (2020). Blockchain-enabled resource management and sharing for 6G communications. *Digital Communications and Networks, 6*(3), 261–269.

Yu, Y., Li, Y., Tian, J., & Liu, J. (2018). Blockchain-based solutions to security and privacy issues in the Internet of Things. *IEEE Wireless Communications, 25*(6), 12–18.

Zaghloul, E., Li, T., Mutka, M. W., & Ren, J. (2020). Bitcoin and blockchain: Security and privacy. *IEEE Internet of Things Journal, 7*(10), 10288–10313.

Zheng, Z., Xie, S., Dai, H., Chen, X., & Wang, H. (2017, June). *An overview of blockchain technology: Architecture, consensus, and future trends* [Congress]. Big Data (BigData Congress), IEEE International Congress, pp. 557–564.

# 2 Software-Defined Networking: Evolution, Open Issues, and Challenges

*Sukhvinder Singh and Preeti Gupta*

## CONTENTS

2.1 Introduction ..................................................................................................31
2.2 Software-Defined Network Architecture ......................................................32
2.3 Software-Defined Networks: Bottom-Up Scenario ......................................33
    2.3.1 Layer I: Network Infrastructure ........................................................35
    2.3.2 Layer II: Southbound Interfaces ........................................................35
    2.3.3 Layer III: Network Hypervisors ........................................................35
    2.3.4 Layer IV: Controller (NOS) ...............................................................37
    2.3.5 Layer V: Northbound Interfaces ........................................................37
    2.3.6 Layer VI: Language-Based Virtualization .........................................37
    2.3.7 Layer VII: Programming Language ...................................................38
    2.3.8 Layer VIII: Network Application .......................................................39
2.4 SDN and Cloud Integration ..........................................................................41
2.5 Present Implementation and Challenges ......................................................41
2.6 SDN-Based Cloud Network ..........................................................................42
2.7 The Current State of SDN Implementation ..................................................42
2.8 Security Challenges in SDN .........................................................................42
    2.8.1 Switch-Level Security Challenges .....................................................43
    2.8.2 Controller-Level Security Challenges ................................................43
    2.8.3 Channel-Level Security Challenges ...................................................43
2.9 SDN and Society 5.0 – Futuristic Approach .................................................45
2.10 Conclusion and Future Scope .......................................................................45
References .............................................................................................................46

## 2.1 INTRODUCTION

Computer networking has gone through many phases of development from resource-sharing Arpanet in the early 1970s to the present era of automation-based networking. What distinguishes SDN from traditional networks is the latter's communication protocols and the interconnection between the resources based on wired and wireless technologies are more complex and harder to manage; while the former's transport protocols and distributed controls inside the routers and switches implement flow of information from source to destination around the world, with standardization requirements using the specifications of network policies being met by different vendors who have different APIs and vendor-specific commands implemented in the SDN. Given that, despite a wide range of technological adaption, the traditional networks are complex to manage (Benson et al., 2009), with network environments being prone to faults, load changes, and security breach issues, the fact that these networks compel network engineers and administrators to usually have to reconfigure and alternatively manage the specific vendor-oriented device, explicitly, is a drawback. Automatic mechanisms are virtually not found in legacy networks, so that complexity is added for innovation to take place and improvement to catch on vertically integrated, proprietary network devices having a close nature.

An SDN is an evolving network continuously trying to overcome the shortcomings of the traditional networks by suggesting ways to separate the control plane (carrying signaling traffic and routing) from the data plane (carrying user traffic). The term SDN was first coined at Stanford University, Stanford, CA, USA (Greene, 2009) and represents the idea and work around OpenFlow (McKeown et al., 2008). As the vertical integration of the network is intercepted by separating control logic from the routers and switches making them simple forwarding devices, there is simultaneously the control logic implementing a centralized system called centralized control that can concentrate on the configuration of different network devices without bothering about which vendors they belong to. A single centralized control takes over networks functions of the network (re)configuration, evaluation, programming logic, and policy enforcement. Many projects are contributing and adapting different ways to make the network more scalable and flexible, and networks are becoming more responsive to the organizations and their customer's needs due to SDN implementation.

## 2.2 SOFTWARE-DEFINED NETWORK ARCHITECTURE

A highly programmable switch infrastructure that separates the control plane from the data plane while providing a unified well-defined application programming interface (API) to communicate with the switches, and to control the communication flow, is what SDN is all about. Enabling the network to control the communication by the software rather than hardware, the Control layer is a key component of this infrastructure that is responsible for managing the flow of information and control functions or multiple network applications, such that the central controller coordinates the configuration of such a framework and the flow rules by submitting the configuration request and conveying the flow rule. The controller on the control layer, in such a paradigm, manages southbound API toward the infrastructure layer and northbound API towards business application, which are discussed in the upcoming section in detail. SDN poses, both, an opportunity and a challenge to the network security community to understand and analyze the flow of data. A conceptual approach can be seen in Figure 2.1.

# Evolving Software-Defined Networking

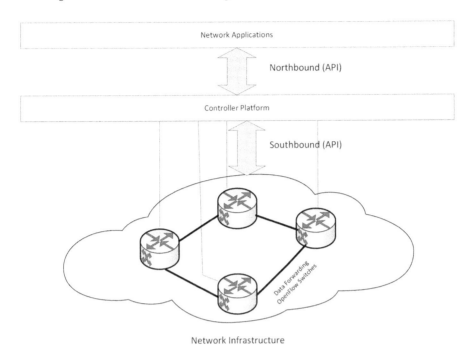

**FIGURE 2.1** SDN Architecture (Simplified View).

One of the useful and notable examples of SDN API is the OpenFlow protocol, which contains the flow tables defining rules that match the traffic, and accomplishes actions such as dropping, forwarding, alternation, etc. OpenFlow performs like a switch, firewall, or a router by taking instructions from the controller, while letting a communication protocol in SDN environments to connect directly to network devices in forwarding plane, and separates hardware and firmware. As a separate software controller, OpenFlow (OF) is the well-known first software-defined networking (SDN) standards that can interact with switches and routers (physical and virtual hypervisor-based), adaptable to changing business requirements.

## 2.3 SOFTWARE-DEFINED NETWORKS: BOTTOM-UP SCENARIO

The composition of different layers of SDN architecture is depicted in Figure 2.2. Each layer defining the specific functionality such as the southbound (connecting forwarding devices) API, northbound (connecting Network applications) API, Network Operating System (NOS)s, and network applications is always present in an SDN deployment, while other specified functions may be present only in a particular deployment, such as language based virtualization or hypervisor. The following section will describe each layer and the properties of the trifold perspective of SDN, while presenting a plane-oriented system design architecture, respectively. This section will introduce each layer of the SDN.

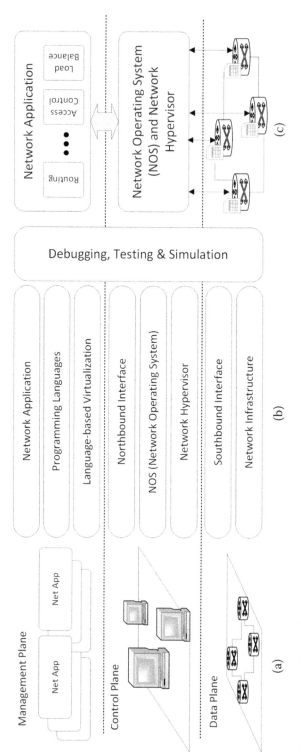

**FIGURE 2.2** SDN (a) Planes, (b) Layers (c) Architecture (System Design).

### 2.3.1 Layer I: Network Infrastructure

The infrastructure of SDN is similar to the traditional network consisting of switches, routers, and middlebox appliances except that they act as forwarding devices without embedded control or proprietary software to make decisions. A logically centralized control system takes over the network intelligence as depicted in Figure 2.2(c) using NOS and applications. These applications on the top (conceptually) ensure the configuration, compatibility over communication, and interoperability among heterogenous forwarding devices, difficult in legacy network with closed and proprietary interfaces.

OpenFlow is the extensive design of SDN data plane central controller; however, Protocol-Oblivious Forwarding (POF) (Song, 2013) is also being pushed parallelly and leveraged to further enhance the network programmability.

In the OpenFlow device, the flow table defines a path through a sequence to handle a packet. The lookup process starts when the packet arrives and matches in the tables of the pipeline. A flow rule is defined with combined different matching fields, as illustrated in Figure 2.3, to handle the packets; if no rule is defined, the packet is discarded. However, a default rule is a setup to send the packet to the controller. Flow rules follow a natural sequence number in the flow table according to the priority.

### 2.3.2 Layer II: Southbound Interfaces

Southbound interfaces (APIs) offer the bridge between the controller and the forwarding device. These APIs separate the control plane and data plane, however, leaving them connected physically or virtually to the forwarding elements. OpenFlow (McKeown et al., 2008) is a widely accepted southbound standard in SDN and it enables an OpenFlow enabled device to communicate with the controller using a common specification with the data plane. Providing event-based messaging on the port or link, OpenFlow changes the forwarding device-generated statistic collection by the controller, and the packet-in message to the forwarding device when they don't know where to forward the new incoming flow. There are other APIs than OpenFlow which are initiated as the southbound interface such as ForCES (Doria et al., 2010), Open vSwitch Database (Pfaff & Davie, 2013), OpFlex (Smith et al., 2014), and OpenState (Bianchi et al., 2014).

### 2.3.3 Layer III: Network Hypervisors

Network Hypervisors set up a virtual network environment to decouple from the underlying physical network and is the common overarching technology in current computer systems. Allowing the sharing of physical network resources using distinct virtual machines, in cloud technology, for example, IaaS (Infrastructure as a Service) lets each user to have his own space to access resources such as computing and storage allocated by the virtual environment wherein the provided physical infrastructure could be shared as resources and allocated on demand. One of the interesting facts about such infrastructures is migration, where the virtual machine

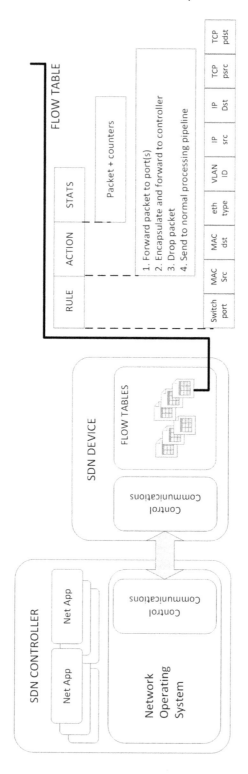

**FIGURE 2.3** Open Flow-Enabled SDN Devices.

can be migrated to another physical server. Even with the advancement in virtualization technologies, still, the network is configured statically at its locations.

SDN and new tunneling techniques such as VXLAN (Zhao et al., 2017) and NVGRE (Sridharan, 2011) could be the glimmer of hope to change the situation by providing on-demand setup of a virtual network.

### 2.3.4 LAYER IV: CONTROLLER (NOS)

Networking specifications and operations have been managed and configured by the closed set of device-specific instructions NOSs (e.g. JunOS and Cisco IOS), but with common functionality and abstracting device-specific characteristics being still absent in the network. SDN promises to enable the management and solving of the problem of the network through the central controller provided by the NOS (Gude et al., 2008).

A NOS (or a controller) is a critical element of SDN architecture supporting applications (control logic) for producing a network set up by configuring the network devices according to the policies proposed by the network operator.

### 2.3.5 LAYER V: NORTHBOUND INTERFACES

Northbound interfaces allow network applications to connect to the underlying network device through NOS, and is still an open issue. However, the southbound interface (OpenFlow) is a widely accepted proposal. Being software implemented systems, not hardware as in the case of southbound APIs, the implementations of Northbound APIs are commonly the widely adapted lead driver, while standards emerge later (Guis, 2012). The future role of northbound interfaces and the issue on how to implement these in SDN are in discussion (Dix, 2013), and these APIs encourage application portability and interoperability as keys to the northbound APIs in control platforms, wherein the POSIX standard (Gibbons et al., 2015) that guarantees the controller and programming language independence can be compared to a northbound interface; while, NOSIX (Yu et al., 2014) which is not a general-purpose API acts as a driver for southbound API, and defines the portable flow model application. It may be a part of the common abstraction layer in the control platform.

Existing controllers such as OpenDaylight (Eftimie & Borcoci, 2020), NOX (Gude et al., 2008), RYU (Asadollahi et al., 2018) define northbound APIs to connect to the network applications and has its own detailed application and specification. Moreover, programming languages provide powerful higher-level abstractions and mechanisms to ease software application development with control function details and data behavior, for example, transparent data plane fault tolerance.

### 2.3.6 LAYER VI: LANGUAGE-BASED VIRTUALIZATION

The capability of expressing modularity and carrying out different levels of abstraction is the essential characteristic of virtualization. For example, virtualization offers a different view of the physical network, wherein a module can be seen as a single switch but while having several different underlying forwarding devices, that simplifies the job of the application developer to control the forwarding rules as the need to reckon about the switch count or their sequences is eliminated. In this way, a big network simply can be set up, developed, and deployed. For instance, Pyretic offers a higher-level abstraction of network topology, and it is the concept implementing a network of objects to incorporate the concept of abstraction, wherein these objects specifically define the network topology and policies but hide the information and enable the required services. Static slicing is another kind of language-based virtualization, where the network slicing is done based on the network's application layer definition by the compiler, but given the compiler's output is a specific monolithic that defined the configuration commands and the slicing information of the network, the hypervisor is not required to dynamically configure the network. For specific requirements, static slicing of deployment is achieved. An example of static slicing is splendid isolation (Gutz et al., 2012).

### 2.3.7 Layer VII: Programming Language

Network programmability in SDN starts in the low level machine language such as OpenFlow and continues to higher-level abstraction to programming the controller using higher-level programming language (Koponen et al., 2014). Assembly-like machine language such as POX (Bholebawa & Dalal, 2018) and OpenFlow's network modules are devoted to low-level details rather than problem-solving ones. These types of languages mimic the forwarding devices while higher-level language creates a higher-level abstraction to ease the task of programming forwarding devices and, also, enables problem focus and a more productive environment to speed up the development and innovation while, also, addressing the challenges of lower-level abstraction. It also promotes reusable modularization in the control plane and fosters the development of network virtualization.

To achieve a diverse goal and overcome the issue of lower-level instruction, higher-level programming language have been anticipated, such as:

- Providing higher levels of abstraction to maintain easy-to-understand network policies and management tasks.
- Preventing device-specific configurations and dependencies among the network.
- Task decoupling (e.g. routing, traffic engineering, access control, security).
- Solving forwarding rules, to avoid conflicting or incomplete rules to prevent triggering switch events in an automated way.
- Addressing race conditions.
- Enhancement in the techniques to resolve conflict.
- Reduction in latency in the new flow processing.

The programming language also supports other specialized abstractions such as monitoring (Muhizi et al., 2017). Several programming languages have been proposed for SDN, even as FML (Field Manipulation Language) (Hinrichs et al.,

2009), Nettle (Voellmy et al., 2011), and Procera (Voellmy et al., 2012) being reactive and functional, will cause policies application written in these languages to be event-triggered based on reactive action (for example connecting or loading new hosts to the network). Another example of this type of programming language is frenetic (Foster et al., 2011), based on purely function abstraction sets and which solved many of NOX/OpenFlow-based problems, is known to work on packet-level abstraction policies by dealing with rules overlapping and offering sequential modules structure.

### 2.3.8 Layer VIII: Network Application

To command the forwarding devices, control logic is implemented in the data plane by the network applications, which are the brain of the network controlling the flow in underlying network devices. For example, to define the path from point X to point Y to move a packet, an application-based topology decides the path to use and trains the controller to mount flow rule in all forwarding devices from X to Y.

SDNs may be deployed in any conventional network environment such as home networks, enterprise networks, and data centers. Given the variation of networks has a broad range of network applications, that are still developing, the functionality, such as load balancing, routing, policy setups, and security enforcements is being performed by the present network application while exploring more innovative methods to make it more efficient and agile such as machine learning, reducing power consumption, and also Quality of Services (QoS) enforcement and reliable functionalities of the data plane.

Applications of SDN can be divided into these five categories:

- Measurement and monitoring.
- Agility and wireless.
- Traffic engineering.
- Datacenter networking.
- Security and dependability.

In order to manage network, the measurement and monitoring behavior and the status of the network components assume importance, but with the advent of SDN, the entire process of measurement and monitoring has become more flexible and accessible. Improvement of OpenFlow based SDN is considered to optimize the process of monitoring by reducing the overburden on control plane by sampling and estimation techniques such as deterministic packet sample techniques, two-staged bloom filters, and traffic monitoring. Bandwidth requirement can also be monitored by SDN application which can further help in matching the latency and bandwidth requirements. Other monitoring frameworks such as Payless help flexible monitoring by SDN without overburdening the control plane by providing different mechanism and capabilities in the real-time scenario.

SDN applications provide an easy deployment opportunity to wireless and cellular networks, because such networks have suboptimal use in the current distributed control plane environment for managing the limited spectrum, allocating radio resources, implementing handover mechanisms, managing interference, and performing efficient load balancing between cells. Realization of features such as simplified administration, seamless subscriber mobility, easy management of heterogeneous network technology, and QoS which are hard-to-implement in the wireless communication network is being made possible by the programmable stack layer of the network. For example, OpenRadio provides a software-implemented layer to abstract the wireless protocol from the hardware. SoftRAN is another example that allows the operator to enhance the algorithm for better handover and transmit power control.

Many applications such as Quality of Service (QoS) for SDN, Plug-n-server, SIM-PLE, FlowQoS, and Middlepipes propose to engineer the traffic to optimize power consumption and for load balancing. They also offer to aggregate network utilization and other generic enhancement techniques. With load balancing being the first concept envisioned by the SDN application, it's SDN again which simplifies different services running inside the network by taking appropriate action whenever a new server is installed by the network while considering both the computing capacity of the server and network load. Other application-purpose routing and application-aware networking for data streaming are put in use by providing packets scheduler.

To support a virtualized application in a complex setting with enhanced agility and cost effective infrastructure, we are witness to changes in the field of Datacenter networking. SDN applications can help data-centric infrastructures to deal with their significant challenges such as computing, networking, and storage, even as small- to large-scale cloud data centers strive to offer high and flexible bandwidth with low latency, a high level of resilience, and an intelligent network. There are still many open issues to be tackled in SDN use in data centers due to legacy networks' complexity and inflexibility, but SDN can change such irksome situations by applying different techniques to solve issues such as optimized network resource management to overcome the problem of idle resources or inadequate resources vis-a-vis demand, migration of data to overcome problems of static networks, failure avoidance to maximize system availability, fast deployment to overcome issues of proprietary softwares like user training, optimized resource utilization to overcome problems of wastage, troubleshooting to overcome problems of software-hardware alignment, administrating to overcome the issue of control and maintenance, minimizing latency to overcome the problem of data input-output delays, and controller operating cost reduction.

Due to rapid digitization, network security breaches are on the rise, with persistent attacks and threats making it crucial to develop an environment capable enough to handle security attacks effectively. Security and dependability techniques such as access control and multi-path are emerging in the framework of SDNs, where the first technique involves enhancing the security of the network using SDNs, and the second, the SDN's security itself.

Due to the huge demand for network resources, it is now mandatory to have efficient network management. Given the quantity of data and the flow of information that are increasing fast in the telecommunication network, the scalability issue and allocating of resources cannot be resolved by the traditional classification methods and network management. SDN has an effective way to handle such issues, that is, by its design, SDN makes it possible to analyze the data and allocate network resources, and to effectively manage it too.

## 2.4 SDN AND CLOUD INTEGRATION

Cloud computing is one of the widely accepted paradigm shifts built around the core idea of computing such as on-demand resources, agility, scaling, and establishment based on usage. The diverse models like Infrastructure as a Service (IaaS), Platform as a Service (PaaS), Network as a Service (NaaS), and Software as a Service (SaaS), are providing services and are still evolving everyday to fill the gaps and concerns voiced by the industries and standards bodies (Mell & Grance, 2011). For example, some of the concerns are: implementation of virtual network inside IaaS using existing technology, the possibility of implementing SDN (Greene, 2009), standardization, different topologies' involvement, and procedures across multiple data centers, and so on.

SDN is one of the appealing platforms to the tenants by its providing a software controller to implement control logic rather than putting it on the physical switches, so that each tenant's control logic can be implemented on the software controller instead of in physical switches. OpenFlow (McKeown et al., 2008) provides an open API to perform packet forwarding based on flow rules tables, statics generation, and notify changes in topologies.

## 2.5 PRESENT IMPLEMENTATION AND CHALLENGES

The existing cloud structure is based on the "one size fits all" model to enable clouds to operate in various environments of the cloud. The forwarding protocols, network topology, and security policies come together as all the model's requirements; however, it avoids the optimization of usage and proper management of clouds. Cloud users should have the capacity to specify the bandwidth requirement of applications in the cloud to ensure the performance is comparable to an 'on premise' implementation. Applications with many tiers have each their own requirements of specification about the bandwidth to fulfill the transaction within the acceptable time frame. However, the security appliances are implemented for the secure communication and protection of the data from the different attacks by the enterprises, alongside other vulnerabilities as load balancing, caching, and application acceleration. The regulatory policy impacting the configuration of each switch and router apart, the Protocols, changing requirements, L2 spamming tree protocol (STP), along with vendor specification commands make it challenging at the cloud network scale to operate. Hypervisor and network appliance are typically implemented based on the static configuration of a physical network that makes location dependency constraints. Datacenter connectivity to make a vision "one cloud" is another challenge (Azodolmolky et al., 2013).

## 2.6 SDN-BASED CLOUD NETWORK

As discussed, SDN is one of the trending infrastructures where forwarding functionality is separated from the network control functionality and directly programmed using the higher-level abstraction programming language. The migration of control into an accessible computing device is logically centralized to be "abstract" for applications and services in network infrastructure.

The significance of a logically centralized controller could be understood from the detailed view of network availability and cloud resources that will allow the cloud appliances to adequately resource the data center. The level of services and bandwidth will be provided by a link. A high-level informative description of an SDN based cloud contains:

- A cloud infrastructure containing OpenFlow-enabled backbone control nodes to connect the data center to the enterprise.
- SDN-based centralized control to define the network policies and flow rules in the cloud.
- WAN network virtualization applications.
- A core node (OpenFlow) to switch the traffic between control nodes.
- A hybrid procedure and arrangement of cloud by using software to manage the cloud provider, enterprises, workflow inside the cloud, and resource management of the compute/storage in data centers.

## 2.7 THE CURRENT STATE OF SDN IMPLEMENTATION

SDN started as academic (McKeown et al., 2008) experiments; however, today it has become more popular in IT industries to manage their networks. Google has deployed an SDN connected to its data center assisting the company to enhance operational efficiency and reducing the cost of the operation at the global level. Many other companies are also adopting and researching more on SDN, like NSX of VMware also works on a virtualization platform, a commercial solution to deliver fully-functional software-based networking independent to underlying network devices, which is based on the SDN principle. Open Network Foundation (ONF) (Nunes et al., 2014), OpenDaylight (Eftimie & Borcoci, 2020), RYU ingenuities (Asadollahi et al., 2018) are the important indications of SDN from an Industrial perspective. Moreover, many vendors are working to provide service delivery automation and intelligent service by incorporating cloud based-SDN implementation to enable on-demand bandwidth procedures as per tenant's needs.

## 2.8 SECURITY CHALLENGES IN SDN

OpenFlow is one of the most known implementations of the SDN; however, decoupling of data plane from control plane and securing it become the major concern (Shaghaghi et al., 2020). Attributes such as programmability and software-centric controlling concept make it more prone towards the security issues and challenges. For example, Denial of Service (DoS) attack due to the centralized controller and flow-table limitations, or the trust between controllers and network devices, and network applications and controllers are making the IT experts think. The configuration anomalies and program-oriented flaws can make the SDN environment vulnerable and produce many security challenges, such as the ones identified for OpenFlow-based SDN within its assets: switch level security challenges, controller level security challenges, and security challenges at channel level.

### 2.8.1 Switch-Level Security Challenges

SDN switch level vulnerability, where the switch is compromised by targeting flow tables access, has to do with switches containing information associated with routing, network management, and access control that can destabilize the network (Li et al., 2016). An attacker who utilizes the switch to gain unauthorized access physically or virtually to compromise the SDN-connected hosts and disrupt the network communication is a real threat.

For example, let an attacker try to insert a malicious flow table to compromise the security mechanism through flooding numerous useless rules; as a result, the normal flow rules are unable to store, and perform regular communication. In such a scenario, arbitrary responses occur to a normal request. Such an example is depicted in Figure 2.4(a). At the SDN switch level, CIA (Confidentiality, Integrity, Availability) aspects may be compromised by malicious input actions such as XSS (Cross-Site Scripting), spoofing, scanning, and so on.

### 2.8.2 Controller-Level Security Challenges

The SDN controller level vulnerability, where attackers utilize the centralized controller to compromise the Confidentiality, Integrity, and Availability, is known to adversely affect the CIA triad of SDN. As SDN decouples the data plane from the control plane to initiate a centralized controller that deals with all incoming traffic, it may itself become the key bottleneck for several attacks such as denial-of-service (DoS) and flooding attacks. For example, the controller obtains the forwarding rules when it receives unknown packets, so that by gaining access to the controller, an unauthorized person can cause the unavailability of response when the requests arrive from the data plane. An example is depicted in Figure 2.4 (b).

Malicious action and intrusions such as hijacking, spoofing, scanning, and so on may apply at the controller level entities.

### 2.8.3 Channel-Level Security Challenges

The channels related to vulnerability, where the attacker utilizes or gains access to channel communication, when affected, will cause communication between the components and the administration to falter. In OpenFlow-based SDN there is no trust mechanism existing between the centralized controller and the data plane to verify secure communication, which becomes the loophole in the security infrastructure for the intruder to compromise the security of by becoming the man-in-middle exploiting the communication. Figure 2.4 (c) depicts the

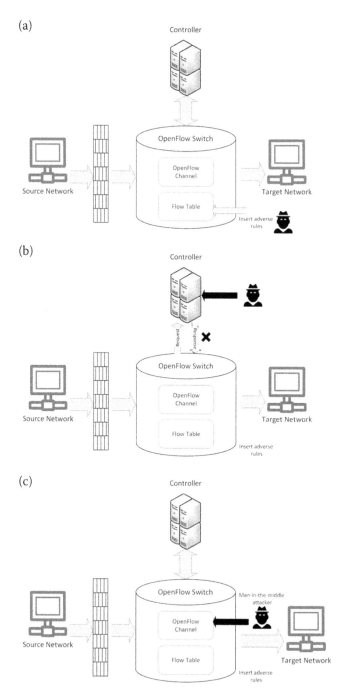

**FIGURE 2.4** (a) Switch Level Security Challenges: Inserting Malicious Rules. (b) Controller Level Security Challenges: DoS Attacks. (c) Channel Level Security Challenges.

man-in-middle-attack to the channel to hijack the messages between network parties who believe they are communicating within the trusted channel (Li et al., 2016). Various attacks may be triggered, for example, the man-in-middle-attack, repudiation, monitoring, and others, at the channel level to disrupt the communication.

## 2.9 SDN AND SOCIETY 5.0 – FUTURISTIC APPROACH

Society 5.0 aims to resolve social challenges by incorporating the fourth industrial revolution (IoT, AI, Big Data, and Robotics). It aims to look towards a system that highly integrates virtual space and real space i.e. cyberspace and physical space. As IoT plays an important role in the definition of Society 5.0, it is fruitful to identify the role of SDN in dealing with challenges in IoT implementation, with one of the biggest challenges before network administrators for dealing with IoT being having the ability to collect data and later conduct analysis for facilitating a positive user experience in real-time. However, with its capabilities, SDN is able to redirect traffic automatically when needed, significantly improving the IoT applications. Through the adaptation of virtual networks, provisioning of storage and computing resources can be done efficiently leading to instant delivery for analyzing data.

## 2.10 CONCLUSION AND FUTURE SCOPE

The chapter systematically reviews the state of the art of SDN from the perspective of evolution, need, architectural framework, and open issues and challenges. The limitations of traditional legacy networks that led to the development of SDN architectural framework are also chalked out. A simplified view of SDN architecture is presented identifying the functionality of layers, data flow, SDN API, protocols, and similar other features. While a bottom-up scenario presents and elaborates the different layers of SDN infrastructure, a layer-wise discussion is carried out on Network Infrastructure, Southbound Interfaces, Network Hypervisors, Controller, Northbound Interfaces, Language-Based Virtualization, Programming Language and Network Application. Towards the latter half of the chapter, the role of SDN in cloud integration, along with the open challenges, has been discussed. The current state of SDN implementation brings into discussion the areas where SDN implementation is effectively achieved. Further, the security challenges raised by separation of the data plane from the control plane in SDN are dissected. Paving a path towards futuristic approaches, SDN's role towards contributing in establishing Society 5.0 has also been identified in the chapter.

Though the SDN framework prevents security breaches due to its ability at providing symmetric and centralized controller, still as identified here, it can be subjected to new vulnerabilities as the central controller can be a single point of failure bringing the whole network down. In future endeavors, the role of techniques like machine learning and deep learning for implementation of flow-based anomaly detection system for securing the controller can be studied.

## REFERENCES

Asadollahi, S., Goswami, B., & Sameer, M. (2018, February). *Ryu controller's scalability experiment on software defined networks* [Conference session]. Current Trends in Advanced Computing (ICCTAC), International Conference, pp. 1–5.

Azodolmolky, S., Wieder, P., & Yahyapour, R. (2013, June). *SDN-based cloud computing networking* [Conference session]. Transparent Optical Networks (ICTON), 15th International Conference, IEEE, pp. 1–4.

Benson, T., Akella, A., & Maltz, D. A. (2009). *Unraveling the complexity of network management* (pp. 335–348). NSDI.

Bholebawa, I. Z., & Dalal, U. D. (2018). Performance analysis of SDN/OpenFlow controllers: POX versus floodlight. *Wireless Personal Communications*, *98*(2), 1679–1699.

Bianchi, G., Bonola, M., Capone, A., & Cascone, C. (2014). OpenState: Programming platform-independent stateful OpenFlow applications inside the switch. *ACM SIGCOMM Computer Communication Review*, *44*, 44–51. 10.1145/2602204.2602211.

Dix, J. (2013). Clarifying the role of software-defined networking northbound APIs. *Network World*. http://www.networkworld.com/news/2013/050213-sherwood-269366.html.

Doria, A., Salim, J. H., Haas, R., Khosravi, H. M., Wang, W., Dong, L., ... & Halpern, J. M. (2010). Forwarding and control element separation (ForCES) protocol specification. *RFC*, *5810*, 1–124.

Eftimie, A., & Borcoci, E. (2020, June). *SDN controller implementation using OpenDayLight: Experiments* [Conference session]. Communications (COMM), 13th International Conference, IEEE, pp. 477–481.

Foster, N., Harrison, R., Freedman, M. J., Monsanto, C., Rexford, J., Story, A., & Walker, D. (2011). Frenetic: A network programming language. *ACM Sigplan Notices*, *46*(9), 279–291.

Gibbons, J. W., & Agah, A. (2015). Modelling content lifespan in online social networks using data mining. *International Journal of Web Based Communities*, *11*(3–4), 234–263.

Greene, K. (2009). TR10: Software-defined networking. Technology Review (MIT).

Gude, N., Koponen, T., Pettit, J., Pfaff, B., Casado, M., McKeown, N., & Shenker, S. (2008). NOX: Towards an operating system for networks. *ACM SIGCOMM Computer Communication Review*, *38*(3), 105–110.

Guis, I. (2012). The SDN gold rush to the northbound API. https://www.sdxcentral.com/articles/contributed/the-sdn-gold-rush-to-the-northbound-api/2012/11/

Gutz, S., Story, A., Schlesinger, C., & Foster, N. (2012, August). *Splendid isolation: A slice abstraction for software-defined networks* [Workshop]. Hot Topics in Software Defined Networks, First Workshop, pp. 79–84.

Hinrichs, T. L., Gude, N. S., Casado, M., Mitchell, J. C., & Shenker, S. (2009, August). *Practical declarative network management* [Workshop]. Research on Enterprise Networking, 1st ACM Workshop, pp. 1–10.

Koponen, T., Amidon, K., Balland, P., Casado, M., Chanda, A., Fulton, B., & Zhang, R. (2014). Network virtualization in multi-tenant datacenters [Symposium]. Networked Systems Design and Implementation ({NSDI} 14), 11th {USENIX} Symposium, pp. 203–216.

Li, W., Meng, W., & Kwok, L. F. (2016). A survey on openFlow-based software defined Networks: Security challenges and countermeasures. *Journal of Network and Computer Applications*, *68*, 126–139.

McKeown, N., Anderson, T., Balakrishnan, H., Parulkar, G., Peterson, L., Rexford, J., & Turner, J. (2008). OpenFlow: Enabling innovation in campus networks. *ACM SIGCOMM Computer Communication Review*, *38*(2), 69–74.

Mell, P., & Grance, T. (2011). The NIST definition of cloud computing.
Muhizi, S., Shamshin, G., Muthanna, A., Kirichek, R., Vladyko, A., & Koucheryavy, A. (2017). *Analysis and performance evaluation of SDN queue model* [Conference session]. Wired/Wireless Internet Communication, International Conference, Springer, Cham, pp. 26–37.
Nunes, B. A. A., Mendonca, M., Nguyen, X. N., Obraczka, K., & Turletti, T. (2014). A survey of software-defined networking: Past, present, and future of programmable networks. *IEEE Communications Surveys & Tutorials, 16*(3), 1617–1634.
Pfaff, B., & Davie, B. (2013). The open vswitch database management protocol. *Internet Requests for Comments, RFC Editor, RFC*, 7047.
Shaghaghi, A., Kaafar, M. A., Buyya, R., & Jha, S. (2020). Software-defined network (SDN) data plane security: Issues, solutions, and future directions. *Handbook of Computer Networks and Cyber Security*, pp. 341–387.
Smith, M., Dvorkin, M., Laribi, Y., Pandey, V., Garg, P., & Weidenbache, N. (2014). OpFlex control protocol, internet draft, internet engineering task force.
Song, H. (2013). *Protocol-oblivious forwarding: Unleash the power of SDN through a future-proof forwarding plane* [Workshop]. Hot Topics in Software Defined Networking, Second ACM SIGCOMM Workshop, pp. 127–132.
Sridharan, M. (2011). NVGRE: Network virtualization using generic routing encapsulation. draft-sridharan-virtualization-nvgre-00. txt.
Voellmy, A., & Hudak, P. (2011). Nettle: Taking the sting out of programming network routers [Symposium]. Practical Aspects of Declarative Languages, International Symposium, Springer, Berlin, Heidelberg, pp. 235–249.
Voellmy, A., Kim, H., & Feamster, N. (2012). *Procera: A language for high-level reactive network control* [Workshop]. Hot Topics in Software Defined Networks, First Workshop, pp. 43–48.
Yu, M., Wundsam, A., & Raju, M. (2014). NOSIX: A lightweight portability layer for the SDN OS. *ACM SIGCOMM Computer Communication Review, 44*(2), 28–35.
Zhao, Z., Hong, F., & Li, R. (2017). SDN Based VxLAN optimization in cloud computing networks. *IEEE Access, 5*, 23312–23319. 10.1109/ACCESS.2017.2762362.

# 3 Performance Enhancement in Cloud Computing Using Efficient Resource Scheduling

*Prashant Lakkadwala and Priyesh Kanungo*

## CONTENTS

3.1 Introduction ..................................................................................................49
3.2 Literature Survey ...........................................................................................50
3.3 Proposed Research ........................................................................................52
3.4 Simulation Setup ...........................................................................................57
3.5 Results Analysis ............................................................................................59
3.6 Conclusions ...................................................................................................64
3.7 Limitation and Future Work .........................................................................65
References ..............................................................................................................65

## 3.1 INTRODUCTION

With the need for computational resources growing exponentially in tandem with the advent of different kinds of computational and communication devices, a number of large and small-scale applications are serving their clients 24 hours, and they will offer new challenges and computational opportunities related to computational workloads. To deal with computational complexities, the cloud offers a cost-effective solution (Hashem et al., 2015), where cloud computing is a new way of computation that serves medical, communication, agriculture, and many more applications, and the cloud, a new perspective on computing offering the software and hardware resources on sharing and on-demand basis to enhance computational productivity. The cloud not only offers computational resources, but it also provides storage and data transfer solutions (Botta et al., 2016). These resources are efficient, scalable, and remotely consumable.

While offering a number of benefits to their clients, the cloud service providers (CSP) are worried about the infrastructure running cost being financially and computationally draining. In this context, to reduce the infrastructure running cost, some essential techniques such as resource management, regulated resource consumption, and provisioning have been used. On the other hand, the mismanagement of resources (low or higher) can exceed the computational and outfitted cost of the server. This phenomenon can negatively impact the performance of the cloud. Thus, the proposed work aimed to investigate and spell out ways to enhance the resource utilization of cloud servers by scheduling approaches (Dharani & Kalaiarasu, 2016). With this aim in mind, different experimental states of affairs have been measured for demonstrating resource scheduling and management (Calheiros et al., 2011), not to mention the efforts to optimize the performance of cloud servers in terms of resource exploitation using the two variants of scheduling algorithms. Deliberating on the proposed resource scheduling techniques as promising for enhancing resource preservation and minimization of cloud operational cost, this chapter also describes the simulation of methodologies for memory management, and, finally, conducts a study on the performance of the extended priority-based algorithm vis-a-vis the priority-based algorithm.

## 3.2 LITERATURE SURVEY

This section highlights the valuable contributions to and research work within the domain of resource scheduling and provisioning offering new directions for contemplation and valuable opportunities for extending the proposed study to new horizons.

Though the cloud asset planning relies upon the QoS necessities of an application, its acquiring a proficient and best approach towards fulfilling this aim is fundamental. Singh and Chana (2016a) offer an assessment of asset allocation and management procedures by incorporating procedures dependent on 110 research articles, where the resource planning methods have various classes, according to their sorts and advantages, tools, aim, and strategies, leading to the assessment helping with arriving at calculations with the choice of the most spectacular of these.

Zuo et al. (2017) propose a procedure, they named it MOSACO, which is dependent on multi-objective scheduling that advances the pool of resources and limits task execution time and expenses, while also improving QoS and upgrading the benefit of providers. The framework demonstrates task culmination time, deadline violations, cost, and level of resource utilization, while illustrating the report with comparable techniques.

QoS necessity of jobs ensues from the provisioning of suitable assets. With the disclosure of best Task-Asset pair still an advancement issue, these methods work for productive provisioning. Singh and Chana (2016a) investigate cloud asset provisioning, in which the task is sorted into different provisioning strategies, and QoS incorporates asset management, provisioning, advancement, components, and correlations, advantages, and issues.

Asset scheduling and energy conservation are critical issues in the cloud. Furthermore, it is hard to analyze the performance and power utilization of provisioning approaches. Consequently, simulations are utilized for assessment. Lin et al. (2017) have modified the CloudSim for resource planning and energy utilization model for a precise valuation, in which there are six mixes of allocation techniques illustrated, alongwith diverse methods demonstrating different energy cost.

The energy-saving of cloud servers is crucial for cloud providers, while to control cloud running expenses by consolidating cloud services into a limited number of servers is another concern. The latter is a goal in cloud asset management that motivates improvements in dynamic decision making in asset allocation and finding equilibrium between consolidation and replication. Gill et al. (2019) propose a cuckoo optimization-based asset scheduling strategy known as CRUZE that groups and executes task on cloud assets and improves energy-conservation, and, the results of CRUZE deployment shows that it is skilled in reducing energy utilization and improves quality and CPU usage.

Resource scheduling in cloud computing has been done for provisioning of resources, and is well planned according to the prerequisites of QoS, which says it is obligatory to anticipate and approve resource provisioning before planning. An asset provisioning and planning strategy has been introduced by Singh and Chana (2016a). In it, the tasks are re-grouped depending on the QoS prerequisites, and assets are provisioned prior to planning. Lastly, the model has been estimated using simulation, and results show that it is effective and productive.

Given the VM planning, in the light of sale auction framework, is offered by Kong et al. (2016), therefore to start with, the sequencing of the customers' bids is performed, followed by the client groups being screened and applicably VM asset being allocated. Last, the cost is determined by the average payments and competitive payments.

Utilizing cloud radio access network (C-RAN) with the mobile edge cloud computing (MEC), a mobile service provider can effectively deal with the traffic and upgrade the capabilities of gadgets. Given energy consumption is an issue for MSP, Wang et al. (2018) present a structure for the power-performance tradeoff of MSP utilizing planned network and computational assets, in which the resource scheduling is figured as a stochastic issue and network plans, an optimization technique. The standard strategy accepts that work demands have fixed lengths and can be done within the boundaries of a decision making interval. Furthermore, it isn't reasonable for dynamic or variable length jobs, or to broaden the Lyapunov procedure and give the VariedLen for decision making. The calculation can arrive at time normal benefits near the ideal while maintaining system stability and low congestion.

Resource planning is a cycle of the allocation of resources over time to act upon an obligatory task and decision making. Madni et al. (2019) plan a hybrid gradient descent cuckoo search (HGDCS) dependent on gradient descent and cuckoo search for handling the asset scheduling problem. The employment of this asset allocation algorithm contrasts the makespan, throughput, load balancing, and improvement of existing allocation vis-a-vis the HGDCS-proposed one.

Network planning issues are networks under the impact of multi-objective optimization; however, it manages single-objective frameworks which need to be tackled as multi-objective issues. Subsequently, meta-heuristic calculations can be valuable. Madni et al. (2019) present a MOCSO calculation for resource scheduling issues, where the essence of network planning is to diminish the expense and upgrade the performance of networks. Given network planning is tackling issues in IaaS, an examination of networks helps identify impacts of network planning and results show that it performs better.

## 3.3 PROPOSED RESEARCH

The proposed work is motivated by the aim to investigate the cloud infrastructure performance factors which influence the different job scheduling techniques; therefore, to design an improved scheduling technique to manage resources effectively is the goal. This section offers an understanding of the approaches for enhancing the cloud infrastructure performance.

A. System Overview

Cloud computing has become popular due to resource scalability and enormous computing power as various online applications are ported to the cloud environment, and the cloud effectively handling provisioning the required resources according to the application's requirements. This is possible due to service level agreements (SLA) (Hameed et al., 2014). If the cloud server is overloaded with client requests, then the server can borrow the resources from the other infrastructure setup; whereas, when the server is underloaded, it outsources the resources to other overloaded infrastructures (Manvi & Shyam, 2013). In this manner, the server arranges the computational and storage resources for their clients. But when the resource use is ideal, then its running cost becomes an issue for the cloud infrastructure provider.

In this context, to maximize the profit to the service providers, the scheduling technique is proposed for investigation, wherein it is assumed that required resources are the primary memory of the VMs (Virtual Machines), and the focus of the investigation is on managing the resources and scheduling them to measure the impact of scheduling in computational resources i.e. memory (Agarwal & Wenisch, 2017). Thus initially, different resource scheduling techniques (Kalra & Singh, 2015; Zhan et al., 2015) are evaluated and, then, two enhanced scheduling approaches are presented. First, a priority-based resource scheduling technique is proposed that helps to manage resources efficiently, and secondly, the priority-based task scheduling technique is extended to improve and accept the dynamically changing workload. Both the introduced techniques are implemented using a simulation tool (CloudSim) to analyze the impact on memory utilization of the cloud infrastructure use. The experiments with the dynamically changing workload are carried out to deal with the diverse workloads.

B. Aim and Objective

The cloud systems are hugely affected due to sudden growth of traffic or workload when it is reasonable to queue the processes. Due to this waiting time of the processes, their workload too is growing. Therefore, processes are queued in waiting state (Gholami et al., 2016). To solve such kinds of problems, it is required to migrate the process from one resource to other (Singh et al., 2018) and with the need to adjust the demand using local resource management strategies. During process migration, bulk data transfer takes place (Wu et al., 2015). Thus, we identified two key issues:

1. Dynamically changing workload, and
2. Frequent process migration

Thus the proposed work include two kinds of solutions, first, the local resource management, and second, design cost function to make decisions for workload or data migration (Chen & Wu, 2016). Additionally, the work also includes a method which identifies duplicate processes to preserve expensive resources which can substantially improve the resource availability (Sharma et al., 2019). In this chapter, the work is focused on finding the solution for the first problem by developing a local resource management technique using two methods:

1. First, to create a priority based list to schedule the jobs,
2. And second, to extend the previous solution by two steps,
    a. First, create three types of priority lists,
    b. Second, use the priority list and prepare final job scheduling sequence list.

This section has been providing a basic concept of the required scheduling model. The next section involves the proposed methodology.

C. Methodology

Introducing the proposed technique for improving the cloud resources preservation and scheduling, we assumed two kinds of experimental cloud infrastructures. The required configuration for simulation of the proposed strategy has been illustrated in Figures 3.1 and 3.2.

Figure 3.1 shows a single data center and Figure 3.2 shows a resource scheduling based on multiple cloud data centers. In single datacenter based model, a number of clients are connected to the scheduler for submitting their task, but the clients don't submit their request directly to cloud infrastructure, and herein, a broker is required. A broker is a third party entity, who is responsible for

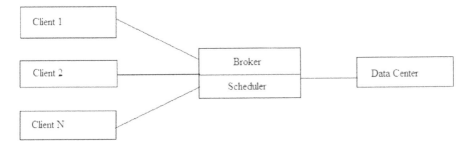

**FIGURE 3.1**   Single Data Center.

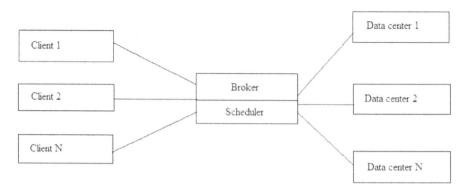

**FIGURE 3.2** Multiple Sources of Request and Multiple Data Centers.

collecting the jobs from clients, and then scheduling the jobs for action by datacenters (Sharkh et al., 2016). It may be possible that a broker is connected with multiple datacenters; however, that is not an issue since the broker has a scheduler to schedule the job with single or multiple datacenters. In Figure 3.2, broker is connected with multiple clients as well as multiple datacenters (Calheiros et al., 2012). In this chapter, we consider Figure 3.1 which is based on single datacenter-based experimental cloud infrastructure.

    a. Priority-based Task Execution

According to the cloud infrastructure of Figure 3.1, the system includes a number of users or clients, and a single datacenter built with a number hosts consisting of a number of virtual machines (VMs) (Mosa & Sakellariou, 2017). The VMs have been used for processing the user's request, based on the broker having programmed the client jobs with datacenter resources. After execution of the jobs, the datacenter returns results of execution to the broker and the broker responds to the client (Pascual & Alonso, 2008). With the clients submitting their appeal to the broker in raw format, it is up to the broker to convert it into a special format known as cloudlets, and to plan the job's execution by the cloudlets positioned in a queue maintained by scheduler. However, the cloudlet has three main attributes, viz., Process length, File input size, and File output size. Let, this scheduler's queue is $Q = \{j_1, j_2, ..., j_n\}$, where $j_n$ is a cloudlet. Additionally, we need to manage these jobs using a priority-based scheduling algorithm. The review of literature on different types of Algorithms in resource scheduling is summarized in Table 3.1.

    The algorithm begins with the evaluation of Q, the process queue available with the scheduler. On the basis of evaluation, the scheduling algorithm sorts all the processes in the queue on the basis of their length (Abdulhamid et al., 2016), after which the minimum (min) and maximum (max) length of job is used to calculate a threshold, equal to the mean value of the min and the max job length. Later, the threshold value is used to compare all the cloudlets in the queue and assign a priority label (i.e. low and high) and to sort the labeled cloudlets in an ascending order and to submit the cloudlets to the datacenter.

## TABLE 3.1
### Literature Review Summary

| Ref. | Paper Type | Direction | Algorithms | Problem Type |
| --- | --- | --- | --- | --- |
| Singh and Chana (2016a) | Survey | Resource scheduling | – | – |
| Zuo et al. (2017) | Research | Resource scheduling | MOSACO | Optimization |
| Singh and Chana (2016b), Agarwal and Wenisch (2017) | Survey | Resource scheduling | | Optimization |
| Lin et al. (2017) | Simulation | Energy and resource | Different algorithms | Measuring power consumption |
| Gill et al. (2019) | Research | Energy and resource consolidation | CRUZE | Optimization |
| Singh and Chana (2016c) | Research | Provisioning and scheduling | k-means clustering | Cluster based optimization |
| Kong et al. (2016) | Research | VM scheduling | Minimum cost | Cost minimization |
| Wang et al. (2018) | Research | VM scheduling | VariedLen algorithm | Optimization |
| Madni et al. (2019) | Research | Resource scheduling | HGDCS | Optimization |
| Madni et al. (2019) | Research | Resource scheduling | MOCSO | Multi-objective problems |

b. Dynamically Varying Workload

The cloud infrastructure for dynamic workload scheduling is illustrated in Figure 3.2. In order to simulate this solution, first, we need to generate a sequence of dynamically changing workloads for which a random function having the option of accepting two inputs: a minimum and a maximum value, and producing a random number, must be implemented. Using this function, and the min and max values, we generate jobs of random length (Chaudhry & Khan, 2015). Now, in order to schedule the generated dynamic workload optimally, we introduced an extended priority-based job scheduling algorithm, discussed in Tables 3.2 and 3.3.

The extended priority based scheduling technique is designed in two parts: in the first part, the algorithm, described in Table 3.2, is assigning a priority to the each cloudlet, and in the second part, the algorithm, as shown in Table 3.3, is used to schedule processes. The aim of the first algorithm (priority assignment) is to process the cloudlets queuing to the cloud broker to create three priority lists (i.e. low priority, high priority and medium priority). To do this, a fixed window size $J_{max}$ is prepared, where $J_{max}$ is the maximum cloudlet length, to select all the cloudlets within the scope of this window. The algorithm selects a cloudlet for scheduling when it has a length higher than 33% of the maximum cloudlet length, or accreting length outside the low priority window; or, a length higher than the window comprising cloudlets above 33% but below 66% of the maximum cloudlet length, or accreting length outside the medium

## TABLE 3.2
## Priority Assignment

Input: process queue $Q = \{j_1, j_2, ....., j_n\}$
Output: priority based queue $Q_p$
Process:
1. $Q_n = Read\ Process\ Queue(Q)$
2. $J_{min} = Get\ Minimum(Q_n)$
3. $J_{max} = GetMaximum(Q_n)$
4. $J_{threshold} = \frac{J_{min} + J_{max}}{2}$
5. $for(i = 1; i \le n; j++)$
    a. if $(j_i \le J_{threshold})$
        i. $Q_p = j_i.\ AddLabel(lowPriority)$
    b. Else
        i. $Q_p = j_i.\ AddLabel(highPriority)$
    c. End if
6. End for
7. Return $Q_p$

## TABLE 3.3
## Schedule Process

Input: process queue $Q = \{j_1, j_2, ....., j_n\}$
Output: priority assigned job list
$JList = \{JListLow,\ JListMid,\ JlistHigh\}$
Process:
1. $Q_n = ReadProcessQueue(Q)$
2. $J_{max} = GetMaxLength(Q_n)$
3. $for(i = 1; i \le n; i++)$
    a. if $(J_i \le (J_{max} * 0.33))$
        i. $JListLow.\ Add(J_i)$
    b. if $((J_{max} * 33) > J_i \le (J_{max} * 0.66))$
        i. $JListMid.\ Add(J_i)$
    c. if $((J_{max} * 0.66) > J_i)$
        i. $JListHigh.\ Add(J_i)$
    d. end if
4. End for
5. Return $JList = \{JListLow,\ JListMid,\ JlistHigh\}$

priority window; or, a length higher than the window comprising cloudlets above 66% and below 100% of the maximum cloudlet length, or accreting length outside the higher priority window. The algorithm (Table 3.2) shows the priority assignment of all the cloudlets in a window. Based on the length of the jobs, the algorithm assigns priority levels (i.e. low, high and medium), and these categories of jobs are kept in three separate lists, where the job lists are used to prepare the job scheduled list.

Table 3.3 provides the working of the extended priority based cloudlet scheduling algorithm which gives a scheduled list of cloudlets. According to this scheduled cloudlet lists, there is assigning of the resources. The scheduling algorithm (Table 3.3) is used for dealing with the dynamically changing workload. According to the given algorithm, the cloud scheduler accepts the priority Cloudlet lists (i.e. low, medium, and high) and returns a scheduled cloudlet list. Picking two cloudlet lists at a time: the cloudlet list is picked associated with small cloudlets, or low priority cloudlet list; and the second cloudlet list includes the larger length of cloudlets list, or high priority job list. Using these cloudlet lists, one smallest and one highest length of cloudlets is selected. The selected cloudlets are placed on a list called as scheduled cloudlet list. When a cloudlet list is finished, the medium priority cloudlet list is used. The direction of accessing cloudlet from the medium priority cloudlet depends on the lists elements having finished recently. For instance, if low priority cloudlet list is finished first, then the cloudlets are accessed from top of the list; otherwise, it is accessed from the bottom. In this way, first the low and higher length jobs are organized in a sequence for processing and, then, the medium length of cloudlets are processed. This approach is used to simulate scheduling of the dynamic workload in cloud datacenter.

The discussed algorithms for priority based resource allocation and dynamic workload scheduling algorithm are implemented in next section using CloudSim, which is a discrete event simulator to simulate the cloud processes. The next section discusses the required simulation configuration.

## 3.4 SIMULATION SETUP

The proposed two different cloud job scheduling techniques based on priority based job scheduling and dynamically changing workload scheduling techniques are demonstrated in this section. Thus the section includes the simulation scenarios and the setup of the CloudSim tool.

A. Simulation Setup

In order to demonstrate the priority based and dynamically changing workload on server, we need to configure the cloud simulation using the different parameters. Table 3.4 shows the required parameters and its values for appropriate configuration.

B. Simulation Scenario

In this experiment we have tried to demonstrate the working of a priority-based job scheduling and dynamically changing workload scheduling, in which both the scheduling techniques are implemented here using the CloudSim tool and JAVA technology. Thus to carry out experiments for performance evaluation of both the scheduling strategies, the following two simulation scenarios are followed for demonstration.

## TABLE 3.4
## Scheduled Job List

Input: priority assigned job list $JList = \{JListLow, JListMid, JlistHigh\}$
Output: scheduled job list SList
Process:
1. $L = getLength(JListLow)$
2. $M = getLength(JListMid)$
3. $H = getLength(JListHigh)$
4. $T = L + M + H$
5. $for (i = 1; i < T; i++)$
   a.  *if* $(flag == 0 \, \&\& \, L \, != 0)$
       i.   $SList.Add - (JListLow.FirstElement)$
      ii.   $JListLow.Remove - (FirstElement)$
      iii.  $L = L-1$
      iv.  *if* $(H != 0)$
          1.  $flag2 = 0$
      v.   Else
          1.  $flag1 = 0$
      vi.  End if
      vii.  $flag = 1$
   b.  *else If* $(flag1 == 0 \, \&\& \, M != 0)$
      i.   *if* $(L == 0)$
          1.  $SList.Add - (JListMid.FirstElement)$
          2.  $JListMid.Remove - (FirstElement)$
          3.  $M = M - 1$
      ii.  *else if* $(H == 0)$
          1.  $SList.Add - (JListMid.LastElement)$
          2.  $JListMid.Remove - (LastElement)$
          3.  $M = M - 1$
      iii.  End if
   c.  *else if* $(flag2 == 0 \, \&\& \, H != 0)$
      i.   $SList.Add (JListHigh.Last - Element)$
      ii.   $JListHigh.Remove - (LastElement)$
      iii.  $H = H-1$
      iv.  $Flag2 = 1$
      v.   *if* $(L != 0)$
          1.  $Flag = 0$
      vi.  Else
          1.  $Flag1 = 0$
          i.  End if
   d.  End if
6. End for
7. Return SList

Performance Enhancement in Cloud Computing

1. **Priority based scheduling:** This scenario is implemented to explain the potential of priority-based job scheduling algorithm as described in Table 3.4. With this purpose in mind, the cloud is configured using the above-defined parameters, and then the space shared scheduling technique is modified to incorporate the proposed cloudlet assignment policy. Using the developed strategy, the cloudlet is submitted to the cloud infrastructure and their performance is recorded.
2. **Dynamically changing workload scheduling:** This scenario demonstrates the extended version of priority-based job scheduling. The concept of this algorithm is discussed in Tables 3.2 and 3.3, which is implemented with the help of CloudSim. The algorithm is capable of dealing with random or dynamic length of jobs, which is produced on the basis of a random function. The demonstration includes the priority assignment of the dynamic jobs and their scheduling.

## 3.5 RESULTS ANALYSIS

The result analysis of the proposed scheduling algorithms offers an explanation of job scheduling where there are two different scheduling approaches implemented, and which are compared based on various performance parameters like response time, processing cost, and memory usage.

A. Response Time

The time required to process a cloudlet is computed as processing time, where the response time is measured by summing waiting time and execution time. The following function can be used for measuring response time:

$$response\ time = execution\ time + waiting\ time$$

Further, to measure the mean response time, the following function is used:

$$mean\ response\ time = \frac{1}{N} \sum_{i=1}^{N} response\ time_i$$

Where, N is the number of jobs for scheduling, and *response time$_i$* is the $i^{th}$ job's response time.

The comparison of the performance of both the scheduling algorithms is shown in Figure 3.3. That is, a line graph which demonstrates priority-based scheduling performance uses a blue plotline and the dynamically changing workload scheduling performance uses a red plotline. In this line graph, the number of cloudlets to be process is shown on the X axis, and the time taken to process cloudlet is shown on the Y axis. The results in terms of response time have been measured with different size of job windows. According to the measured performance of proposed scheduling algorithms, the dynamic workload scheduling technique demonstrates

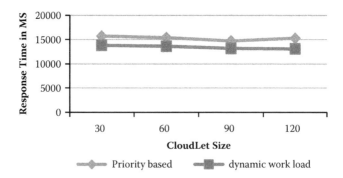

**FIGURE 3.3** Response Time of Cloudlets.

less response time as compared to the initial priority-based workload scheduling approach. Therefore, the extended priority scheduling technique is superior to the classical or previously proposed priority-based scheduling technique.

B. Processing Cost

The cloud servers are dealing with a large volume of jobs and processing of large jobs have an impact on processing cost, where the processing cost is an essential parameter of jobs scheduling in the cloud. The total cost of processing the cloudlets using infrastructure resources is termed as processing cost. The mean processing cost of cloudlets is measured using the following formula:

$$mean\ processing\ cost = \frac{1}{N} \sum_{i=1}^{N} cloudlet_i \cdot proceessing\ cost \qquad (3.1)$$

The comparative processing cost of both the approaches is shown in Figure 3.4. That is, a line graph representation of processing cost obtained during experiments is plotted with measurement of processing cost doesn't include the waiting time

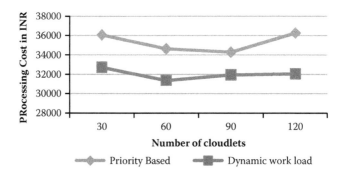

**FIGURE 3.4** Mean Processing Cost.

cost. For the experimental scenarios (cloudlets 30, 60, 90, and 120) randomly generated cloudlets are used within a predefined range. In this line graph, X axis shows the number of cloudlets available for experimentation, and the Y axis shows total processing cost in INR. Here, we assumed that per unit processing is 3 INR. It is evident from the results that the proposed scheduling technique for scheduling the dynamic length of jobs requires less cost as compared to classical priority-based scheduling technique.

C. Mean Memory Usage

In job scheduling techniques, the scheduled jobs are assigned to the VMs (Virtual Machines). The VMs are responsible for the execution of jobs. Therefore, VM resources are used for the processing of jobs. The average memory used in VMs for all the submitted jobs is called mean memory usage, calculated on the basis of following function:

$$Mean\ memory = \frac{1}{N} \sum_{i=1}^{N} VM_i \cdot Memory\ Usage \qquad (3.2)$$

The memory usage during execution of processes in VMs is measured and explained in this section. Figure 3.5 illustrates the affiliation among the number of cloudlets and memory used. That is, a line graph where X axis is evidence for the number of cloudlets used in the experimentation and Y axis shows memory used by the given processes is plotted. The results show that the dynamic load-based scheduling approach reduces the data-centers' resource requirements in terms of memory usage.

D. Mean Memory Cost

The cost of memory required to process submitted jobs is termed as mean memory cost, and for all the VM's memory usages, it is computed, where cost as per both the algorithms is measured using the following formula:

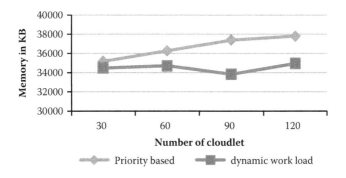

FIGURE 3.5  Mean Memory Usages.

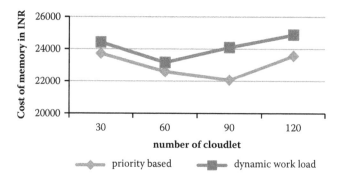

**FIGURE 3.6** Memory Cost.

$$\text{Mean memory cost} = \frac{1}{N} \sum_{i=1}^{N} VM_i \cdot \text{Memory Usage} * \text{memory cost} \quad (3.3)$$

Figure 3.6 shows the bond connecting quantity of cloudlets and the cost of memory. That is, a line graph representation of attained trial results with increasing size of cloudlets used is plotted. The X axis of this line graph is the number of cloudlets submitted for experiment and Y axis shows the cost of memory in terms of INR. As revealed in Figure 3.6, the dynamic work load-based scheduling algorithm is costlier than priority-based scheduling but gives better result.

E. Scheduling Time

Once the previous sections have delineated on network performance after applying scheduling approaches on the cloud, this section provides the evaluation of algorithms for their own performance in terms of time required for scheduling. This parameter computes the time required for scheduling of the resource, where the time consumed is also known as time complexity or time consumption of scheduling algorithms. It is calculated using the following formula:

$$\text{time consumed} = \text{end time} - \text{start time} \quad (3.4)$$

Figure 3.7 compares the performance of both the implemented algorithms (i.e. priority-based scheduling and extended priority-based scheduling). Processing time of the proposed algorithms is represented using line graph. The X axis of this graph represents the number of cloudlets to be processed during simulation and Y axis shows the time required for scheduling. The time consumption of the algorithms is measured here in terms of milliseconds. As shown in graph, the dynamically work load-based scheduling algorithm is expensive in terms of time with respect to priority-based technique.

# Performance Enhancement in Cloud Computing

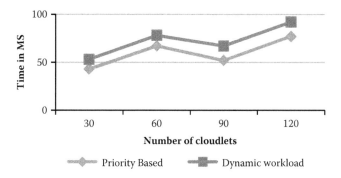

**FIGURE 3.7** Time Consumption

### F. Memory Utilization of Algorithm

The execution of an algorithm needs some amount of data to be held in the main memory, where the memory space is termed as the memory usage of the algorithm. The following formula is used for computing the memory usage:

$$memory\ usage = total\ space - free\ space \qquad (3.5)$$

Figure 3.8 shows the comparative performance of both the approaches of resource scheduling. According to the performance, the priority-based technique consumes less amount of memory with respect to the dynamic work load-based scheduling. As the dynamic work load-based scheduling works on two different sequences, it needs to hold an additional amount of data over the main memory.

Both the algorithms are implemented using CloudSim simulator, with each algorithm's performance measured and compared. Based on conducted experiments, the consequences of the algorithms are summarized in Table 3.5. According to the performance of both the resource scheduling techniques, both the algorithms are working well, and offer advantages by reducing computational resources

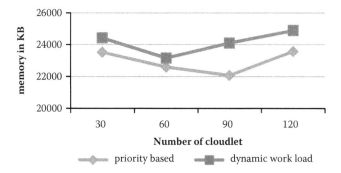

**FIGURE 3.8** Memory Usages

**TABLE 3.5**
**Simulation Setup**

| S. No. | Parameters | Values |
|---|---|---|
| 1 | Number of Cloudlet | 30, 60, 90, 120 |
| 2 | Host Bandwidth | 10000Gbit/s |
| 3 | Host Storage | 1000000 GB |
| 4 | VM Image Size | 10000 |
| 5 | Host RAM | 2048 |
| 6 | Virtual machine RAM | 512 |
| 7 | Number of CPU | 1 |

**TABLE 3.6**
**Performance Summary**

| S. No. | Parameters | Priority-based Scheduling | Dynamic Work Load Scheduling |
|---|---|---|---|
| 1 | Memory usages | Low | High |
| 2 | Time usages | Low | High |
| 3 | Cloudlet Processing time | High | Low |
| 4 | Cloudlet processing cost | High | Low |
| 5 | Cloudlet memory usages | High | Low |
| 6 | Cloudlet memory cost | High | Low |

consumption. The extended priority-based job scheduling is efficient and effective for managing the datacenter resources. On the other hand, it is a little bit more expensive compared to the priority-based scheduling due to larger running time and memory requirements. The extended priority-based scheduling algorithm is also effective for managing dynamic workloads. Based on the experiments, the performance summary is prepared and shown in Table 3.6.

## 3.6 CONCLUSIONS

The cloud is a popular technology; and, it is increasingly acceptable for various computing tasks. That in itself makes clouds popular due to their scalable resources and on-demand availability, which require managing a huge workload. In such conditions, mismanagement of resources can increase the cloud infrastructure running cost, and so, we need efficient and effective scheduling techniques for handling diversity in the process execution. The aim of this chapter was to introduce a CloudSim simulation for describing the process of resource allocation and obtaining the impact of job scheduling in memory resource utilization.

The two resource scheduling techniques have been described and their simulation carried out in this chapter. The first method is developed for scheduling jobs on a priority basis, where the algorithm computes a threshold for categorizing the jobs according to their length and by assigning priority, with the threshold computed dynamically based on workload for scheduling. In the second algorithm, the previously introduced algorithm is extended to improve task performance. Therefore, three priority-based lists are prepared by using the processes' length-based threshold. Further, these priority-based lists are used to generate the scheduled process list.

According to our findings during the system design and experimentation, the segmentation of the large process queue can help in efficient scheduling. The priority assignment offers the advantage to decide which cloudlet is needed to be processed first. The small segmentation of cloudlets, and alternate assignment of low and higher length cloudlets, also help in round robin-based scheduling techniques. In addition, the experiment's results demonstrate the employment of priority-based scheduling techniques as helpful in optimizing the performance of datacenter by reducing cost, and waiting time of cloudlet processing.

## 3.7  LIMITATION AND FUTURE WORK

The cloud's performance has been improved by including some of the changes in the scheduling strategy. But advances in some of the necessary features can improve the productivity of cloud more, whereby motivation ensuing from the current introduced work is forwarded to the near future in the hope of further research in the direction of the following extensions of the work:

1. The current proposed techniques offer the solution for dynamically appeared workload scheduling solutions; in next endeavors, we are investigating the bulk data transfer and workload scheduling.
2. The future work also extends the current work in the direction of managing process migration and communication overhead relevant issues.
3. Elimination of duplicate processes to optimize the performance of computational system.

## REFERENCES

Agarwal, N., & Wenisch, T. F. (2017). Thermostat: Application-transparent page management for two-tiered main memory. In *Proceedings of the Twenty-Second International Conference on Architectural Support for Programming Languages and Operating Systems (ASPLOS '17)* (pp. 631–644), April 8–12, Xi'an, China. © ACM. ISBN 978-1-4503-4465-4/17/04.

Abdulhamid, S. M., Madni, S. H. H., & Abdullahi, M. (2018). Fault tolerance aware scheduling technique for cloud computing environment using dynamic clustering algorithm. *Neural Computing & Applications*, 29(1), 279–293). doi: 10.1007/s00521-016-2448-8.

Botta, A. et al. (2016). Integration of cloud computing and internet of things: A survey. *Future Generation Computer Systems*, 56, 684–700.

Chaudhry, I. A., & Khan, A. A. (2015). A research survey: Review of flexible job shop scheduling techniques. *International Transactions in Operational Research, 00*, 1–41. 10.1111/itor.12199.

Calheiros, R. N. et al. (2011). CloudSim: A toolkit for modeling and simulation of cloud computing environments and evaluation of resource provisioning algorithms. *Software – Practice and Experience, 41*, 23–50.

Calheiros, R. N. et al. (2012). A coordinator for scaling elastic applications across multiple clouds. *Future Generation Computer Systems, 28*, 1350–1362.

Chen, Y., & Wu, J. (2016). *High network utilization load balancing scheme for data centers* [Conference session]. IEEE Global Communications Conference. 10.1109/GLOCOM.2016.7841872.

Dharani, R., & Kalaiarasu, M. (2016). Efficient resource allocation and scheduling in cloud computing environment. *International Journal of Research in Computer Applications and Robotics, 4*(3), 48–55.

Gholami, M. F. et al. (2016). Cloud migration process—A survey, evaluation framework, and open challenges. *The Journal of Systems and Software, 120*, 31–69.

Gill, S. S., Garraghan, P., Stankovski, V., Casale, G., Thulasiram, R. K., Ghosh, S. K., ... & Buyya, R. (2019). Holistic resource management for sustainable and reliable cloud computing: An innovative solution to global challenge. *Journal of Systems and Software, 155*, 104–129.

Hameed, A. et al. (2014). A survey and taxonomy on energy efficient resource allocation techniques for cloud computing systems. In *Computing*. Springer-Verlag. 10.1007/s00607-014-0407-8.

Hashem, I. A. T. et al. (2015). The rise of "big data" on cloud computing: Review and open research issues. *Information Systems, 47*, 98–115.

Kalra, M., & Singh, S. (2015). A review of metaheuristic scheduling techniques in cloud computing. *Egyptian Informatics Journal, 16*, 275–295.

Kong, W., Lei, Y., & Ma, J. (2016). Virtual machine resource scheduling algorithm for cloud computing based on auction mechanism. *Optik, 127*(12), 5099–5104.

Lin, W. et al. (2017). Multi-resource scheduling and power simulation for cloud computing. *Information Sciences, 397–398*, 168–186.

Manvi, S. K. S., & Shyam, G. K. (2013). Resource management for Infrastructure as a Service (IaaS) in cloud computing: A survey. *Journal of Network and Computer Applications, 41*, 424–440.

Madni, S. H. H. et al. (2019a). Hybrid gradient descent cuckoo search (HGDCS) algorithm for resource scheduling in IaaS cloud computing environment. *Cluster Computing, 22*(1), 301–334.

Madni, S. H. H., Abd Latiff, M. S., & Ali, J. (2019b). Multi-objective-oriented cuckoo search optimization-based resource scheduling algorithm for clouds. *Arabian Journal for Science and Engineering, 44*(4), 3385–3602.

Mosa, A., & Sakellariou, R. (2017). Virtual machine consolidation for cloud data centers using parameter-based adaptive allocation. *ECBS '17, Cyprus ACM*, 978-1-4503-4843-0/17/08.

Pascual, J. A., & Alonso, J. M. (2008). Effects of job placement on scheduling performance. *Actas de las XIX Jornadas de Paralelismo*, 393–398, ISBN: 978-84-8021-676-0.

Sharkh, M. A. et al. (2016). Building a cloud on earth: A study of cloud computing data center simulators. *Computer Networks, 108*, 78–96.

Sharma, P. et al. (2019). Resource deflation: A new approach for transient resource reclamation. *EuroSys* 19, March 25–28, ACM, ISBN 978-1-4503-6281-8/19/03.

Singh, S. et al. (2018). Comparative study of existing data scheduling approaches and role of cloud in VANET. *Environment. Procedia Computer Science, 125*, 925–934.

Singh, S., & Chana, I. (2016a). A survey on resource scheduling in cloud computing: issues and challenges. *Journal of Grid Computing, 14*, 217–264. 10.1007/s10723-015-9359-2.

Singh, S., & Chana, I. (2016b). Cloud resource provisioning: survey, status and future research directions. *Knowledge Information System.* London: Springer-Verlag. 10.1007/s10115-016-0922-3.

Singh, S., & Chana, I. (2016c). Resource provisioning and scheduling in clouds: QoS perspective. *Journal of Supercomputing, 72*, 926–960. DOI: 10.1007/s11227-016-1626-x.

Wang, X. et al. (2018). Dynamic resource scheduling in mobile edge cloud with cloud radio access network. *IEEE Transactions on Parallel Distributed and Systems, 29*(11), IEEE. 10.1109/TPDS.2018.2832124.

Wu, Y. et al. (2015). Orchestrating bulk data transfers across geo-distributed datacenters. *IEEE Transactions on Cloud Computing*, 2168–7161, IEEE.

Zhan, Z. H. et al. (2015). Cloud computing resource scheduling and a survey of its evolutionary approaches. *ACM Computing Surveys, 47*(4), 63:1–63:33.

Zuo, L. et al. (2017). A multi-objective hybrid cloud resource scheduling method based on deadline and cost constraints. *Special Section on Emerging Trends, Issues, and Challenges in Energy -Efficient Cloud Computing, 5*(2017), 2169–3536, IEEE.

# 4 Hyper-Personalized Recommendation Systems: A Systematic Literature Mapping

*Bijendra Tyagi and Dr. Vishal Bhatnagar*

## CONTENTS

4.1 Introduction ........................................................................................... 69
    4.1.1 Key Building Blocks of Hyper-Personalization ............................. 70
4.2 Literature Review .................................................................................. 71
4.3 Research Method ................................................................................... 77
    4.3.1 Objectives ....................................................................................... 77
    4.3.2 Methodology ................................................................................... 78
4.4 Discussion .............................................................................................. 79
    4.4.1 Personalization vs. Hyper-Personalization ..................................... 79
    4.4.2 Pitfalls of Traditional Approaches .................................................. 79
    4.4.3 Why Hyper-Personalization? .......................................................... 81
    4.4.4 Hyper-Personalization Framework ................................................. 81
4.5 Results .................................................................................................... 83
    4.5.1 Classification on the Basis of Research Papers Used ..................... 83
    4.5.2 Focus Areas ..................................................................................... 84
    4.5.3 Experimental Analysis of Campaign Designing ............................ 85
4.6 Conclusion ............................................................................................. 86
References ...................................................................................................... 86

## 4.1 INTRODUCTION

One of the major factors for any industry or business to grow and survive in this competitive world is its customers. Given organizations are adopting different strategies to attract and keep a hold on their existing customers, thus, a common strategy is to personalize products or services to meet needs of their customers, even as today's customers are comparatively more smart and better informed than before, who no longer want ready-made/conventional services or products.

Hyper-personalization is defined as the process of using data to provide more personalized and specific products, services, and content, wherein the key element is to interact on a one-to-one basis with users/individuals, and not with the segments

**FIGURE 4.1** Key Building Blocks.

or groups they belong to. It is basically the next stage of digital marketing constructing emails that change content on the basis of customer's location and the time of accessing the email.

In this paper, we comprehensively collected the literature associated to personalization from academic databases including journals like Information sciences, Expert systems with Applications, Computers in Human Behavior, and others, and reviewed 20 international journal articles published from 2010 to 2019.

Real time data or user's personal data, like the past interaction of the user with the search engine (queries that were used or documents that were clicked in the past), have been represented as useful in order to recommend the personalized search results as per the users' information need.

### 4.1.1 Key Building Blocks of Hyper-Personalization

Below are the building blocks to achieve hyper-personalization as Figure 4.1 shows:

- ◦. **Extensive customer personas** – Organizations should collect customer data from all possible channels and create an extensive customer persona. Regularly updating these personas with the latest insights will help marketers/advertisers target customers with the most appropriate content, experiences, and products that dynamically match needs of the customer.
- ◦. **Marketing automation platforms and robust data** – Customer data platforms are responsible for analyzing and converting the data collected from various online channels into customer insights. A strong platform utilizes these customer insights to generate highly personalized marketing campaigns and achieve better engagement of customer with products.
- ◦. **Customer first strategy** – A marketing plan should be created such that it treats every customer as a unique individual and not as a segment. Providing

relevant content predictively at the most appropriate time and at the right point of customer's journey will definitely improve emotional engagement of customer.

## 4.2 LITERATURE REVIEW

| Approach | Research paper | Authors/Year | Findings |
|---|---|---|---|
| Collaborative filtering | An approach for combining results of Recommender system techniques based on collaborative filtering | Edjalma Queirz da Silva, Celso G. Camilo-Junior, Luiz Mario L. Pascoal, Thierson C. Rosa / 2016 | The authors presented an approach to bind the results of RS techniques by the use of Collaborative filtering. The item was to recommend to target users not tried so far by calculating the similarities between existing users who are already using these products. The main problem targeted here was to figure out the best combination of expectations of user and sufficient items to be recommended. |
| | | | Invire was the system which was proposed in this study, and it used to produce list of items that should be recommended to customers on the basis of the Search algorithm's result. |
| | | | Every system has its own pros and cons, similarly there were some limitations of this approach as well: |
| | | | • Dependent on base recommendation approach chosen initially. |
| | | | • Not suitable to be utilized in small database or initial stage applications (Silva et al., 2016). |
| | Multimedia Recommendation with Item- and Component-Level Attention | Jingyuan Chen, Hanwang Zhang, Xiangnan He∗, Liqiang Nie, Wei Liu, Tat-Seng Chua / 2017 | This study proposed how implicit feedback (item and component level) in Multimedia recommendation can be improved by Collaborative filtering based attention mechanisms. |
| | | | The main problem which was targeted by this study was that the majority of existing systems based on CF were not suitable for multimedia recommendation as they ignored an important factor i.e. implicitness in user's interaction with multimedia content. There resides 2 types of implicitness with respect to multimedia content: |

*(Continued)*

| Approach | Research paper | Authors/Year | Findings |
|---|---|---|---|
| | | | • Item – level: the preferences of user for the item like videos, pictures etc. are not known.<br>• Component – level: the preferences of user for the specific components (like frames in video) inside an item are not known.<br>The attention model introduced here was a neural network using two main modules: component level module which used to gather information about multimedia items, and the second was item level module which was used to arrange the preferences of these items. (Chen et al., 2017). |
| | Collaborative Memory Network for Recommendation Systems | Travis Ebesu, Bin Shen, Yi Fang /2018 | The authors here focused on neighborhood- or memory-based approach. They presented an extensive architecture to combine the two main classes of CF models with the help of Collaborative Memory Networks (CMN). They introduced two components: memory component and neighborhood component (which was based on neural attention mechanism).<br>The CMN defined in this approach used to work on three states of memory:<br>• An internal memory specific to user.<br>• A memory specific to item/product.<br>• The collective neighborhood state (Ebesu et al., 2018). |
| **Hybrid approach - collaborative filtering and deep learning** | Collaborative Filtering and Deep Learning Based Recommendation System For Cold Start Items | Jian Wei, Jianhua He, Kai Chen, Yi Zhou, Zuoyin Tang / 2017 | This paper presented two recommendation systems here for the purpose to solve the problem CCS and ICS for the new products. CCS and ICS can be explained as:<br>• Complete cold start (CCS) problem in which rating records are not all available.<br>• Incomplete cold start (ICS) problem in which only few rating records are accessible for a small group of new products or users.<br>Sparsity and Cold start were the problems faced by these models and to overcome these problems an integrated approach using CS and ML algorithms was implemented with the aim of improving the performance of recommendation basically for the CS items. |

| Approach | Research paper | Authors/Year | Findings |
|---|---|---|---|
| | | | It was concluded that due to very few ratings received for the items, the item-based model ICS was not able to make better recommendations, and thus CCS item-based model was better in such cases (Wei et al., 2017). |
| | Music Recommendation using Collaborative Filtering and Deep Learning | Anand Neil Arnold, Vairamuthu S. / 2019 | This study proposed an approach to use collaborative filtering and deep learning for presenting music recommendation system. The target here was to improve the recommendation system by collaborating deep learning with filtering techniques. The hybrid RS examined song's album art in order to figure out unique labels using YOLO which is one of the fastest algorithm used for object detection, and has been proven as extremely good for object detection in real-time. |
| | | | Authors took a sample of 1000 songs with different attributes as user id, song id, title, count, release, artist's name, year of release, and others. They applied a custom recommender which determined the recommendation score and arranged these songs based on the calculated score. After that using the recommendation ranks, the songs were recommended to the users. Firstly they determined the number of times the song was listened to and then mapped them based on the count. Once the count was generated, they determined the recommendation score and arranged them in ascending order. The next part, basically the deep learning part, led to the recognition of album art with the help of object detection algorithms. In this study, the algorithm for object detection "(YOLO) you only look once" was used. |
| | | | As every approach has its limitations, similarly, the limitation here was the faulty accuracy of recommendations. The system did not recommend songs accurately to the listeners on the basis of user history and preferences (Neil & Vairamuthu, 2019). |

*(Continued)*

| Approach | Research paper | Authors/Year | Findings |
|---|---|---|---|
| **Personalization of web** | Location: The Impact of Geolocation on Web Search Personalization | Chloe Kliman-Silver, Aniko Hannak, David Lazer, Christo Wilson, Alan Mislove / 2015 | The authors explained the concept of Personalization and geolocation. Filter Bubble Effect (where some algorithms for personalization conclude the useful information to be unnecessary and prevent the users from accessing it) was mainly focused by the model proposed here. The aim here was to analyze the impact of personalization techniques based on location in the results from Google. The authors observed that the variation in search results depends on physical distance and it expands as distance increases. Following observations were drawn from the study:<br>1. Algorithm was not activated to same degree by all types of user queries.<br>2. All user queries did not require location based. personalization like searching for CCD did not need Maps.<br>3. Noisy search results were obtained by Google. (Kliman-Silver et al., 2015). |
| | Toward Semantic Web Using Personalization Techniques | Sinan Diwan/ 2016 | The authors here used personalization techniques to customize web servers in a way to make demented web pages looks like it has been composed semantically. The author explained that personalization had gone through different phases. Authors introduced a personalization schema to personalize web pages through the participation of users. In this study, the Authors described that Semantic Web is the next future of the web where all the contents will be connected together based on meanings, and the idea proposed here was to create extensible mark-up language (XML) based descriptors, one for each web page being fetched from the repository, at server side, and provide user the ability to mark-up contents on current web page upon his/her experience. It was concluded that:<br>• XML carriers should be constructed which can virtually regulate non- |

| Approach | Research paper | Authors/Year | Findings |
|---|---|---|---|
| | | | semantic constructed web pages to gather virtual semantic content. |
| | | | • Web servers should have a theoretical model to perceive search engine queries to develop or upgrade current knowledge base, which later on might be used as an intelligent reference model to reply to these queries. |
| | | | This XML-based descriptor will be used for every search query originating from a search engine (e.g. google, yahoo), and upon its fields, the web server can effectually construct connections based on the meaning (Diwan, 2016). |
| | Advancing E-Commerce Personalization: Process Framework and Case Study | Maurits Kaptein, Petri Parvinen / 2015 | This paper represented the framework and case study for an e-commerce personalization. The problem focused here was high growth of real-time data points in online marketing due to enhancement in computing power. |
| | | | The aim of this study was to: |
| | | | a. Categorization of personalization attempts. |
| | | | b. Development of ideal personalization approaches. |
| | | | Consumer behavior and technological methods were explained as the two main components to be focused on for any successful personalization approach. Three assumptions regarding consumer psychology were presented: |
| | | | • The various types of content like product categories and price have effect on the bottom line. |
| | | | • Bottom line effects may vary between consumers. |
| | | | • The effects should be relatively balanced between individuals. |
| | | | The conclusion here was that this framework can help to select suitable content with the focus on product selection for personalization (Kaptein & Parvenin, 2015). |
| **E-learning** | Generalized metrics for the analysis of | Fathi Essalmi, Leila Jemni Ben Ayed, Mohamed | The authors proposed an approach for learning objects (involved in course) based recommending personalization |

(*Continued*)

| Approach | Research paper | Authors/Year | Findings |
|---|---|---|---|
| | E-learning personalization strategies | Jemni, Sabine Graf, Kinshuk / 2015 | techniques. The main question that was focused here was -: how to forecast personalization techniques that are well suited for particular courses? The motivation behind this research or the main problem that was targeted through this study was the level of difficulty in using several different personalization elements to personalize each course for a learner. Let's say, if we need to examine 20 parameters of personalization for personalizing any course, and make an assumption that each of the personalization parameters contains four different learner's characteristics, then the professor who is answerable for the course has to prepare 60 (=20*4) different learning scenarios. So in such a case, the process of evaluation will consume more time and can be the reason for decreasing the motivation of learner. Thus it becomes very difficult to answer this question: "what are the personalization parameters that should be considered in the learner profile?" and the answer comes from this study (Essalmi et al., 2015). |
| | Recommender System Based on Web Usage Mining for Personalized E-learning Platforms | Sunil and M. N. Doja / 2017 | This study represented an e-learning platform for learning assets recommendation using web-usage mining for learners, and provided them an efficient platform and personalized services. The proposed system was able to mine through stored server log files and database of back office of e-learning platforms, such as the categorization of the users, and the accumulation and categorization of the web pages often visited by the learners. The study revealed that web-usage mining based analysis of the quality courses can surely help e-learning models to: 1. Improve structure of website. 2. Classify the learners. 3. Group the web pages in order to reach required results. |

| Approach | Research paper | Authors/Year | Findings |
|---|---|---|---|
| | | | 4. Provide specific learners the recommendation of learning resources (Sunil & Doja, 2017). |
| Gaussian Process and Machine Learning | Perception-based personalization of Hearing Aids using Gaussian process and active learning | Jens Brehm Bagger Nielsen, Jakob Nielsen, Jan Larsen / 2015 | The authors of this paper proposed another approach which described Hearing Aids Personalization based on Active learning and Gaussian processes. This study introduced a hearing-aid interactive personalization technique which obtains a setting for an individual of the hearing aids from direct inbuilt feedback of users. An interactive HA personalization system (IHAPS) based on machine learning described here enhanced various parameters which are directly based on aural perception of user rather than on verbal unclear description. With the involvement of user process of listening to and comparing HA settings, the Proposed System enabled the user to perceive his sound preference. IHAPS is based on visualized interactive loop. The loop basically contains following parts:<br>• Modeling<br>• Active learning, and<br>• User-interaction<br>The Authors concluded that IHAPS certainly may be suitable for both, the client to tweak the hearing aids and for the hearing-care professional. In order to reach there, the replicability of an individual setting is required to be studied further (Nielsen et al., 2015). |

## 4.3 RESEARCH METHOD

### 4.3.1 Objectives

The objective of this study is to consolidate research findings around the aim of locating different trends and major approaches that can be observed in the area of hyper-personalization and their evolution over time, while elaborating on the influence of various technologies and industries on the evolution of the phenomenon of hyper-personalization.

This is going to provide a clear view on how different components and mechanism of hyper-personalization have been generated and what are the primary aspects of such systems. The objectives are defined in the form of questions given below and extensive analysis was done to figure out their answers:

- Question 1: What can be depicted as the primary components of personalization, speaking on the basis of research literature?
- Question 2: How the domain of hyper-personalization has evolved over time, speaking on the basis of research literature?
- Question 3: How the technological transformations have affected the growth of personalized systems?
- Question 4: Which industries can be observed to have governed/regulated hyper-personalization?

Question 1 is targeted at collecting useful information on typical constituents of personalization and how this domain has changed over time.

Question 2 and Question 3 will basically focus on the trends and evolution of trends in the domain of hyper-personalized systems.

Question 4 will be focusing on different industries, which have been regulating the activities and thoughts surrounding personalized recommendation research.

### 4.3.2 Methodology

Our goal here is to present a review of techniques and approaches explained in different researches done on Hyper-personalization. To provide an overall catalogue of the academic literature on personalization techniques, the following conference databases and online journals were consulted: IEEE, Elsevier, Springer.

Each article was reviewed carefully and was then classified on the basis of journals and year of publication, as shown in later part of this paper.

The methodology used for this study is depicted in the flowchart shown below in Figure 4.2:

1. The data from different sources such as customer's personal data and behavioral data is collected in order to create hyper-personalized environment for the user. The collected data is then transformed in large volumes from one format to another to make it suitable for storing and processing together and, then finally, the unprocessed data is loaded to or processed data downloaded from data marts and data warehouses.
2. Next, the main step is to create hyper-personalized content by learning the customer's requirements through the data collected in data warehouse from different sources. Different steps like translation, localization, and content management are performed in order to fulfill customer's needs mapping to preferances.
3. The hyper-personalized content created in step 2 is then analyzed with the help of different processes like clustering, classification, and association for

# Hyper-Personalized Recommendation Systems

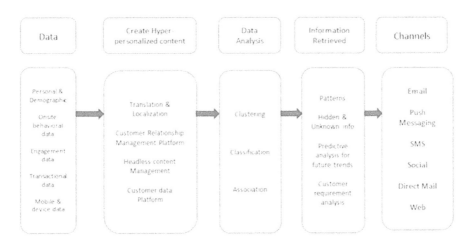

**FIGURE 4.2** Research Methodology.

extraction of useful information from the data and to take decisions on the basis of corresponding analysis.

4. Different hidden and unknown patterns are discovered on the basis of analysis and future trends are predicted in order to re-use same information for similar groups of users.
5. Finally, the discovered well suited subset of data is delivered to users through various channels, like personalized emails, SMS, or push messages in order to attract customers to products or services.

## 4.4 DISCUSSION

### 4.4.1 PERSONALIZATION VS. HYPER-PERSONALIZATION

Personalization helped marketers sell their products and services with the aim of fulfilling the needs and preferences of each individual, because personalization, by itself, is not enough for customers in today's scenario of customizing customer experience to meet with their preferences. Hyper-personalization is the next step to the traditional personalization techniques, where the table below represents the differences between the two approaches on the basis of various attributes as depicted in Table 4.1

### 4.4.2 PITFALLS OF TRADITIONAL APPROACHES

- Organizations using personalized conversation generally incur higher expenditure because it involves the usage of both time and money to arrange content which can attract various groups. In place of purchasing the traditional hoardings, organizations will now be spending money on tools, and various ads and content, to retrieve customer information and distribute information to them because simple brochures or pamphlets are no longer

**TABLE 4.1**
**Personalization vs. Hyper-Personalization**

| Attributes | Personalization | Hyper-Personalization |
|---|---|---|
| DATA | Personal and transactional information such as name, organization, purchase history etc. is elicited by traditional personalization approaches. | Behavioral and real-time data is elicited by Hyper-personalization such as browsing behavior, engagement data, and in-application behavior. |
| TECHNIQUES | This approach may simply use profiling to make certain assumptions about the customer on the basis of their characteristics. | It uses more complex and evolved techniques like web usage mining, machine learning, etc. |
| RESULTS | It might create slight or no customer interest for the product's brand. | It ensures more meaningful interactions and personalized communication with customers which helps in gaining customer loyalty. |
| EXAMPLE | A brand may advertise summer clothing to customers who bought similar products during previous season. | Promoting the similar items with ads based on real time data points like location, time, date, and payment method of the last purchase. |
| APPLICATION | Traditional e-commerce websites | Amazon which creates a user profile to sell products. |

enough to deal with customers of today looking to fulfill their needs and get product recommendations as per their preferences.
- Only search history and cookies will not work, as a few failures at reaching preferred products can be frustrating for users/customers. For example, a personalized email containing recipient's name will not attract the customer's attention if they do not find the content of the email to be as per their interest.
- Relevance of content plays a very important role in personalization which is a lacuna in traditional techniques and, ultimately, leads to distancing in customer loyalty. Let's take an example of an individual who was looking for information on French revolution who may personally be interested in African history, so that if the individual starts getting recommendations from Amazon on French revolution even when he or she no longer has any curiosity in the topic, it will lead to the customer moving away from the Amazon website.
- Organizations have to call for a lot more specific authorization requests while collecting personal data. Given the customer's privacy is an additional factor which should be considered while designing personalized experience, the continuous barrage of advertisement can sometimes make users feel like they are being tracked by someone or their privacy is getting breached.
- There's no use in continuously advertising goods or services which are only purchased periodically, such as major appliances or home renovations. Wasting a customer's time with irrelevant advertisements will definitely cause harm to a brand's trust – which once gone is difficult to get back. The marketers or

organizations should take into consideration that the products cannot be recommended always on the basis of history of user's behavior since their needs and goals might change over time and static recommendations will not work in such a scenario.

### 4.4.3 WHY HYPER-PERSONALIZATION?

Hyper-personalization has become important for marketers to sell their products and services to smart customers who are no longer interested in useless advertisements or generic approaches. Apart from this, there are facts which should be factored into consideration by marketers while designing a personalized experience for today's customers. Some of them are listed below:

- The message that we send as marketer has 8 seconds to catch and hold the online channel user's attention. To get observed, our communication should stand out and has to be clutter-breaking.
- Engagement of channel user with content has reduced by 60% and information overload is making consumers absent-minded.
- Out of total customers, 75% will be more likely to buy from someone who provides personalized offerings according to the preferences of individual. Retrieved from https://newsroom.accenture.com/industries/retail/us-consumers-want-more-personalized-retail-experience-and-control-over-personal-information-accenture-survey-shows.htm

Figure 4.3 depicts the needs for Hyper-personalization

### 4.4.4 HYPER-PERSONALIZATION FRAMEWORK

Hyper-personalized framework helps organizations take advantage of omni-channel data to generate real-time personalized customer experiences. Given customers use multiple digital devices or gadgets, such as tablets, mobile phones, laptops, or other technological devices, thus, the trail left by the customer can be tracked by tracking

FIGURE 4.3  Hyper-Personalization.

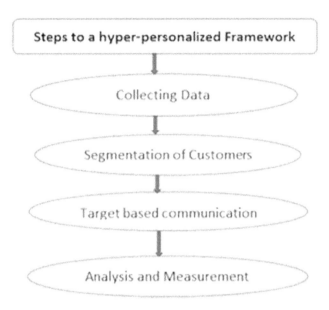

**FIGURE 4.4** Hyper-Personalization Framework.

the device-based interaction signs helping businesses collect loads of customer information on their lifestyle and online behavior. Employing this data to generate personalized experiences across the buyer's journey is essential to put into effect the benefits of hyper-personalization as shown in Figure 4.4.

- **Collecting Data** – The first and most important step towards a hyper-personalized framework is data collection because without data there is no understanding by the organization of its customers. Traditional personalization methods make it very difficult to understand the customer as evident from customers in today's world, who think that brands are not able to understand them as individuals.
- **Segmentation of Customers** – The data collected will help the organization gain an understanding of customer's interaction with their brand, and to construct Hyper-Personalized messages and experiences for them. This process of hyper-personalization is incomplete without segmentation of customers, which is the process of grouping customers into various subsets on the basis of differentiated factors such as satisfaction, location, average spend, demographic, and interaction history with brand. If we combine data and segmentation, we can get loyal and engaged customers.
- **Target based Communications** – Once the customer segmentation is done, hyper-personalized communication can be initiated. Selecting the right channel along with the right time is essential for the success because the chances of conversation will be more if selectively targeted ads and relevant communications are generated.

- **Analysis and Measurement** – It is not enough to only run a targeted campaign. Measuring the success of campaign is important. Once the metrics of a campaign have been figured out which are directly associated to the revenue, it can easily be replicated for sustained results.

## 4.5 RESULTS

On the basis of this study, we figured out the answers to all the four questions based on which the complete research was done. The research was aimed at analyzing personalization techniques along with their pros and cons.

- Online channel user's data, collected through various sources and the tools/techniques which utilize this data in order to study the user behavior, can be considered as the primary components of personalization. Understanding the channel user's behavior, preferences, and needs is the most important step to sell personalized products or services and gain high revenue.
- Privacy plays a vital role today and channel users are not ready to accept personalization at the cost of their personal details falling in the hands of others. With the evolution in requirements and behaviors of customer, there is a need to enhance traditional patterns and create hyper-personalized environment which considers each channel user as unique.
- Introduction of artificial intelligence and machine learning techniques have enhanced the data utilization process to efficiently analyze large volumes of data and draw patterns to generate recommendations which can fulfill the needs of customer.
- Various industries like e-commerce, e-learning, healthcare, banking, or another sector can be considered to have utilized hyper-personalized products for their users successfully and earned higher customer satisfaction and loyalty as compared to fallouts from use of non-personalized services.

Different research papers that were studied in order to understand the concept of personalization were classified on the basis of journals and year of publication.

### 4.5.1 Classification on the Basis of Research Papers Used

The articles were analyzed by approach, year of publication, and the total number of articles in each journal. The graphs below represent the distribution:

a. Distribution by Publication Year – Different articles published between 2009 and 2020 were studied for this research and, on the basis of this classification, it was observed that 2016 was the year when maximum number of articles on personalization were published which depicted how personalization has helped various industries grow as per Figure 4.5.
b. Distribution by Journals – The articles published by different authors in different journals as depicted in Figure 4.6.

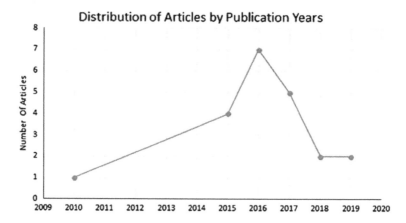

**FIGURE 4.5**  Distribution of Articles by Publication Years.

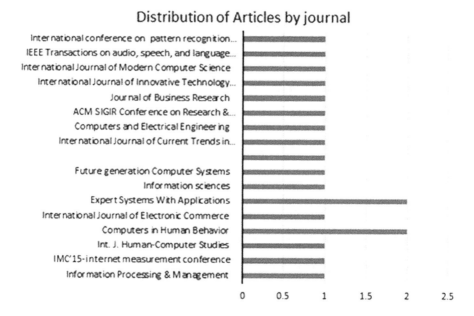

**FIGURE 4.6**  Distribution of Articles by Journal.

### 4.5.2 Focus Areas

**Content Delivery**

Process of hyper-personalization is not confined to how consumers are targeted by marketers and how messages are designed for them. The need of today's end user is to get engaged with the most concise and relevant content in innovative and expressive ways such as virtual reality (VR), stories on social media, videos, or some other medium. Thus we need to analyze and study further on how to leverage these techniques in a hyper-personalized system.

## Data Analysis

The data is in huge volumes and processing/analyzing a data-point while ignoring another is hard to do since it's difficult to choose between millions of datapoints, thus resulting in wrong prediction or over-personalized products.

Attribute analysis and event sequence analysis should be embedded in data analysis process to form patterns based on channel user's past experience with website/product.

## Reaching Consumers

Artificial Intelligence based chatbot services can help consumers in navigating to different products according to their needs and to roam about the website. This application can be utilized to take care of all plausible concerns of the consumer after he or she purchases the organization's product, which means the need to enable efficient after-sales support.

## Data Security

Organizations need a lot more specific authorization requests while collecting personal data. Not only do they need to be clear about how to collect, process, and store data, but also have the proper mindset to manage transparency and consent which if properly done will give the individual control over how their data is getting processed, by whom, and for what purpose. Thus, hyper-personalization is achievable within the framework of data privacy.

### 4.5.3 EXPERIMENTAL ANALYSIS OF CAMPAIGN DESIGNING

On the basis of the study, we were able to figure out various parameters that are essential for any hyper-personalized network to work and earn revenue. This can be explained with the help of an example based on the e-commerce industry given below:

- **Trigger point**: Idle cart (user doesn't proceed with purchase after adding product to cart).
- **User based attributes:** Name, membership status, message medium chosen.
- **Behavioral attributes:** Products viewed.
- **Data points on the basis of past purchase:** Favorite day of purchase, discount coupon applied, device/gadget used for purchase.

A personalized message (containing personalized token on the basis of data) is created with the help of a marketing cloud platform for advertisement of discount applied on product the customer viewed and added to the cart (if the discount was applied in past purchase).

Later, a message can be sent to the device/gadget used for past purchase and a supplementary offer of faster delivery can be triggered if the user belongs to premium category(like prime accounts on Amazon). Analyzing the historical data, the communication channel can be figured out which had the highest engagement in the past and incorporated in new campaign. On checking purchase data of channel user, the message can be triggered on a particular day of the week driven by prospects of results as brought out by better results in the past. Thus, it becomes a perfectly hyper-personalized campaign.

## 4.6 CONCLUSION

In this paper, we give a brief overview about various approaches used by different researchers in order to model a system which can personalize products or services for their users like E-learning platforms, hearing aids settings, hyper-personalized goods based on user's DNA, music or movie recommendation systems, and Semantic web with the help of multiple techniques like collaborative filtering, machine learning, deep Learning and neural networks, and a lot of other different mechanisms. The aim here was to figure out the basis of each approach along with the main problem that was targeted by different researchers during this time period. We discussed the working of some of the proposed mechanisms along with their limitations.

Further knowledge needs to be gleaned and research needs to be conducted in this domain, particularly in the implementation of approaches defined in this study in order to fulfill the requirement of today's customers. The future work, which can be extended based on further investigations into the relationships between various subsystems, is in order which can help organizations to achieve a hyper-personalized environment to gain new customers insights.

## REFERENCES

Chen, J., Zhang, H., He*, X., Nie, L., Liu, W., & Chua, T. (2017). Attentive collaborative filtering: Multimedia recommendation with item-and component-level attention. *ACM Conference, SIGIR'*, *17*, 335–344.

Diwan, S. (2016). Towards semantic web using personalization techniques. *International Journal of Applied Engineering Research (IJAER)*, *11*(4), 2451–2453.

Ebesu, T., Shen, B., & Fang, Y. (2018). *Collaborative memory network for recommendation systems* [Conference session]. Research & Development in Information Retrieval, ACM SIGIR Conference, pp. 515–524.

Essalmi, F., Ayed, L., Jemni, M., Graf, S., & Kinshuk (2015). Generalized metrics for the analysis of E-learning personalization strategies. *Computers in Human Behaviour*, *48*(C), 310–322.

Kaptein, M., & Parvinen, P. (2015, March). Advancing E-commerce personalization: Process framework and case study. *International Journal of Electronic Commerce*, *19*(3), 7–33.

Kliman-Silver, C., Hannak, A., Lazer, D., Wilson, C., & Mislove, A. (2015). *The impact of geolocation on web search personalization* [Conference session]. IMC'15-internet Measurement Conference, pp. 121–127. 10.1145/2815675.2815714

Neil, A., & Vairamuthu, S. (2019). Music recommendation using collaborative filtering and deep learning. *International Journal of Innovative Technology and Exploring Engineering (IJITEE)*, *8*(7), 964–968.

Nielsen, J. B. B., Nielsen, J., & Larsen, J.(2015). Perception-based personalization of hearing aids using Gaussian processes and active learning. *IEEE Transactions on Audio, Speech, and Language Processing*, *23*(1), 162–173.

Silva, E., Celso, G., Junior, C., Pascoal, L., & Rosa, T. (2016). An evolutionary approach for combining results of recommender systems techniques based on collaborative filtering. *Expert Systems With Applications*, *53*, 204–218.

Sunil, & Doja, M. N. (2017). Recommender system based on web usage mining for personalized e-learning platforms. *International Journal of Modern Computer Science*, *5*(3), 48–53.

Wei, J., He, J., Chen, K., Zhou, Y., & Tang, Z. (2017). Collaborative filtering and deep learning based recommendation system for cold start items. *Expert Systems With Applications*, *69*, 29–39.

# 5 Evolutionary Computational Technique for Segmentation of Bilingual Roman & Gurmukhi Handwritten Script

*Gurpreet Singh and Manoj Sachan*

## CONTENTS

5.1 Introduction ..................................................................................................87
5.2 Literature Review .........................................................................................89
5.3 Framework for Segmentation Process of Bilingual Hrs ..............................92
5.4 Word Level Segmentation ............................................................................93
5.5 Pre-Processing ..............................................................................................95
      5.5.1 Duplicate Points Removal from Stroke Data (DPR) ......................97
      5.5.2 Missing Points Identification from Input Stroke (MPI) .................98
      5.5.3 Normalization and Centering of Strokes (NCS) .............................98
5.6 Script Identification .....................................................................................99
5.7 Stroke Level Segmentation ........................................................................100
5.8 Results and Discussion ..............................................................................103
5.9 Conclusion .................................................................................................105
5.10 Future Scope ..............................................................................................106
References ............................................................................................................106

## 5.1 INTRODUCTION

The efficiency of handwriting recognition applications always reflects the intelligent systems handling the complex computations (De Jong, 2019; Xue et al., 2016). A number of evolutionary computational models has been tried by the researchers to create these intelligent systems. To build the applications, the extraction of different Region of Interests (ROI's) is one of the necessary steps for any pattern recognition application, and the process of extraction for ROI is known as segmentation (Akouaydi et al., 2017; Wigington et al., 2018). Segmentation is responsible for the identification of some meaningful patterns from the raw data, and in the case of automatic handwriting recognition, the segmentation process is based on the idea of making meaning of the pixels involved in the formation of strokes or characters or even words (Babbel.com, n.d.; Sachan et al., 2011; Verma, 2003). But, in case of online handwriting recognition applications, pinpointing the information about those points which are touched by the digital pen over the surface of any digitizer during handwriting stroke formation, i.e. between pen up and pen down events, is considered as a process of segmentation and identification of a more refined unit "stroke". A single extracted stroke during online handwriting process is equivalent to the binary form of output in case of segmentation process under offline handwriting recognition environment("Best OCR and handwriting dataset for machine learning applications", n.d.; "Roman Recognizer", n.d.; Blumenstein et al., 2002; Chim et al., 1999).

Once pixels forming handwriting are identified, the identification of other meaningful units such as text lines, words, characters, and sub parts of characters, etc are next. From the handwritten text, the text lines are extracted by using the concept of horizontal projection. Figure 5.1 shows the expected output from horizontal projection approach (Gattal et al., 2016; Marti & Bunke, 2001; Pesch et al., 2012; Razzak et al., 2009; Wshah et al., 2009).

From Figure 5.1, we can see that during the identification of upper boundary and the lower boundary of a complete line of text, we need the dimensions of the whole of the writing area from the digitizer's surface. The topmost co-ordinates of every stroke drawn vis-a-vis the lowest value of the horizontal axis have to be recorded to draw the upper boundary of the complete text line, and in a similar manner, to get the lower boundary of the different text lines, the lowest coordinates of every stroke drawn with respect to the highest value of the horizontal axis have to be recorded. Thus, the topmost value helps in the identification of lower boundary of the different text lines existing in the handwriting sample supplied as input. There is a need to maintain two different value sets to store the highest and the lowest values, respectively, for each stroke and one derived attribute i.e. the height of each stroke along with these two values. The combination of these values helped in the identification of text line boundaries.

Next, the segmentation of complete handwritten words is needed. Figure 5.2 shows the output of vertical projection technique, which can extract a complete

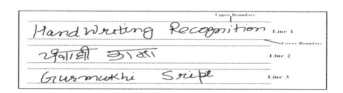

**FIGURE 5.1** Horizontal Projection for Text Line Segmentation.

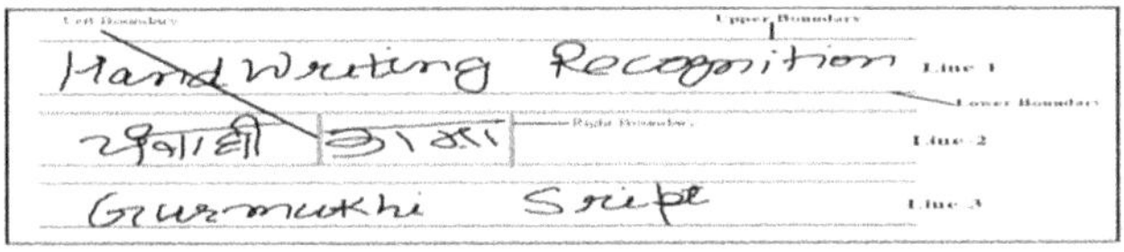

# Evolutionary Segmentation Technique

**FIGURE 5.2** Vertical Projection for Complete Word Identification.

word by identifying the left and right outer lines of the word, when a word is represented by a complete rectangle observed in a similar way as line segmentation approach. Apart from vertical projection technique, in case of online handwriting recognition where only stroke data is concerned, the word level segmentation is carried out on the basis of recording of connected strokes information alongwith the high and low values of vertical axis for each stroke. A derived parameter about the width of each stroke needs to be recorded to accomplish the task of word segmentation. Further, the character extraction process uses the knowledge base approaches existing with different scripts. Some of the factors affecting the process of the segmentation of individual characters are the consideration of monolingual or bilingual or even multilingual systems handling the problem of intermixed text.

## 5.2 LITERATURE REVIEW

With the aim of understanding the work already done in the area of segmentation of handwritten data, the main focus of this section of the chapter is to collect the state of the art information of work done at what depth by other researchers. Following are some of the selective and relevant research works in the field of Online and Offline handwriting recognition, addressing the issue of segmentation.

Samanta et al. (2015) worked on an online handwriting recognition system which didn't incorporate the initial idea of the script used to write words as an important factor in handwriting recognition. Authors tested the performance of the system against three different scripts: Roman for English language, Arabic, and Bangla. Proposing different segmentation schemes in the category of curve analysis at various discrete points, local minima point calculation, and ANGseg analysis, the authors captured the directional movement of digital pen. In comparison, local minima based segmentation provided better results as per their observations.

Teja and Namboodiri (2013) presented an online handwriting recognition system for Indian language "Malayalam", wherein they considered the concept of bulky strokes formation and formation of meaningful script symbols from these bulky strokes. The technique for maximising the curvature data was used by the author to get the idea about various end points, and cubic spline technique was used to expound on the idea about the structure of the strokes. An accuracy level of approximately 95% was observed with the help of support vector machine mechanism.

Sachan et al. (2011) presented a segmentation technique for online Gurmukhi handwriting recognition system that needed, along with the handwritten strokes information, information on pen pressure exerted by the writer on the surface of digitizer, the number of pen down/ pen up events, and the timing of the formation of each stroke. To handle the situation of over segmentation, they introduced the concept of merging of sub strokes, even as the performance of the technique was measured with the closed vocabulary system having the dataset of 2150 words having the overall accuracy as 86.40% with the implementation of merging of sub strokes algorithm. A lacuna of the proposed segmentation algorithm was in its not showing desired segmentation results when words with slanted strokes were provided as input. Number of times the proposed algorithm mistakenly considered " [ " and " { " as different headline strokes were counted, as well as how this also led to the incorrect identification of words.

Verma (2003) proposed segmentation approach addressed the complexity of cursive nature of handwriting style, wherein he worked on English language using Roman script. The rule-based model was proposed by considering coding information of contour present in handwriting samples, and from contour information, some segmentation points were selected and validated by using various parameters like average character size, type of left character availability, etc. For the purpose of classification, Artificial Neural Network (ANN) was used to recognize both character and non-character types of segmented area, while with the handwriting sample, the computation about baseline area was performed. Then, the rule-based approach was used to train the network for further classification, where the approach's main focus was on the detection or identification of loops present in the sample, to get an idea about the Hat shape contours and their removal, removal of false segmentation points using average width of characters, and missing points detection. He followed the bench mark dataset CEDAR to train the proposed system and observed the classification accuracy of 99.67% by transition feature and 100% while considering the training module. At the time of testing, recognition accuracy achieved was 86.84%.

Zimmermann and Bunke (2002) proposed a method of segmentation based on Hidden Markov Model (HMM) classifier, while working on IAM dataset containing handwriting samples of English text. The level of complexity considered was the cursive style of handwriting, and their findings added two more fields as boundary box at word level and ground truth for every word having information about punctuation characters. The overall accuracy achieved by their system was recorded as 98%. In analyzing the approach, they found 417 complete pages of text and about 25000 words from these pages had been addressed for boundary detection. To deal with the concept of word boundary box detection, they firstly converted the input image to its equivalent Binary form; then, connected component analysis was used to find the smaller connected components, with these components processed by the HMM classifier to identify various word classes.

Bluche (2016) proposed neural network based technique (MDLSTM-RNN) Multi-Dimensional Long Short Term Memory Recurrent Neural Network, which is used to process offline handwritten paragraphs to generate equivalent digital output, carried out segmentation at the level of complete lines over the input text presented in the images. For training and testing part of his proposed system, the handwriting databases of "Rimes" and IAM were used, and the observed error rates revealed by the softmax decoder from observing 150 dbi images were 8.4% and 4.9% for IAM and "Rimes" databases, respectively. The overall impact of this research work reflected 20% and 50% improvement over IAM and "Rimes" databases, respectively. Valid for implementation under a single paragraph only, the proposed concept was not workable if the input image represents a complete page with multiple paragraphs.

Sadri et al. (2004) working on a segmentation technique for the extraction of handwritten numerals present in the input text image, considered the skeleton tracing approach for segmentation by focusing on both foreground and background features of the image. Traversing each connection point of differently connected components, their analysis was carried out in observing multi directional movement of the components. For detecting various background features, a vertical projection approach was used; while the overall segmentation was made possible on the result of performance of both foreground and background features. They considered NIST SD19 database, for the purpose of training and testing, and also proposed whether to record the recognition feedback or the performance of overall system for each input to construct some recognition scores as new parameter in the database, so that the new parameter may help in future implementations.

Wshah et al. (2009) worked on the segmentation of offline Arabic script under automatic handwriting recognition system, and proposed the solution of segmentation for complete Arabic words containing maximum of three characters. Contributing to the arrival of lexicon for these Arabic words to deal with the complexity level of the script due to the existence of large number of words, the authors proposed both chain code method and skeleton formation approach to handle the problem under consideration. Overall they achieved 93% accuracy for the segmentation process. The work was done with material from the databases DARPA and IFTN.

Marti and Bunke (2001) to carry out segmentation and recognition of words, proposed the extraction of complete lines from text images and their classification, and, by addressing the issues of writer independent systems, they took the path of unconstrained segmentation. With the difficulty of picking a suitable threshold for the purpose of segmentation, as different thresholds depended upon plausibly existing inter and intra word distance, where these kind of distances may vary in case of writer independent and unconstrained systems, the authors found that the choice of low or high values of threshold may result in information loss. They claimed 95.56% accuracy of their proposed algorithm for line segmentation and overall 73.45% recognition results observed at the time of word level text classification.

Table 5.1 imparts the thought that most researchers are working on the concept of monolingual scripts, so that only one script across all these monolingual systems is used to write the handwritten text. At the time of segmentation, rules applicable for a single language and its script are considered by the researchers. For segmentation under bilingual or multilingual environment, the main barrier is to identify the script or the language of the part of the text under consideration.

## TABLE 5.1
## Summary of Literature Reviewed

| Authors | Script/Language | Technique |
|---|---|---|
| Samanta et al. (2015) | English/Arabic/ Bangla | MinSeg and ANGSeg |
| Teja and Namboodiri (2013) | Malayalam | Ballistic strokes |
| Sachan et al. (2011) | Gurmukhi | SVM, kNN |
| Zimmermann and Bunke (2002) | IAM dataset English | HMM |
| Bluche (2016) | English | MDLSTM-RNN |
| Sadri et al. (2004) | English | Skeleton Tracing |
| Wshah et al. (2009) | Arabic | Lexicon Generation |
| Marti and Bunke (2001) | English | Intra word distance |

## 5.3 FRAMEWORK FOR SEGMENTATION PROCESS OF BILINGUAL HRS

As the bilingual system under consideration accepts online handwritten text (Kicinger et al., 2005), and since by its bilingual nature, the written text contains words from Roman and Gurmukhi scripts, therefore to convert the handwritten data to its equivalent digital form, an important step after pre-processing is segmentation. Different levels of segmentation have to be performed to get the digital data from handwritten samples with the help of evolutionary computational techniques. Figure 5.3 highlights these different levels, wherein the first step to perform during segmentation is to segment at the level of complete words out of bilingual intermixed text, and the next is the segmentation at the level of unique strokes representing a single word, which helps to provide the valuable information about the script of the word under consideration.

If script identification process directly depends on the results of segmentation phase at the level of strokes, then we can proceed with the recognition engine of the respective script for the extraction of refined information about the presence of different characters in the segmented word.

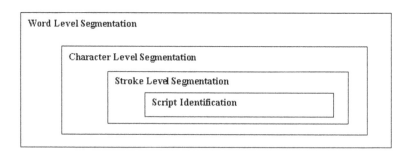

**FIGURE 5.3** Hierarchy of Segmentation Process in Online Handwriting Recognition Application.

## 5.4 WORD LEVEL SEGMENTATION

If word level segmentation has been performed based on the connectivity of different strokes with one another, and if the connectivity can be confirmed based on the existence of common stroke points between different strokes and the appearance of one stroke inside the boundary of any other stroke, then the minimum and maximum values of x and y coordinate points can be recorded for each stroke by considering its different stroke points. These coordinates help in the process of predicting the boundary of the stroke. Different sets containing a number of connected strokes having been identified, every set represents a complete handwritten word. Algorithm 5.1 is used to segment different words from the online text given as input to the system, which makes the base criteria of the segmentation to be different words as the presence of space between two consecutive words and the identification of a stroke which never show any kind of connectivity with the previously identified sets of the complete words. Algorithm 5.1 takes input as the set of strokes generated by the writer during the writing process. Initially, first stroke is considered as the first segmented word.

---

### ALGORITHM 5.1  WORD LEVEL SEGMENTATION (S, W)

*// Where S is a set to k input strokes and W represents the superset represented by different m set of complete Words*

Step 1.
Set $W=NULL$, $W(S_1)=1$, $i=2$, $j=2$;

Step 2.
*Repeat While K strokes in S
Loop*
Set $J:=k$
*Repeat While $j>=1$ or $Common(j,j-1)==True$
Loop*
*If $Common(j,j-1)==True$
Then*
Set $W(S_j)=i-1$
Set $j=j-1$ and $i=i-1$;
*Else*
$W(S_j)=i$
*End if*
Set $J=J-1$
*End Loop
End Loop*

Step 3.
*Segmented Words Set*

---

The value $W(S_1) = 1$ represents the assignment of word label 1 to first stroke $S_1$. Then for the rest of the strokes, the connectivity relation as common points or overlapping of boundary boxes with exactly its previous stroke has been calculated by using the procedure Common(). If any kind of relationship of the current stroke exists, particularly with its previous stroke, then the word segmentation level assigned to the previous stroke is reflected with the current stroke also. In the end, different subsets of the strokes with a common word segmentation level value are treated as the words belonging to the handwriting sample under consideration.

The following nomenclature is used to represent the segmentation process at complete word level when a bilingual text $T_i$ is considered.

$$W_{seg} = \frac{S_{true}}{S_{false}} \times 100 \tag{5.1}$$

$W_{seg}$ is used to express the percentage of correct segmentation of handwritten words from bilingual text $T_i$. $S_{true}$ and $S_{false}$ represents the total number of correctly segmented words and total number of falsely segmented words respectively for bilingual text $T_i$. Some samples have been considered to show the effectiveness of segmentation process. The results obtained from these samples are presented with the help of following examples.

Figure 5.4(a) represents a handwriting sample containing nine complete words of bilingual Gurmukhi-Roman text. The word level segmentation results of the same are represented in Figure 5.4(b). The eighth word of this sample, which is actually a Gurmukhi script word shows over segmentation. The reason behind this over segmentation of the data is found to be the distortion produced during the generation of writing sample by the writer. The cause for such kind of distorted strokes may be the speed of the writer, may be the quality of the surface of digitizer, or the quality of the stylus used.

Figure 5.5 (a) represents another handwritten bilingual online sample consisting of a total of ten words, where each word has been written using different number of strokes. The application under consideration has been able to segment all the words exactly as per the requirement against the handwritten sample. The result of segmentation is reflected by Figure 5.5 (b).

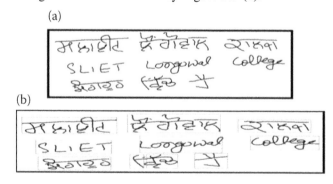

**FIGURE 5.4**  a: Bilingual Handwritten Sample-1, b: Word Level Segmentation Sample-1.

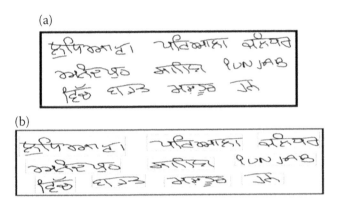

**FIGURE 5.5** a: Bilingual Handwritten Sample-2. b: Word Level Segmentation Sample-2.

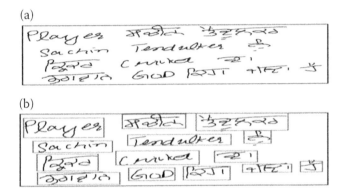

**FIGURE 5.6** a: Bilingual Handwritten Sample-3. b: Word Level Segmentation Sample-3.

Figure 5.6 (a) shows a handwritten sample of fourteen different words. The system has been able to correctly segment all the words present in the sample. This is also presented in Figure 5.6 (b). Similarly, the performance of the approach proposed for word segmentation for bilingual online handwritten text has been tested over many samples during the testing phase.

## 5.5 PRE-PROCESSING

As already mentioned, the quality of the input strokes mainly depends upon the type of digitizer and stylus used, while other factors like awareness among the writers about the use of stylus and digitizer screen, writing speed of the individual writer etc. can also affect the quality of the captured strokes. Input having some kind of noise that, if directly supplied to the higher phases like segmentation and classification, gives rise to an expectation of erroneous results in the overall process to appear can be dealt with by singling out the kind of noise present in the input strokes, or pre-processing steps. Some of the factors for noise detected are the presence of redundant stroke points, or the absence of some specific stroke points, which may affect the overall prediction about the stroke category and the normalization of stroke points to deal with the variety of writing surfaces having different dimensions. The presence of duplicate points always increases the processing overhead for the overall process and the removal of redundant points is considered as one of the basic steps in all kind of handwriting recognition applications. Many researchers proposed algorithms mainly for the purpose of redundant point removals based on the language of the script under consideration (Akouaydi et al., 2017; Bluche, 2016; Elleuch et al., 2016; Marti & Bunke, 2001; Naeem Ayyaz et al., 2012; Sadri et al., 2004). This chapter also describes the algorithm implemented for the removal of redundant or duplicate points from the bilingual (Gurmukhi/Roman) handwritten data, where the other main concern during the pre-processing phase is to identify or recover all those missed out points which may contribute to the extraction of the exact information about the shape or the structure of the strokes under consideration. Some of the reasons behind the missing out of this information are:

- The poor quality of the hardware.
- High writing speed of writers.
- Writing on the edges of the digitizer's surface.
- The use of passive stylus technology.
- Poor writing style or no proper command over the script under consideration.

While checking the quality of hardware used for capturing the handwritten strokes, it can easily be observed that, if a passive stylus having low quality is used to collect handwriting samples then there is a possibility to miss out on some important coordinate points during the generation of different strokes. The passive stylus is considered incapable to provide smooth handwriting strokes. Secondly, stylus apart, the hardware quality is often affected by the use of digitizers where the writing surface is different from the display surface, given these devices mainly come with USB support which hardware environment is a disadvantage to the writer who cannot watch the output screen and the movement of stylus on writing pad simultaneously. This kind of conscious environment may lead to uneven handwriting and the missing out of some important information at the time of stroke creation. Next, a very high writing speed of various writers may also cause the missing out of some important information about the shape or structure of stroke under consideration. Sometimes, the writers are holding a passive stylus and with the writing speed of writer being very fast, the stroke points capturing process of the passive stylus may result in multiple sub strokes rather than the formation of a single large stroke. This kind of distorted information affects the overall recognition process and may generate some strokes which might not fall under the category of defined stroke classes. Fourthly, when the writers start writing from the top or from any specific corner or edge of the digitizer's surface, which, during the phase of writing flows, does not help the writer who may have to lift the pen up in case of exact boundaries of writing pad causing distorted strokes. Thus, the writer may miss out some important stroke points or produce some unknown sub stroke in place of a single bigger stroke. Lastly, missing points can cause concern, in case of stroke formation, when the familiarity level of writer with the language and script under

# Evolutionary Segmentation Technique

consideration or with the digital pen based writing environment is little, with the result that instead of writing a single unique stroke the writer may produce many sub strokes. These sub strokes may or may not fall under the basic strokes categories. All of these can be considered the reasons for improving the quality of input supplied to the online handwriting recognition system, and to deal with these issues of noise and to improve the quality of the overall input, the following pre-processing algorithms are implemented for the online handwriting recognition system for bilingual (Gurmukhi & Roman) environment:

- Duplicate Point Removal from stroke data (DPR).
- Missing Points Identification from input stroke (MPI).
- Normalization & Centering of stroke data (NCS).

### 5.5.1 Duplicate Points Removal from Stroke Data (DPR)

This pre-processing step is necessary to avoid the overhead produced with the introduction of repetitive stroke points within a single stroke. Many times the main cause of this is considered to be the familiarity level of the writer with digital writing environment or a not that good quality of the hardware equipment used for the purpose of writing. Algorithm 5.2 provides a solution for removal of these unnecessary data points.

---

**ALGORITHM 5.2 DUPLICATE POINTS REMOVAL(S, XVAL[], YVAL[])**

// 'S' is a stroke and two empty arrays Xval[] and Yval[] to hold X and Y values of each coordinate of the stroke respectively.

Step 1.
*Consider each stoke point of stroke "S"*

Step 2.
*Compare current stroke point with the data stored in Parallel arrays Xval [] and Yval[] for respective index values.*

Step 3.
*Add point in the parallel array if no previous match found*
*Else*
*Discard the point*

Step 4.
*Repeat step 2 and 3 for each stroke point of stroke S.*

Step 5.
*Stop*

---

This algorithm worked over two vectors Xval[] and Yval[]. These vectors are used to store the 2-D coordinate information of the points touched by the writer on the surface of digitizer during the stroke formation. As per the essence of the algorithm, before considering any new entry these vectors consider the current values under consideration with already stored data at respective indexes of parallel arrays.

### 5.5.2 Missing Points Identification from Input Stroke (MPI)

Algorithm 5.3 is used to find the missing points during the generation of stroke data. The main concept used to identify the required missing points is the use of threshold. Here, a distance threshold is maintained to identify the missing point between two consecutive stroke points which helps in improving the quality of input data.

### 5.5.3 Normalization and Centering of Strokes (NCS)

This algorithm is used for the purpose of generating strokes from the relative plane to the absolute plane of the digitizer's writing surface with specific dimensions.

---

**ALGORITHM 5.3 *MISSING POINTS IDENTIFICATION (S, XVAL[], YVAL[],N)***

*// S represents the stroke under consideration, n is the total number of points under stroke S and Xval[] & Yval[]are considered to be parallel arrays containing the respective coordinates information of each stroke point.*

Step 1.
*Consider every point of stroke "S"*

Step 2.
*Store the distance between each consecutive stoke point in Distance[] array.*

Step 3.
*Fid the Average distance to analyze Distance threshold*

Step 4.
*Scan and find the consecutive stroke points having distance greater than the double of distance threshold.*

Step 5.
*if any pair of points found then create new stroke point using interpolation.*

Step 6.
*Stop*

Evolutionary Segmentation Technique

---

**ALGORITHM 5.4   NORMALIZATION AND CENTERING OF STROKES (S, XVAL[], YVAL[])**

/ S is the input stroke under consideration and Xval[] and Yval[] are the parallel linear arrays used to hold the coordinate information of each point contained by stroke S.
/

Step 1.
Find Min and Max values of X and Y components of stroke points.

Step 2.
Assign new X and Y values to each stroke point as per the absolute plane on the scale of origin (0,0).

Step 3.
Stop.

---

Algorithm NCS (Normalization and Centering of Strokes) is used for this purpose. The main reason for the requirement of this algorithm is the availability of digitizer's surface in variety of sizes.

As per the size of the surface, each hardware device have different contraction level of pixels under the writing area. So, there is a requirement for a normalized plane, which gives the capability to write some machine independent code. To express the captured strokes under the normalized plane Algorithm 5.4 NCS is introduced.

## 5.6   SCRIPT IDENTIFICATION

A critical phase of the bilingual online handwriting recognition system, script identification's biggest challenge is the identification of the script associated with the segmented word. In this work, an algorithm was proposed for the segmentation of complete words from the stroke data provided as an input by different writers, where information on these segmented words is provided as an input to the script identification phase. Considered important because of the unique nature of different scripts as per their own character sets and rules of writing, the script identification process associated with the segmented word performs segmentation at the level of different strokes as the first step. That is, the total number of strokes with the detailed information about their respective set of stroke points are the main findings of the stroke level segmentation for each identified word. The second step under script identification process is the categorization of each stroke as one of three different categories, that is "Roman", "Gurmukhi" or "Common". The detailed description about these categories has been provided in the next part of this chapter. The shape and structural features of strokes have been explored to find the suitable category of a each stroke. For this, at the level of feature extraction; DZIC

**TABLE 5.2**
**Roman Strokes**

| S_ID | Stroke | S_ID | Stroke | S_ID | Stroke | S_ID | Stroke |
|---|---|---|---|---|---|---|---|
| 1 | / | 11 | ∧ | 21 | G | 31 | ⊣ |
| 2 | < | 12 | L | 22 | M | 32 | N |
| 3 | V | 13 | O | 23 | P | 33 | S |
| 4 | y | 14 | Z | 24 | a | 34 | ℓ |
| 5 | b | 15 | ι | 25 | d | 35 | e |
| 6 | B | 16 | g | 26 | R | 36 | h |
| 7 | J | 17 | ℓ | 27 | n | 37 | p |
| 8 | q | 18 | ℳ | 28 | r | 38 | ß |
| 9 | u | 19 | v | 29 | w | | |
| 10 | x | 20 | y | 30 | 3 | | |

(Directional Zone Identification Code) and SVTC (Stroke Vector Trajectory Code) algorithms have been implemented, where these algorithms have been used to collect information about the shape and the structure of the stroke under consideration. The classification phase is the last phase with results of the different strokes as identified for the bilingual online handwriting recognition system of Gurmukhi and Roman scripts. For the classification of these strokes, different classifiers have been used such as HMM, SVM, MLP and DNM. The last part of this chapter shows the script identification results for the different intermixed samples of Punjabi and English words.

## 5.7 STROKE LEVEL SEGMENTATION

Dealing with the identification of handwriting strokes, and as the system under consideration is an online system for handwriting recognition, stroke level segmentation provides a complete input unit known as stroke by recording each pen down followed by pen up event. Further, for handling the input in the form of intermixed text of Gurmukhi and Roman scripts, a total of 84 different stroke classes has been observed. These classes are divided into following three different categories:

- Roman Strokes.
- Gurmukhi Strokes.
- Common Strokes.

# Evolutionary Segmentation Technique

Roman strokes are the unique strokes which belong to Roman script only. A total of 38 unique Roman stroke classes have been identified during the implementation of bilingual handwriting recognition system. These Roman stroke classes are represented by Table 5.2. This table also reflects the unique stroke id's that are assigned to each stroke class. Similarly, Gurmukhi strokes are the unique strokes that belongs to Gurmukhi script only. A total of 34 unique Gurmukhi stroke classes have been observed for the bilingual handwriting recognition system under consideration. These Gurmukhi stroke classes are represented by Table 5.3. During the implementation process, it has been observed that, any particular word written under bilingual environment, either contain all Roman strokes or all Gurmukhi strokes. Another important finding is that both Gurmukhi and Roman scripts contain some similar strokes.

Identified as "Common" strokes, the strokes in Table 5.4 represent the twelve common stroke classes observed during the formation of input text supplied to bilingual handwriting recognition system. These common strokes are used to complete the description of some characters from the character set of both (Roman, Gurmukhi) scripts.

## TABLE 5.3
### Gurmukhi Strokes

| S_ID | Stroke | S_ID | Stroke | S_ID | Stroke | S_ID | Stroke |
|---|---|---|---|---|---|---|---|
| 39 |  | 49 |  | 59 |  | 69 |  |
| 40 |  | 50 |  | 60 |  | 70 |  |
| 41 |  | 51 |  | 61 |  | 71 |  |
| 42 |  | 52 |  | 62 |  | 72 |  |
| 43 |  | 53 |  | 63 |  |  |  |
| 44 |  | 54 |  | 64 |  |  |  |
| 45 |  | 55 |  | 65 |  |  |  |
| 46 |  | 56 |  | 66 |  |  |  |
| 47 |  | 57 |  | 67 |  |  |  |
| 48 |  | 58 |  | 68 |  |  |  |

## TABLE 5.4
### Common Strokes of Bilingual (Gurmukhi + Roman) System

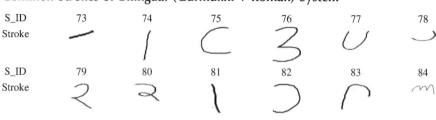

Table 5.5 represents all categories of strokes along with the data reflecting the segmentation accuracies achieved by the proposed system during the identification and recognition phase of bilingual system. Category "R" represents Roman strokes, "G" represents the Gurmukhi strokes and "C" represents the common strokes. Figure 5.7 graphically represents these results. Overall 98.19% stroke

## TABLE 5.5
### Stroke Segmentation Accuracy Analysis

| Sample No. | Total Words Present | Correct Detection | False Detection | %age |
|---|---|---|---|---|
| 1 | 9 | 8 | 1 | 88.88 |
| 2 | 10 | 10 | 0 | 100 |
| 3 | 7 | 7 | 0 | 100 |
| 4 | 9 | 8 | 1 | 88.88 |
| 5 | 14 | 14 | 0 | 100 |
| 6 | 10 | 10 | 0 | 100 |
| 7 | 9 | 9 | 0 | 100 |
| 8 | 17 | 16 | 1 | 94.12 |
| 9 | 15 | 15 | 0 | 100 |
| 10 | 13 | 12 | 1 | 92.31 |
| 11 | 15 | 15 | 0 | 100 |
| 12 | 12 | 12 | 0 | 100 |
| 13 | 9 | 9 | 0 | 100 |
| 14 | 6 | 6 | 0 | 100 |
| 15 | 18 | 18 | 0 | 100 |
| 16 | 17 | 16 | 1 | 94.12 |
| 17 | 8 | 8 | 0 | 100 |
| 18 | 10 | 10 | 0 | 100 |
| 19 | 14 | 14 | 0 | 100 |
| 20 | 10 | 10 | 0 | 100 |

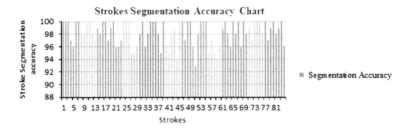

**FIGURE 5.7** Stroke Segmentation Accuracy.

segmentation accuracy has been observed during the testing phase of the system under consideration. As the dealing is with the Online handwriting recognition systems, thus it is possible to fetch the stroke data directly by using "INK Overlay" class in "Visual Studio" environment. The INK Overlay class is used to fetch the digital ink generated by the pen up and pen down events in case of digitizer and stylus environment. The identification of different categories of the strokes for the bilingual system of Gurmukhi and Roman scripts is based on the different experimental observations. All these mentioned categories of strokes collectively are well capable to express any character present in the character set of Gurmukhi or Roman script.

## 5.8 RESULTS AND DISCUSSION

The word level segmentation results obtained from the above samples and few more tested samples are represented in Table 5.6. However, Table 5.5 and Figure 5.8 represent the overall word segmentation results.

The results are observed to be about 97.09% accurate. Some of the reasons behind over segmentation or under segmentation are as follows:

- Competency of writing in the selected scripts.
- Overlapping of strokes while writing.
- Generation of distorted strokes due to the high speed of writers.
- Quality of Digitizer or Stylus.
- Familiarity of writer with the digitizer and stylus environment.

The accuracy of stroke segmentation process is known to approach perfection. But, sometimes the writer writes on the boundary areas of the writing surface, causing a single stroke to be misinterpreted as the existence of multiple strokes, which leads to the false declaration of stroke class. About 35 stroke classes, out of the total of 84, show 100% stroke segmentation results and other stroke classes also represent a good level of segmentation accuracy i.e. approximately 95%. The segmentation accuracy of this system has been observed to be improving with the familiarity level of writer increasing with digitizer and stylus.

**TABLE 5.6**
**Word Segmentation Results Using Different Samples**

| S_ID | S_Type | %age | S_ID | S_Type | %age | S_ID | S_Type | %age |
|---|---|---|---|---|---|---|---|---|
| 1 | R | 100 | 29 | R | 96 | 57 | G | 97 |
| 2 | R | 100 | 30 | R | 98 | 58 | G | 95 |
| 3 | R | 100 | 31 | R | 100 | 59 | G | 94 |
| 4 | R | 97 | 32 | R | 96 | 60 | G | 96 |
| 5 | R | 96 | 33 | R | 98 | 61 | G | 99 |
| 6 | R | 100 | 34 | R | 100 | 62 | G | 100 |
| 7 | R | 100 | 35 | R | 100 | 63 | G | 98 |
| 8 | R | 98 | 36 | R | 100 | 64 | G | 96 |
| 9 | R | 98 | 37 | R | 98 | 65 | G | 100 |
| 10 | R | 100 | 38 | R | 95 | 66 | G | 98 |
| 11 | R | 100 | 39 | G | 100 | 67 | G | 100 |
| 12 | R | 98 | 40 | G | 96 | 68 | G | 96 |
| 13 | R | 97 | 41 | G | 95 | 69 | G | 100 |
| 14 | R | 99 | 42 | G | 98 | 70 | G | 98 |
| 15 | R | 98 | 43 | G | 99 | 71 | G | 99 |
| 16 | R | 100 | 44 | G | 94 | 72 | G | 97 |
| 17 | R | 100 | 45 | G | 100 | 73 | C | 100 |
| 18 | R | 97 | 46 | G | 100 | 74 | C | 100 |
| 19 | R | 99 | 47 | G | 97 | 75 | C | 100 |
| 20 | R | 100 | 48 | G | 100 | 76 | C | 98 |
| 21 | R | 96 | 49 | G | 100 | 77 | C | 100 |
| 22 | R | 96 | 50 | G | 96 | 78 | C | 97 |
| 23 | R | 97 | 51 | G | 93 | 79 | C | 99 |
| 24 | R | 100 | 52 | G | 98 | 80 | C | 100 |
| 25 | R | 100 | 53 | G | 100 | 81 | C | 98 |
| 26 | R | 100 | 54 | G | 100 | 82 | C | 99 |
| 27 | R | 95 | 55 | G | 100 | 83 | C | 100 |
| 28 | R | 95 | 56 | G | 95 | 84 | C | 96 |

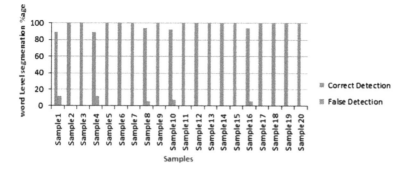

**FIGURE 5.8** Performance of Word Level Segmentation Results.

## 5.9 CONCLUSION

A segmentation method is proposed in this chapter to extract the minor details from the handwritten text, which in this work is an intermixed text, given the text consists of the words written by using English and Punjabi languages. For the purpose of writing, the Gurmukhi script is used for Punjabi language and Roman script is used for English language. The main reason behind picking this bilingual system is the popularity of these languages across the globe. The difficulty level of segmentation of the handwritten data due to the involvement of two different kinds of writing scripts has been handled by dividing the work into three different steps of segmentation: first, to divide the entire handwritten text into segmented words, which requires information on the existence and the formation of strokes according to the respective area of the digitizer; whereas, the information about the height of strokes and the coordinates of strokes' plane helped in the segmentation of complete words from the handwritten text, with the accuracy of word level segmentation observed as 97.09%, even as failure cases due to the overlapping of different strokes during the sample collection or due to the familiarization level of writers with digital pen and digitizer environment were also observed. Second, considering the targeted unit as unique characters helps in the formation of correct digital complete words. Lastly, for stroke segmentation, pen up and pen down events of the digital pen and digitizer environment are recorded. Due to the ability of digitizer surface to record stroke coordinates while writing on its surface, no extra efforts were applied on the implementation of vertical or horizontal projections as applicable in case of offline handwriting recognition systems. The overall stroke level recognition accuracy is observed as 95%. Most of the strokes of the stroke set shows the accurate identification all the time. Three different categories of strokes are considered in this work: "Roman (R)", "Gurmukhi (G)", and "Common (C)". Based on the observation of various characters of the character set of Gurmukhi and Roman scripts, stroke level segmentation identifies its goal. After the segmentation of any complete word from the handwritten text, the idea about the script used to write that complete word i.e. Gurmukhi or Roman is based on the presence of different type of strokes in that word. It has been observed that if all strokes of a complete word fall in the category of Gurmukhi strokes, then the script of the entire word has to be considered as Gurmukhi means the word belongs to Punjabi language. Sometimes, if all strokes of a complete word belongs to the stroke category Roman (R), then the overall script associated with the complete word has to be considered as Roman means the language used to write that word must be English. Sometimes, a segmented word contains some of the strokes belonging to the stroke set of both Gurmukhi and Roman scripts, which meant that stroke belongs to the common stroke set of Gurmukhi and Roman scripts. In this situation, the rest of the strokes of the segmented word play an important role because if the rest of the strokes belongs to the category of "R", then the language of the formation of that word under consideration is English; otherwise, rest of all strokes must belong to the category "G" of Gurmukhi script. A significant amount of accuracy has also been observed during script recognition phase.

## 5.10 FUTURE SCOPE

The overall automatic handwriting recognition system under consideration is a bilingual system. Punjabi and English languages are used to write the different samples under consideration. In future, same kind of algorithms with minor modifications can be used for segmentation by considering different combinations of bilingual systems or even for multilingual systems. The concept of identification of different stroke categories can be applied on any multilingual system also for the purpose of identifying the script of any word under consideration.

## REFERENCES

Akouaydi, H., Njah, S., & Alimi, A. M. (2017). Android application for handwriting segmentation using PerTOHS theory. *Ninth International Conference on Machine Vision (ICMV 2016)*, *10341*(March), 103410F. 10.1117/12.2269059.

Babbel.com. (n.d.). Most spoken languages of the world. Retrieved September 19, 2020, from https://www.babbel.com/en/magazine/the-10-most-spoken-languages-in-the-world.

Best OCR and handwriting dataset for machine learning applications. (n.d.). Retrieved March 6, 2020, from https://lionbridge.ai/datasets/15-best-ocr-handwriting-datasets/.

Bluche, T. (2016). Joint line segmentation and transcription for end-to-end handwritten paragraph recognition. *Advances in Neural Information Processing Systems*, *(Nips)*, *29*, 838–846, Chicago.

Blumenstein, M., Cheng, C. K., & Liu, X. Y. (2002). *New preprocessing techniques for handwritten word recognition* [Conference session]. Visualization, Imaging and Image Processing (VIIP 2002), Second IASTED International Conference, ACTA Press, Calgary, pp. 480–484.

Chim, Y. C., Kassim, A. A., & Ibrahim, Y. (1999). *Character Recognition using Statistical Moments*, *17*, 299–307.

De Jong, K. (2019). *Evolutionary computation: A unified approach* [Conference session]. GECCO 2019 Companion - Genetic and Evolutionary Computation Conference Companion, (January 2006), pp. 507–522. 10.1145/3319619.3323379.

Elleuch, M., Maalej, R., & Kherallah, M. (2016). A new design based-SVM of the CNN classifier architecture with dropout for offline Arabic handwritten recognition. *Procedia Computer Science*, *80*, 1712–1723. 10.1016/j.procs.2016.05.512.

Gattal, A., Djeddi, C., Chibani, Y., & Siddiqi, I. (2016). *Isolated handwritten digit recognition using oBIFs and background features* [Workshop]. Document Analysis Systems, DAS 2016, 2th IAPR International Workshop, pp. 305–310. 10.1109/DAS.2016.10.

Kicinger, R., Arciszewski, T., & De Jong, K. (2005). Evolutionary computation and structural design: A survey of the state-of-the-art. *Computers and Structures*, *83*(23–24), 1943–1978. 10.1016/j.compstruc.2005.03.002.

Marti, U. V., & Bunke, H. (2001). *Text line segmentation and word recognition in a system for general writer independent handwriting recognition* [Conference session]. Document Analysis and Recognition, ICDAR, 2001-Janua, International Conference (August), pp. 159–163. 10.1109/ICDAR.2001.953775.

Naeem Ayyaz, M., Javed, I., & Mahmood, W. (2012). Handwritten character recognition using multiclass SVM classification with hybrid feature extraction. *Journal of Engineering & Applied Sciences*, *10*, 57–67.

Pesch, H., Hamdani, M., Forster, J., & Ney, H. (2012). *Analysis of preprocessing techniques for Latin handwriting recognition* [Workshop]. Frontiers in Handwriting Recognition, IWFHR, pp. 280–284. 10.1109/ICFHR.2012.179.

Razzak, M. I., Hussain, S. A., Sher, M., & Khan, Z. S. (2009). Combining offline and online preprocessing for online urdu character recognition. *Lecture Notes in Engineering and Computer Science, 2174*(1), 912–915.

Roman Recognizer. (n.d.). Retrieved February 1, 2020, from https://docs.microsoft.com/en-us/windows/win32/tablet/ink-recognition-sample.

Sachan, M. K., Lehal, G. S., & Jain, V. K. (2011). A novel method to segment online Gurmukhi script. In International Conference on Information Systems for Indian Languages (pp. 1– 8). Springer, Berlin, Heidelberg.

Sadri, J., Suen, C. Y., & Bui, T. D. (2004). *Automatic segmentation of unconstrained handwritten numeral strings* [Workshop]. Frontiers in Handwriting Recognition, IWFHR, (February 2014), pp. 317–322. 10.1109/IWFHR.2004.21.

Samanta, O., Roy, A., Bhattacharya, U., & Parui, S. K. (2015). *Script independent online handwriting recognition* [Conference session]. Document Analysis and Recognition, ICDAR, vol. 2015-Novem, pp. 1251–1255. 10.1109/ICDAR.2015.7333964.

Teja, S. P., & Namboodiri, A. M. (2013). *A ballistic stroke representation of online handwriting for recognition* [Conference session]. Document Analysis and Recognition, ICDAR, pp. 857–861. 10.1109/ICDAR.2013.175.

Verma, B. (2003). A contour code feature based segmentation for handwriting recognition [Conference session]. Document Analysis and Recognition, ICDAR, 2003-Janua (Icdar), pp. 1203–1207. 10.1109/ICDAR.2003.1227848.

Wigington, C., Tensmeyer, C., Davis, B., Barrett, W., Price, B., & Cohen, S. (2018). Start, follow, read: End-to-end full-page handwriting recognition. *Lecture Notes in Computer Science (Including Subseries Lecture Notes in Artificial Intelligence and Lecture Notes in Bioinformatics), 11210 LNCS*, 372–388. 10.1007/978-3-030-01231-1_23.

Wshah, S., Shi, Z., & Govindaraju, V. (2009). *Segmentation of Arabic handwriting based on both contour and skeleton segmentation* [Conference session]. Document Analysis and Recognition, ICDAR, pp. 793–797. 10.1109/ICDAR.2009.152.

Xue, B., Zhang, M., Browne, W. N., & Yao, X. (2016). A survey on evolutionary computation approaches to feature selection. *IEEE Transactions on Evolutionary Computation, 20*(4), 606–626. 10.1109/TEVC.2015.2504420.

Zimmermann, M., & Bunke, H. (2002). Automatic segmentation of the IAM off-line database for handwritten English text. *Proceedings - International Conference on Pattern Recognition, 16*(4), 35–39. 10.1109/icpr.2002.1047394.

# 6 A Metric to Determine the Change Proneness of Software Classes Using GMDH Networks

*Ashu Jain, Dhyanendra Jain, and Dr. Prashant Singh*

## CONTENTS

6.1 Introduction ........................................................................................................ 109
6.2 Literature Review .............................................................................................. 110
6.3 Research Methodology ...................................................................................... 111
6.4 Group Method of Data Handling Polynomial Networks (GMDH) ............ 111
    6.4.1 Structure of a GMDH Network ...................................................... 112
    6.4.2 Algorithm ........................................................................................... 112
6.5 Empirical Data Collection ................................................................................ 114
    6.5.1 Independent and Dependent Variables ......................................... 114
    6.5.2 Machine Learning Algorithms Used ............................................. 114
6.6 Evaluation Metrics Used .................................................................................. 114
6.7 Results ................................................................................................................ 115
    6.7.1 Cross Validation Results ................................................................. 115
    6.7.2 Results of Friedman Test ................................................................ 115
6.8 Class Change Factor (CCF) Metric ................................................................. 115
    6.8.1 Definition of CCF Metric ................................................................ 116
    6.8.2 Validating the CCF Metric .............................................................. 117
    6.8.3 Application of CCF Metric ............................................................. 117
6.9 Conclusion and Future Work .......................................................................... 118
References ................................................................................................................... 118

## 6.1 INTRODUCTION

Software metrics compute the various features of software in quantitative terms when the software is either running or not. Formerly, researchers used to take into account the static metrics which depict only the structural behavior of the software, but nowadays, the researchers are turning more and more towards the dynamic metrics. For the computation of dynamic metrics, the execution of the program code is a must.

Class change proneness is a quality of the product or the likelihood that a software class will change given the arrival of the product is still a prospect. Recognizable proof of change vulnerable classes in software is important for:

1. It will help in the efficient use of resources which helps to pull out the time, and the finance, associated with the maintenance of the software.
2. It can guide the product engineers to take change hindering plan steps in the early periods of SDLC for change prone classes to anticipate presentation of deformities.
3. It will help in contemplating just the change prone classes and performing confirmation exercises in change prone classes to distinguish changes in the underlying periods of programming in progress.
4. These practices guarantee the arrival in market of an easeful, decent quality, and viable software product.

Hence estimation of change prone classes will help the project leaders to save time and money to be spent on the software as many researchers (Aggarwal et al., 2005; Elish and Elish, 2009; Fioravanti & Nesi, 2001; Li & Henry, 1993; Van Koten & Gray, 2006) have predicted the change prone nature of classes using static metrics. In this research paper, use of the dynamic metrics has been depicted to forecast on the vulnerability of change in software classes.

The ML model is formed using the data of OSS SoundHelix. The conduct of GMDH algorithm is also compared with four other ML models (Single Decision Tree, PNN, RBF, and Logistic Regression). In light of the outcomes, another metric has been proposed which can be utilized by analysts and specialists to recognize change prone classes of the product. This metric has been approved on five other open source programming (Art of Illusion, Sweet Home 3D, jTDS, jXLS, and jDrumBox).

The rest of the paper is structured into: section 6.2 summarizes the work related to the current study, which has been already done by the researchers; section 6.3 explains the basic structure of the GMDH network and the algorithm used for building the GMDH network; section 6.4 presents the empirical data collection by summarizing the dependent and independent quantifiers used in the study; section 6.5 explains the measures used for evaluation purpose; section 6.6 compares the performance results of the machine learning algorithms with the GMDH network; section 6.7 presents the new metric proposed and also its validation results; and finally, section 6.8 describes the conclusions made and the future work to take wings using this study.

## 6.2 LITERATURE REVIEW

Presenting an overview of the various studies describing the connection between software metrics and software class change vulnerability, this section starts by mentioning Dagpinar and Jhanke (Dagpinar & Jahnke, 2003) who explored the OO metrics. Aimed at finding which metrics play a significant role in the estimation of maintainability of software product, they observed that size and direct coupling metric had a great impact on software maintainability, whereas, cohesion, inheritance, and indirect/export coupling were not so significant. Jain et al. (2016) explored the use of genetic algorithm as imparting clear-sightedness on the maintainability of software, and concluded that genetic algorithms performed better than the other ML algorithms. Van Koten and Gray (2006) demonstrated the Bayesian framework forecast for a question programming system, where the forecast accuracy of the model is surveyed and found to anticipate reasonableness more unequivocally than did the relapse based models as the results proved. TreeNet model was proposed by Elish and Elish (2009) in the year 2009, and it refined the prediction precision over the recently published OO software maintainability estimation models. Jain and Chug (Jain et al., 2016) analyzed the applicability of the dynamic software metrics for software maintainability prediction. Malhotra and Chug (2012) offered three models – GMDH, Genetic Algorithms (GA), and PNN with Gaussian activation function, and the results showed that GMDH method can be applied to forecast the software maintainability to get more accurate and precise results. Sharma and Chug (2015) empirically legitimized the use of dynamic metrics in comparison to static metrics through machine learning algorithms. Wong and Gokhale (2005) displayed the feature code analysis by the use of dynamic metrics, which provided information on the closeness of two features of the software program. Tahir et al. (2010) proposed the AOP as the technique for collecting the dynamic software metrics. Koru and Liu (2007) identified and characterized the change vulnerable classes in two large OSSs using the tree based models and concluded that the proposed strategy can be effectively used for the prioritization purposes. Many researchers have applied the commonly used machine learning techniques. However, no study has been undertaken to derive the application of Group Method of Data Handling (GMDH) polynomial network to the problem of classification of change prone classes using dynamic metrics.

## 6.3 RESEARCH METHODOLOGY

The research methodology has the following four steps: firstly, the six open source softwares are downloaded from the web and the software metrics extracted. Then, for each dataset of the OSS, Modify metric is calculated as a dependent variable that changes with the event of class change of the software. That is, Modify is set to "yes", if there is a change in a class of the software; otherwise, it is set to "no". Third, once the dataset is finalized, Machine Learning algorithms are applied to train and test the model. Fourthly, the performance of each machine learning model is evaluated, and based on the results, a new metric CCF is also being defined.

## 6.4 GROUP METHOD OF DATA HANDLING POLYNOMIAL NETWORKS (GMDH)

GMDH polynomial systems are extremely predictable systems which depend on the information yield relationship, and is the ideal system for taking care of issues on software, including forecast and distinguishing proof.

GMDH systems are designated as the self-sorting out systems which means the association between the neurons is not settled; rather, the associations are resolved amid the system development in order to advance the system.

### 6.4.1 Structure of a GMDH Network

The essential structure of a GMDH system comprises of four layers: info layer, the second and third layers are the concealed layers, and the yield layer (Figure 6.1).

The primary layer being the input layer shows the contribution to every neuron in the system. The second layer which is the hidden layer takes its contributions from two of the input factors. Thus third layer which is likewise the hidden layer takes the contribution from two neurons of the second hidden layer. This procedure is taken up for each hidden layer, if there are more than two hidden layers. Last, the output layer draws its contribution from the past hidden layer and delivers a single value as the yield of the system (Figure 6.2).

### 6.4.2 Algorithm

The basic purpose of the GMDH algorithm is to establish a relationship between m input variables, x1, x2, x3, ..........., xm with the single output variable y.

For the given n set of observations:
t = number of observations in training set.
c = n – t = number of observations in the testing set.
n = total number of observations.
m = number of input variables.
The basic steps followed for building the network are following:

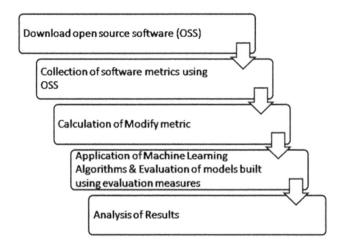

**FIGURE 6.1**  Research Methodology.

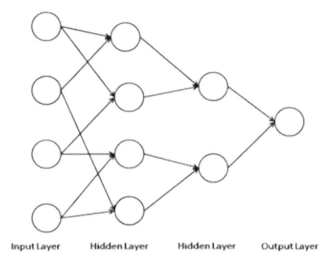

**FIGURE 6.2** Structure of GMDH Network.

1. The first step consists of the construction of new variables: $z_1, z_2, z_3, \ldots, z_{\binom{m}{2}}$.

   From the arrangement of information factors, two factors are taken, at once, to give a slightest square polynomial of the shape

   $$y = A + Bp + Cq + Dp^2 + Eq^2 + Fpq \qquad (6.1)$$

   that will be the best fit observations yi in the preparation set it shaped. These polynomials are put to use for all the $\binom{m}{2}$ combinations.

2. All the $\binom{m}{2}$ polynomials are assessed at the n information focuses and these n qualities will be put away as the segments of the Z cluster. These sections are new factors $z_1, z_2, z_3, \ldots, z_{\binom{m}{2}}$ as far as the info factors are concerned. Only those factors will be retained that can best gauge the output vector y.

3. In the second step, the factors which are not imperative are left out. The segments of Z exhibit, which best predicts the output vector, supplant the sections of X cluster. For this reason, minimum mean squared blunder d is ascertained between the y vector and z vector. At that point, these djs are sorted in the expanding request. The sections which fulfill dj < (M is some recommended number) will supplant the segments of X cluster.

4. From step 2, discover the tiniest estimation of dj and call it MIN. On the off chance that the estimation of MIN is tinier than the past estimation of MIN, rehash steps 1 and 2. In the event that the estimation of MIN is more prominent than the past MIN, then the procedure is halted.

## 6.5 EMPIRICAL DATA COLLECTION

### 6.5.1 Independent and Dependent Variables

Programming measurements are utilized to gauge diverse parts of the programming. These angles can be polymorphism, cohesion, inheritance, size, and so on. We have separated the OO measurements for every one of the six open source programming softwares: jTDS, Art of Illusion, jXLS, orDrumBox, Sweet Home 3D, and SoundHelix.

Dynamic metrics (Jain et al., 2016) are used as independent variables. These metrics are Lack of cohesion of methods (LCOM), Lines of Comment (LOCm),Loose Class Coupling (LCC),Number of Children (NOC), Tight Class Coupling (TCC),Number of Fields (NOF), Number of Assertions per KLOC (NAK), Number of Methods (NOM), Number of Static Fields (NOSF), Number of Static Methods (NOSM), and Number of Test Methods (NTM). Change proneness is utilized as a needy variable, which is the likelihood of progress in the class computed by tallying the quantity of lines of code included, erased, and adjusted. Expansion and erasure of LOC is considered one change while alteration of an LOC is considered two changes. We portray another identifier called MODIFY which was assigned a value of "yes" for all the classes which undergo change and "no" for the unchanged classes in every dataset.

### 6.5.2 Machine Learning Algorithms Used

1. **Single decision tree:** It is a consistent model and known as a parallel tree that shows how the calculation of the dependent variable can be anticipated by utilizing the estimations of an arrangement of indicator factors.
2. **Radial Basis Function (RBF) Neural Network:** Synonymous with the ANN, it produces linear combination of radial basis functions and input neurons as the output of the network.
3. **Probabilistic Neural Networks (PNN):** A system that is forward neural system in light of the Bayesian grouping hypothesis, the PNN system consists of four layers: input layer, shrouded layer, design layer, and output layer.
4. **Logistic Regression:** It uses logistic function to predict the value of the dependent variable.

The performance of the above algorithms has been compared with the execution of GMDH network.

## 6.6 EVALUATION METRICS USED

Following parameters can be used to analyze the results of the algorithms:

1. **Specificity and Sensitivity**: Sensitivity measures the true positive rate. Specificity measures the true negative rate.

2. **Accuracy and Precision**: Accuracy is defined as the percentage of match between actual values and predicted values.
3. **f-measure**: f-measure (f) can be defined as the harmonic mean of precision and recall.
4. **Receiver Operating Characteristic (ROC)**: It is the association drawn between true positive rate and the false positive rate. The productivity of the model is assessed by evaluating the Area under the ROC Curve (AUC).

## 6.7 RESULTS

In this section, the performance of GMDH algorithm is analyzed with other machine learning algorithms (Single decision tree, RBF neural network, PNN, logistic regression, and GMDH network). For building the model, 10-fold cross validation technique is used.

### 6.7.1 Cross Validation Results

The 10-fold cross validation results of all five ML models are given in Table 6.1.

Table 6.1 concludes that GMDH algorithm for the estimation of Change prone classes.

### 6.7.2 Results of Friedman Test

ML models developed are then ranked by using Friedman test (Friedman, 1940) based on their performance. GMDH algorithm results are better than other ML models used and it outperformed all other algorithms with the highest rank of 4.58 (Table 6.2).

## 6.8 CLASS CHANGE FACTOR (CCF) METRIC

A new quantifier "Class Change Factor (CCF)" has been discovered to predict whether a class is vulnerable to change or not. This metric is further validated using five open source datasets.

**TABLE 6.1**
**Comparison of GMDH Network Technique with Other Algorithms Used**

|  | Accuracy | Sensitivity | Specificity | Precision | F-Measure | (AUC, C-Statistic) |
|---|---|---|---|---|---|---|
| Single decision tree | 69.57% | 56.00% | 77.27% | 58.33% | 0.5714 | – |
| GMDH | 73.91% | 52.00% | 86.36% | 68.42% | 0.5909 | 0.720909 |
| PNN | 63.77% | 0.00% | 100.00% | 0.00% | 0.0000 | 0.454545 |
| RBF | 68.12% | 36.00% | 86.36% | 60.00% | 0.4500 | 0.591818 |
| Logistic Regression | 66.67% | 44.00% | 79.55% | 55.00% | 0.4889 | 0.678182 |

**TABLE 6.2**
**Friedman Test Result**

| Algorithms | Mean Rank |
|---|---|
| Single Decision Tree | 3.00 |
| GMDH | 4.58 |
| PNN | 1.83 |
| RBF | 2.92 |
| Logistic regression | 2.67 |

### 6.8.1 Definition of CCF Metric

In view of the after-effects of the SoundHelix dataset utilizing the GMDH calculation, we present Class Change Factor (CCF) metric which can be used to assess those classes on which changes can be performed in the inevitable variant of the product. The CCF metric can be described as follows:

**Name of the metric:** Class Change Factor (CCF)

**Definition of the metric:** For a given class Ci, the change vulnerability can be calculated by the accompanying condition:

$$CCF = 0.264 + 1.012 * [N(1)]^2 - 0.011 * N(7) - 0.008 * [N(7)]^2 + 0.042 * N(1 * N(7)) \tag{6.2}$$

Where

$$N(5) = 7.001 + 14.602 * LOC - 1.899 * (LOC)^2 - 0.834 * NOF - 1.552 * (NOF)^2 + 6.687 * LOC * NOF \tag{6.3}$$

$$N(3) = 34.731 + 132.213 * LOC + 125.507 * (LOC)^2 - 6.431 * N(5) + 0.3152 * [N(5)]^2 - 13.204 * LOC * N(5) \tag{6.4}$$

$$N(1) = 1.818 + 0.211 * LCOM5 - 3.135 * (LCOM5)^2 + 1.961 * N(3) + 0.001 * [N(3)]^2 - 0.955 * LCOM5 * N(3) \tag{6.5}$$

$$N(7) = 1.729 + 4.497 * NOF - 0.735 * (NOF)^2 + 0.902 * N(3) + 0.008 * [N(3)]^2 - 0.284 * N(3) * NOF \tag{6.6}$$

LOC = Linesofcode
NOF = Number of Fields
LCOM5 = Lack of Cohesion of Methods

A class numbered as Ci is assigned the CP value in accordance with the following function:

$$\text{MODIFY}(C_i) = \begin{cases} Yes & if \quad CCF > 0 \\ No & if \quad CCF \leq 0 \end{cases} \quad (6.7)$$

The function defines the MODIFY variable. If the value of CCF quantifier for the given class exceeds zero, then it is prone to change; otherwise, it is not.

### 6.8.2 Validating the CCF Metric

To evaluate the nominated CCF metric, we have used the accuracy, sensitivity, and specificity as the performance evaluators. To perform the validation, five open source software – jXLS, jTDS, Art of Illusion, Sweet home 3D, and orDrumbox have been used. The outcomes have been recapitulated in Table 6.3, which validates the use of CCF metric. It can be concluded from Table 6.3, that all datasets achieved the significant accuracy results. The average accuracy is 81.152%, mean sensitivity value is 0.461 and mean specificity value is 0.77. Hence, CCF can be used as the metric to watch out for the change vulnerability of a class.

### 6.8.3 Application of CCF Metric

Amid the early phases of SDLC, CCF can be useful to programming specialists for its recognizable proof of change prone classes. The significant hotspots for the upgradations in the source code of programming are the visible imperfections and the constantly changing software requirements of the clients. Accordingly, change prone classes of the product delineate the defect prone and weak regions in the product which should be appropriately taken care of before release. Distinguishing proof of change vulnerable classes in the beginning of SDLC can enhance the asset

**TABLE 6.3**
**Validation Result of CCF Metric**

| Datasets | Accuracy | Sensitivity | Specificity |
|---|---|---|---|
| Jtds | 71.42% | 0.24 | 1.00 |
| OrDrumBox | 68.34% | 0.388 | 0.246 |
| Art of Illusion | 95.36% | 1.00 | 0.962 |
| Sweet Home 3D | 73.46% | 0.377 | 0.892 |
| Jxls | 97.18% | 0.3 | 0.75 |

portion implies the asset can be designated to the needy classes which can lessen the wastage of time and cash. Change prone classes should be considered for testing exercises like investigation audits, and so on. This can prompt programming of good quality.

Classes having the CCF value more prominent than zero are vulnerable to change and the classes having CCF value under zero are not vulnerable to change. Programming engineers ought to guarantee that appropriate planning is done while writing the code for such classes, with the goal that they can be proactively refactored in the expected arrivals of the product.

## 6.9 CONCLUSION AND FUTURE WORK

Five ML algorithms – single decision tree, RBF neural network, PNN, logistic regression, and GMDH network, have been figured out using run-time metrics for forecasting the change vulnerability of software class. On the basis of the outcomes, a new quantifier Class Change Factor (CCF) has been presented, which can compute change vulnerability of a software class. This metric is validated by using the dataset of five OSSs. The inferences have been recapitulated in the following points.

1. It is observed that the GMDH serves as the best predictor for software maintainability with 73.91% accuracy and 0.720909 AUC. Hence GMDH could be applied for forecasting the change vulnerability in the initial steps of the SDLC.
2. A metric Class Change Factor (CCF) is proposed for predicting the change vulnerability of a class in early phases of SDLC. The affirmation of CCF (Average Accuracy: 81.152%, Average Sensitivity: 0.461 and Average Specificity: 0.77) shows that CCF can be used as a powerful metric for the estimation of change vulnerable classes.

Our future work focuses on the generalization of the outcomes by reproducing this research on bigger datasets. We can also refine our study by using projects of different coding languages. This study can also be evaluated by applying the class of evolutionary algorithms.

## REFERENCES

Aggarwal, K. K., Singh, Y., Chandra, P., & Puri, M. (2005). Measurement of software maintainability using a fuzzy model. *Journal of Computer Sciences*, *1*(4), 538–542.

Dagpinar, M., & Jahnke, J. H. (2003, November). *Predicting maintainability with object-oriented metrics-an empirical comparison* [Conference session]. Reverse Engineering, 2003. WCRE 2003, 10th Working Conference, IEEE Computer Society, pp. 155–155.

Elish, M. O., & Elish, K. O. (2009, March). *Application of treenet in predicting object-oriented software maintainability: A comparative study* [Conference session]. Software Maintenance and Reengineering, 13th European Conference, IEEE, pp. 69–78.

Fioravanti, F., & Nesi, P. (2001). Estimation and prediction metrics for adaptive maintenance effort of object-oriented systems. *IEEE Transactions on Software Engineering*, *27*(12), 1062–1084.

Friedman, M. (1940). A comparison of alternative tests of significance for the problem of m rankings. *The Annals of Mathematical Statistics, 11*(1), 86–92.

Jain, A., & Chug, A. (2016, August). *Stepping towards dynamic measurement for object oriented software* [Conference session]. Information Processing (IICIP), 1st India International Conference, IEEE, pp. 1–6.

Jain, A., Tarwani, S., & Chug, A. (2016, March). *An empirical investigation of evolutionary algorithm for software maintainability prediction* [Conference session]. Electrical, Electronics and Computer Science (SCEECS), Students' Conference, IEEE, pp. 1–6.

Koru, A. G., & Liu, H. (2007). Identifying and characterizing change-prone classes in two large-scale open-source products. *Journal of Systems and Software, 80*(1), 63–73.

Li, W., & Henry, S. (1993). Object-oriented metrics that predict maintainability. *Journal of Systems and Software, 23*(2), 111–122.

Malhotra, R., & Chug, A. (2012). Software maintainability prediction using machine learning algorithms. *Software Engineering: An International Journal (SeiJ), 2*(2), 19–35.

Sharma, H., & Chug, A. (2015, September). *Dynamic metrics are superior than static metrics in maintainability prediction: An empirical case study* [Conference session]. Reliability, Infocom Technologies and Optimization (ICRITO)(Trends and Future Directions), 4th International Conference, IEEE, pp. 1–6.

Tahir, A., Ahmad, R., & Kasirun, Z. M. (2010). Maintainability dynamic metrics data collection based on aspect-oriented technology. *Malaysian Journal of Computer Science, 23*(3), 177–194.

Van Koten, C., & Gray, A. R. (2006). An application of Bayesian network for predicting object-oriented software maintainability. *Information and Software Technology, 48*(1), 59–67.

Wong, W. E., & Gokhale, S. (2005). Static and dynamic distance metrics for feature-based code analysis. *Journal of Systems and Software, 74*(3), 283–295.

# Section B

*Application of Disruptive Technology in Various Sectors*

# 7 Application of IoT Technology in the Design and Construction of an Android Based Smart Home System

*Adeyemi Abel Ajibesin and Ahmed Tijjani Ishaq*

## CONTENTS

7.1 Introduction ........................................................................................................ 123
7.2 Related Work ...................................................................................................... 125
7.3 The System Model ............................................................................................. 128
7.4 Methodology and Logic Design ....................................................................... 135
7.5 Packaging and Discussion ................................................................................. 139
7.6 Conclusion .......................................................................................................... 140
7.7 Limitations, Future Directions, and Recommendations .............................. 141
References ................................................................................................................... 141

## 7.1 INTRODUCTION

This chapter aims to design and construct an android based smart home system that bypasses conventional switching or control system that not only the physically adept find laborious to operate but also the persons with disability (PWD) find inadequate. If we realize it is usually a challenge for PWD or the aged when left alone to operate home-based switches, which remote wireless controllability can easily overcome, then, to provide one to such people necessitates a smart home. Smart homes are the integration of technology and services through home networking for a better quality of life (Subhamay et al., 2014). As technology grows so also homes are getting smarter, where modern houses are gradually moving away from conventional switches to the centralized control system. The modern system involves wireless technologies and easy-to-operate control switches vis-a-vis the conventional switches which have a lot left to be desired in terms of mobility, for the physically fit if not the elderly and the physically challenged people. Filling this gap is a smart home that is one fitted with automation such as monitoring systems (sensors, actuators, and biomedical monitors) and systems such as lighting, ventilation, and appliances, that are specially wired to enable the home's residents to program, control, and operate remotely.

With automated devices, such as sensors, mounted in the houses, offices, and areas that needed to be covered, some devices can also be worn by an individual. Initially, smart home systems were designed for entertainment and luxury, having advanced settings that needed getting used to, and it was only later realized that there was a need to develop a smart home for special people, who faced many challenges in day to day living. A report released by the World Health Organization (WHO) revealed that there are over seven hundred and eighty-five million people in the world who live with disabilities (WHO Library, 2011). Another survey has also reported that more than one hundred and ten million people in the world are experiencing special cases of disability. In a report by the Population Division of the United Nations, about 10% of the World's population is 60 years and above. It was also estimated that this figure will reach up to 21% by 2050 (United Nations, 2020). This is significant and needed to be considered in Engineering design.

This chapter brings to the physically adept, the PWD, and the elderly or aging people the promise of smart homes. It also considers energy efficiency, comfort, entertainment, and the safety of users. Thus, a system that can make life comfortable for users is required, which can send information regarding the situation at home irrespective of the users' locations (Malik & Bodwade, 2017). This can be in the form of an alert to update the users about the security situation at home, so that an immediate need is met, i.e. to handle an emergency with little or no human intervention, is solved effectively.

In addition to this, smart homes can help save energy. An example is a Z-Wave and ZigBee that can send a command to a device to "sleep" and "wake up" when a reverse command is issued. Another scenario is the ease of electric bulbs that go off when lights are automated or when a person leaves the room. Also, rooms can be heated or cooled depending on its sensing the movement of someone, which shows that the Smart system promises great benefits for the PWD and the elderly living alone (Gibson et al., 2013). Smart homes also offer the advantage to residents to receive notifications, such as the time to take a medication, or sending alerts to the appropriate department such as police or hospital if the user had a fall, especially where there is no immediate assistance. The smart home can help those users, such as an elderly who are easily forgetful, by performing tasks like turning off the oven when the desired temperature is reached.

Researchers have shown tremendous interest in smart systems to further explore their applications and smartphones. For instance, Bluetooth-based home automation using Android smartphones was presented by Piyare and Tazil (2011) and Yan and Shi (2013), in which devices are shown connected to a Bluetooth sub-controller, as well as an in-built component developed by the system is accessed and controlled through a smartphone. It is important to know that Bluetooth technology is limited to a short distance and it is paired with only one device at a time. In the work presented by Muhammad et al. (2017), a Wi-Fi-based approach based on Arduino Ethernet modules were used, and the system's design also consists of an Android compatible Smartphone app, with the only drawback that made an access point (AP) or router to exist between the Arduino Ethernet device and the smartphone a necessity which results in the addition of hardware that in turn increases time delay and cost (Muhammad et al., 2017). As a result, this work considers an Android-based smart home that is developed from scratch as an in-built system, with the design based on an in-built Wi-Fi module (without the need for AP) being flexible and cost-effective. The design process is well organized, articulated, and the reader can easily replicate the design.

As already mentioned, because of the limitations of the conventional switches in terms of giving mobility, especially for the elderly and PWD, this work considers automated devices to mount in houses, offices, and specific areas. Also, wearable devices are useful for this purpose, so that the end-user can control their appliances remotely.

The remainder of this chapter is presented thus: in sections 7.2 and 7.3, the related work and the system model, respectively; sections 7.4 and 7.5 present the research methodology and discussion of packaging and results, respectively; while section 7.6 presents the conclusions reached from research.

## 7.2 RELATED WORK

Not only can the smart home technology be deployed within the home and compound to provide comfort and security, but also the system can help with energy efficiency when a resident can use the home controller to integrate the different levels of home automation. The smart home is ideal for society 5.0 that is promising to accommodate the Fourth Industrial Revolution (Aztiria et al., 2012). If this doesn't present an opportunity for Smart Homes System (SHS) and interdisciplinary research area of computer engineering to be placed at the intersection of computer networks, embedded systems, and applied computing, then it would be wrong to say that it does indeed do it as proved by research and application of research to make smart homes. Recently, research has focused on smart systems and experimentation on various intelligent agent-based computing methods with their application to the community challenges. Many works have been reported for enhancing the performance of SHS (Purohit & Ghosh, 2017; Rajeev, 2013):

- Home Automation System contains a set of home appliances consisting of electric and electronic devices that meet several requirements of functions in the house for the comfort of the householders. Instances of these are washing and cooking machines, refrigerators, thermometers, lighting systems, energy meters, smoke detectors, windows and doors controllers, air conditioners, and sound detectors. These function as sensors, actuators, or both. It should be noted that advanced "smart devices" are developed from time to time. Such advanced systems include smart doors and smart furniture with beds and chairs combined as a system.
- A Control System, which combines the human and software components with control, is one of the works that has been reported to found the system accepting information provided by the sensors and then sending instructions to actuators to achieve certain goals needed by householders.

- Home Automation Network, which ensures that all the SHS components together with the Home Automation System and the Control System are capable of exchanging status and control information.

The Table 7.1 provides an overview of related work in this area, with the authors, their research focus, the types of technology considered, and their main contributions reported. The rapid improvement in computer networking, optimized devices, and increased computational power is all pervasive even as ubiquitous computing motivated smart home research advances in leaps and bounds. Thus, this research work designs and constructs a user-friendly remote control-fitted system that is not only cost- and energy-efficient, but also android- and WiFi technology-based so that it allows the end-user to control appliances such as lighting points without help from anyone.

**TABLE 7.1**
**Summary Table of Literature on Smart Home**

| Authors | Title | Contribution | Technology Used |
|---|---|---|---|
| Kumar and Lee (2014) | Android based smart home system with control via Bluetooth and internet connectivity | A low-cost Smart Living System, which uses Android based User Interface for control of home appliances. Connection to the smart living system can be made from the designed app via Bluetooth or internet connection. It also integrates home security and alert system. | Android Bluetooth WiFi |
| Piyare and Lee (2013) | Smart home-control and monitoring system using smart phone | A low cost and flexible home control and monitoring system using an embedded micro-web server, with IP connectivity, for accessing and controlling devices and appliances remotely using Android based Smart phone app. The proposed system does not require a dedicated server PC with respect to similar systems and offers a novel communication protocol to monitor and control the home environment with more than just the switching functionality. | Android Arduino IoT |
| David et al. (2015) | Design of a home automation system using Arduino | A low cost and flexible home control and environmental monitoring system. To demonstrate the feasibility and effectiveness of this system, devices such as light switches, power plug, temperature sensor, gas sensor, and motion sensors have been integrated with the proposed home control system. | Android Bluetooth WiFi |

## TABLE 7.1 (Continued)
## Summary Table of Literature on Smart Home

| Authors | Title | Contribution | Technology Used |
|---|---|---|---|
| Gupta and Chhabra (2016) | IoT based Smart Home design using power and security management | Design and implementation of an Ethernet-based Smart Home intelligent system for monitoring the electrical energy consumption based upon the real time tracking of the devices at home. | INTEL Development Board Android |
| Baraka et al. (2013) | Low cost arduino/android-based energy-efficient home automation system with smart task scheduling | Used of Home Automation techniques to design and implement a remotely controlled, energy-efficient, and highly scalable Smart Home with basic features that safeguard the residents' comfort and security. The system consists of a house network (sensors and appliance actuators to respectively get information from and control the house environment). As a central controller, we used an Arduino microcontroller that communicates with an Android application, our user interface. | Arduino X10 Zigbee |
| Jabbar et al. (2018) | Design and implementation of IoT-based automation system for smart home | A low-cost Wi-Fi based automation system for Smart Home (SH). It monitors and controls home appliances remotely using Android-based application. The automation system, can easily and efficiently control the electrical appliances via Wi-Fi and Virtuino mobile application. | Android WiFi Arduino |
| Singh et al. (2018) | IoT based smart home automation system using sensor node | Developed a new solution which controls the home situation. The solution uses the sensor and detects the presence or absence of a human object in the housework accordingly. It also provides information about the energy consumed by the house owner regularly in the form of message. | Node MCU Arduino WiFi |
| Tharishny et al. (2016) | Android based smart house control via wireless communication | A simple smart house prototype built to be monitored and controlled through an android application based on smart phone via wireless communication. | Android WiFi Bluetooth |

*(Continued)*

**TABLE 7.1** *(Continued)*
**Summary Table of Literature on Smart Home**

| Authors | Title | Contribution | Technology Used |
|---|---|---|---|
| De Luca et al. (2013) | The use of NFC and Android technologies to enable a KNX-based smart home | Developed and validated an architecture, both hardware and software, able to monitor and manage a KNX-based home automation system through an Android mobile device in an efficient and safe way. Providing more details, a software system able to configure an Android application consistently with the home automation implant was designed and implemented as well as an Android application able to manage the entire home automation system based on the KNX standard. A further Android module, which exploits NFC technology, was developed in order to address the access control issue. A real use case was presented, which demonstrated the effectiveness of the proposed software system. | Android KNX NFC |
| Gunputh et al. (2017) | Design and implementation of a low-cost Arduino-based smart home system | A highly scalable, low-cost, and multi-faceted home automation system based on Arduino microcontroller that is capable of integrating appliance and equipment automation, thermal comfort control, and energy management. | Arduino WiFi |

## 7.3 THE SYSTEM MODEL

The system model and the procedure for the design and construction of the smart home are discussed in this section. The entire system consists of different units: a power supply, sensors, actuators, and the main controller. Figure 7.1 represents the system model of the smart home.

The smart system required an android smartphone with the user application installed. Providing an interface to the users to issue commands for different operations, the android application converts the users' commands into a "GET /" request, sends the message through the Wi-Fi Module to the smart home controller to be processed, when the user's command is received and acted upon. Examples of such requests are situations to open/close the house gate, read the energy consumed by the house, switch ON/OFF load, and read the surrounding temperature. The

# IoT Technology Designed Smart Home System

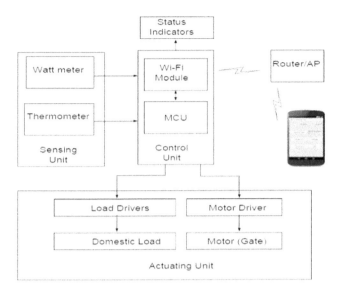

**FIGURE 7.1** The System Model of the Smart Home.

controller on the end of the task be completed will send feedback to the android application through the same link, where the feedback is normally in form of JavaScript Object Notation (JSON). Finally, the users' interface is updated or populated once the feedback is received by the applications.

*Hardware Design*: The circuit diagram for the system used in this project is shown in Figure 7.2, as captured from the Proteus schematic editor. The hardware design involves the design of the circuitry in each of the functional blocks, with details on the design for each of the blocks elaborated below.

1. ***Power Supply Unit Design:*** The power requirement for the system was obtained by summing up all individual power requirements of the subsystem components. This amounted to 30 watts consisting of 12Vdc unregulated at 1A for motor driver, 5Vdc regulated at 700mA for the controller, sensor, and actuator units, and 3.3Vdc at 300mA for Wi-Fi module.
    a. *Unregulated 12Vdc Power Supply Design*: The mains (220Vac) supply was stepped down to 15Vac using a voltage transformer. A bridge rectifier with a maximum current capacity of twice the load current was used to convert the 15Vac into a pulsating DC supply. This pulsating DC is then filtered out using capacitor C1 as illustrated in Figure 7.3.
    b. *Regulated 5v and 3.3v Power Supply*: The 5V regulated power supply was obtained using the circuit diagram in Figure 7.4. Transistor Q1 was chosen to have Ic(sat) greater than IL so that the load can draw up to 1A without heat loss causing the transistor to become warm.

Programmable voltage reference IC TL431 provides the regulation base on the set Vref. Resistor R1 serves as a biasing component Q1 and R2, R3, and RV1 are

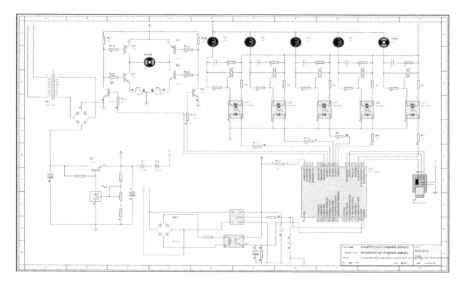

**FIGURE 7.2** Functional Circuit Diagram.

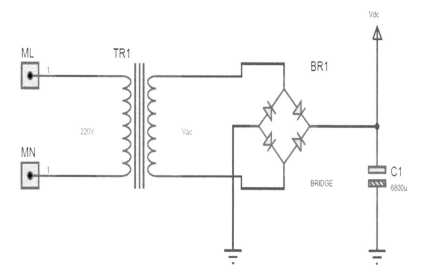

**FIGURE 7.3** Unregulated 12V Power Supply.

used as a voltage divider for the TL431. The regulated 5V output is further filtered using capacitor C2. D1 and D2 reduce the output from 5V to 3.3V.

1. *Sensing Unit Design:* The parameters monitored by this system are temperature, voltage, and current and these are measured using DS18B20, PC817, and ACS785CB-100 sensors, respectively. The following sections discuss the sensors' respective interface to the microcontroller.

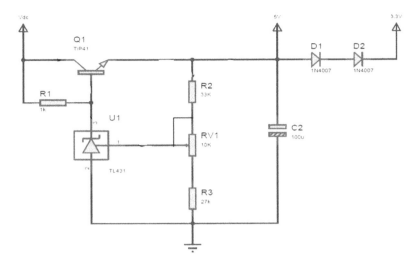

**FIGURE 7.4** Regulated 5V and 3.3V Power Supply Circuit.

a. *Temperature Sensor:* A digital thermometer (DS18B20) was used to read the temperature of the house. Then, the reading is transmitted to the microcontroller. This is performed through one wire communication ($I^2C$) system. The data pin of the DS18B20 IC was connected to pin RE2 of the microcontroller. The circuit diagram of this interface is presented in Figure 7.5. R4 (4.7K) is a pull-up resistor that supplies the data line with sufficient current during its active conversion cycle.

b. *Voltage Sensor:* Optocoupler (PC817) IC was used for voltage sensing because it can work directly with AC supply thereby removing the need for rectification. Refer to Figure 7.6 for how it was configured to work with the microcontroller.

c. *The sensor's output:* This is being fed to a system called Analog to Digital Converter (ADC), which is a channel of the microcontroller. R5 limits the input current for the sensor and R6 converts the output current into voltage for the ADC.

d. *Current Sensor:* The current sensor used is a Hall effect-based linear current sensor IC (ACS758CCB – 100) with a maximum current capacity of 100A and 40mV/A sensitivity. The values of RF, CBYP, and CF were recommended in the datasheet of the IC (Figure 7.7).

2. **Actuator Design**: The actuator unit consists of load drivers and motor drivers. The load drivers were used to put on/off the home lamps and fan while the motor driver which is in form of an H-Bridge configuration operates the house's main gate DC Motor. Figure 7.8 gives the circuit diagram of the load drivers.

TRIACs (Q8025R5) were used for switching the AC loads (fan, lamp1, lamp2, lamp3, and lamp4) while Opto-TRIACs (MOC3041) provide complete

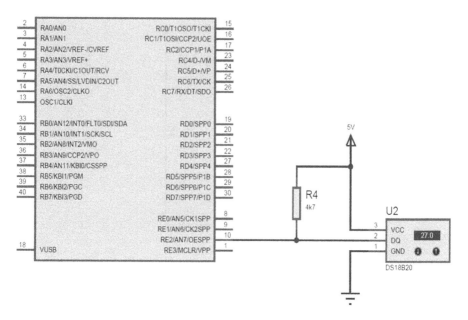

**FIGURE 7.5** Temperature Sensing Circuit.

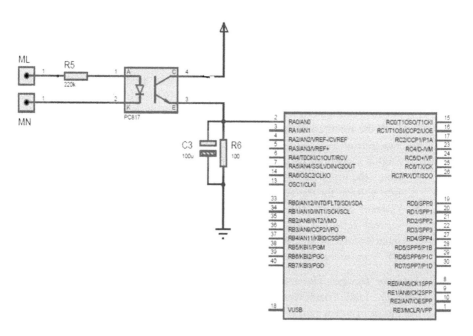

**FIGURE 7.6** Voltage Sensing Circuit.

# IoT Technology Designed Smart Home System    133

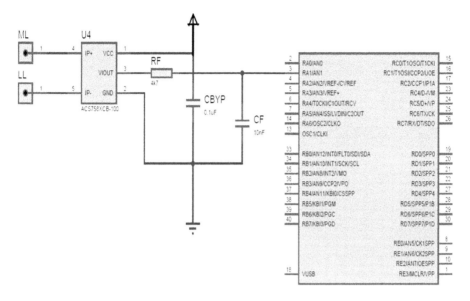

**FIGURE 7.7**  Current Sensing Circuit.

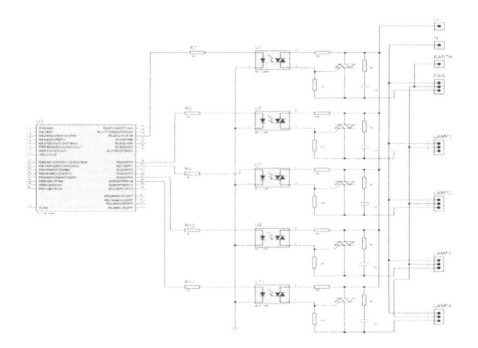

**FIGURE 7.8**  Actuator Circuit Diagram.

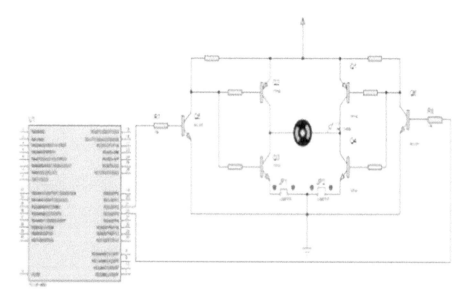

**FIGURE 7.9** Motor Driver Circuit.

isolation between the AC and DC circuits. In this design TRIAC with a current capacity of 25A was used to switch the loads. Snubber network (consisting of R=39 and C=10nF) and the resistors (with values 360 and 330) connecting the Opto-TRIAC with the TRIAC are suggested by the datasheets of the TRIACs. Figure 7.9 illustrates how the motor driver was connected to the microcontroller. The motor drives the house's main gate to either open or close. The open or close state depends on the user input. An interrupt system is considered. Two limit switches were used to achieve this by interrupting the supply when the gate is fully opened or closed.

3. *Controller Design:* The controller unit handles all the processing tasks which include sending and receiving the request, instruction execution, and data processing for the smart home. It consists of a microcontroller and a WiFi Module.
4. *Controller Hardware Design:* PIC Microcontroller family (PIC18F4550) was utilized in this system based on the following system requirements:
    - Universal Synchronous and Asynchronous Receive and Transmit (USART) Communication Module for communicating with the Wi-Fi Module.
    - Analog to Digital Converter (ADC) Module for Current and Voltage reading.
    - Internal Oscillator (8MHz) which eliminate the need for external Crystal.
    - Inter Intergraded Circuit (I2C) or One Wire communication module for communicating with the digital thermometer.
    - Enough I/O ports for connecting more loads (Figure 7.10).

# IoT Technology Designed Smart Home System

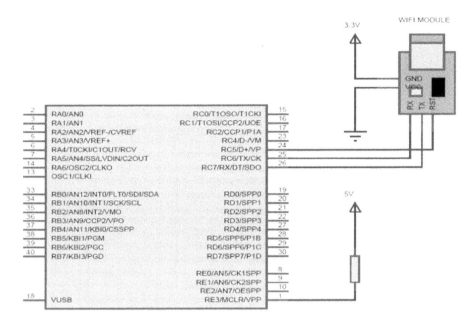

**FIGURE 7.10** Controller Circuit Diagram.

The measured parameters as well as the load's state are transmitted from the microcontroller to the Wi-Fi module via UART communication technology. The Wi-Fi module handles the connection to an android smartphone.

## 7.4 METHODOLOGY AND LOGIC DESIGN

This section discusses the research methodology that is based on software design and the logic for both the MCU and the host android mobile application.

1. *Microcontroller Program:* The controller unit consists of a microcontroller and software. The software, that converts the logic into machine sensible form, comes in different forms. Some of the software that is suitable for microcontroller includes Assembly, C, and BASIC. After writing the program, a compiler converts the code into a Hexadecimal (.hex) file which is then burned into the flash memory of the microcontroller using a programmer. MicroC Integrated Development Environment (IDE) developed by MikroElekronika was used in this project because it comprises a text editor, C compiler, debugger, built-in libraries, and programmer. This tool is considered because of its closeness to machine language and ease of use.

Figure 7.11 represents the program logic of the smart home with the initiation by a microcontroller. That is, the microcontroller triggers the Wi-Fi module to beacon for available smartphones. It authenticates the smartphone and then establishes a connection. The state of loads and sensors, which always be loaded whenever the system is powered on, and after executing a request, is normally stored in EEPROM.

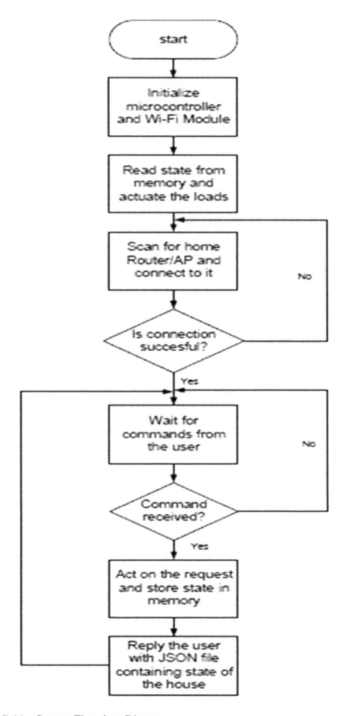

**FIGURE 7.11** System Flowchart Diagram.

2. *Android Application:* Android apps consists of different components, which can be activated separately or together. The individual activity represents different actions to the user's interface. Some activities may be performed independently in the background. Figure 7.12 presents the use case diagram of the smart home app.

An Android Studio is considered for the app. Seek Bar that controls appliances is one of its features. For instance, Toggle Buttons can be used for the appliance such as Lamps and gate control. Figure 7.13 represents the snapshot of the app, which is a user interface.

*Construction Procedure:* This section details the stages involved in the construction, including PCB production and soldering.

1. *PCB Design:* Part of the construction techniques is the use of PCB (Printed Circuit Board) to implement the hardware. The use of PCB reduced the physical dimension and improved the hardware finishing, relative to using the popular Vero board. The PCB layout was designed and routed in Proteus, producing the circuit schematic, PCB layout, and components' layout needed for soldering. The PCB Layout was printed and then transferred onto a copper board using the heating method. Figure 7.14 shows the PCB layouts of the system.

The PCB layout was printed on an artwork paper and transferred onto a copper board using an electric iron. The paper was gently removed from the copper board leaving behind the tracks on the board. Then the board was etched using $H_2O_2$ and

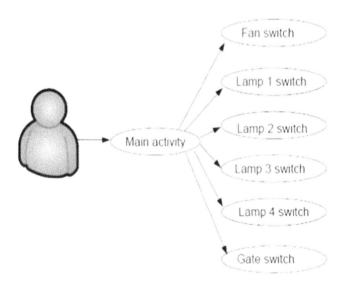

**FIGURE 7.12** Use-Case Diagram.

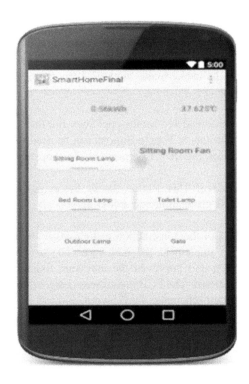

**FIGURE 7.13** Smart Home App Interface.

**FIGURE 7.14** PCB Layout.

HCL in the ratio of 2:1. After etching the board, a thinner solution was used to clean the board. Then the holes were drilled to allow proper insertion of the components for soldering. Figure 7.15 (a) to (c) show the pictures of the board at each stage.

(a)  (b)

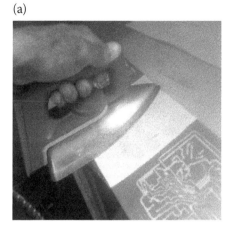
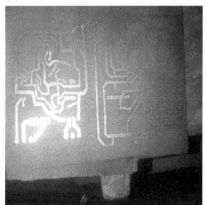

During Transfer     After Etching

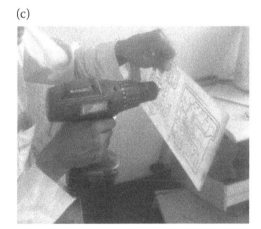

During Drilling

**FIGURE 7.15** PCB Production.

2. ***Component Soldering:*** Soldering was done using soldering Work Station and soldering lead. This process involved the appropriate mounting of various components on the board (PCB) concerning the components' layout. Figure 7.16 shows the pictures of the components side of the finished circuit board.

## 7.5 PACKAGING AND DISCUSSION

***Packaging:*** The smart home system was implemented on a prototyped one-bedroom flat using transparent plastic. Lighting points are installed in the sitting room, bedroom, toilet, and veranda. A ventilation system (fan) was installed in the sitting room. A gear-driven gate was also used as the main gate of the home. Figure 7.17 shows the prototyped home with all the fittings installed.

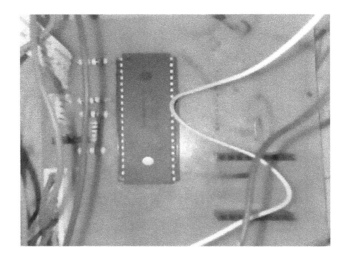

**FIGURE 7.16** Components Side of the Finished Circuit Board.

**FIGURE 7.17** Packaged Hardware.

***Discussion of Results:*** The results obtained after testing showed high precision, reliability, and acceptable tolerance using this smart home system in homes inhabited by PWD or aged persons. The magnitude of the average system error observed for the thermometer is less than 0.6°C on average, which is tolerable and complies with the accuracy of the temperature sensor used.

## 7.6 CONCLUSION

This work has developed an android based platform that enhanced convenience and technical efficiency of a home, given it has the capability to automatically control the home appliances, such as regulate the home temperature, trigger the safety indicator, or update the users on the energy consumption of the home appliances. In this 21st century, things are getting connected, society is changing, and everything becomes smart. A smart card or smart card reader is developed from an ordinary card when a chip is programmed and embedded in the new design. Similarly, fixed-line telephone technology has evolved into mobile phones, and then smartphones, which has become an important tool for society 5.0. Thus, the design and mobile application presented in this chapter can be used to support human development and improve productivity. In particular, the design would serve the PWD and aged people. In providing a platform with significant contributions for the workforce and economy, especially improving the experience of users who are challenged in day to day living by age or disability. Home automation can be extended to include improved convenience for enhancing efficiency in the workplace. Also, the design has limitations in the form of communication issues, for instance, a Wi-Fi module is limited to about 50 meters depending on the environment. The connection will be lost and the user will not be able to control the home's appliances if the radius of connectivity is more than the design range. Mesh technology may be a suitable alternative considering it enjoys a wider coverage. Also, the design required Internet connectivity to enjoy remote connections that were beyond the radius of the in-built Wi-Fi module's reach.

## 7.7 LIMITATIONS, FUTURE DIRECTIONS, AND RECOMMENDATIONS

This work is limited to the design and construction of smart home systems. The future work may consider the actual implementation of the construction in a real-life scenario. Also, future work may consider the use of IOS instead of an android based system and compare the performances. In addition, the LoRa network, which has been proposed as a promising platform for IoT technology may be considered.

Based on our observations, results, and satisfaction metrics displayed in this work, the following recommendations are made for further development:

- The Smart Home can be connected to the internet to provide remote control and data acquisition over the internet.
- More sensors like wearable sensors can be incorporated to help monitor human's well-being such as the health condition of individuals.
- The consumed energy of the house can be sent to the electricity distribution company for the purpose of billing. Therefore, the system performs as the energy metering utility in the house.

## REFERENCES

Aztiria, A., Augusto, J. C., Basagoiti, R., Izaguirre, A., and Cook, D. J. (2012). Discovering frequent user-environment interactions in intelligent environments. Personal and Ubiquitous Computing. *Middlesex University Research Repository*, *1*(16), 91–103.

Baraka, K., Ghobril, M., Malek, S., Kanj, R., & Kayssi, A. (2013, June). *Low cost arduino/ android-based energy-efficient home automation system with smart task scheduling* [Conference session]. Computational Intelligence, Communication Systems and Networks, Fifth International Conference, IEEE, pp. 296–301.

David, N., Chima, A., Ugochukwu, A., & Obinna, E. (2015). Design of a home automation system using Arduino. *International Journal of Scientific & Engineering Research, 6*(6), 795–801.

De Luca, G., Lillo, P., Mainetti, L., Mighali, V., Patrono, L., & Sergi, I. (2013, September). *The use of NFC and Android technologies to enable a KNX-based smart* [Conference session]. Software, Telecommunications and Computer Networks-(SoftCOM 2013), 21st International Conference.

Gibson, M., Gutman, G., Hirst, S., Fitzgerald, K., Fisher, R., & Roush, R. (2013). Expanding the technology safety envelope for older adults to include disaster resilience. In *Technologies for active aging* (pp. 69–93). Boston, MA: Springer.

Gupta, P., & Chhabra, J. (2016, February). *IoT based Smart Home design using power and security management* [Conference session]. Innovation and Challenges in Cyber Security (ICICCS-INBUSH), International Conference, IEEE, pp. 6–10.

Gunputh, S., Murdan, A. P., & Oree, V. (2017, May). *Design and implementation of a low-cost Arduino-based smart home system* [Conference session]. Communication Software and Networks (ICCSN), 9th International Conference, IEEE, pp. 1491–1495.

Jabbar, W. A., Alsibai, M. H., Amran, N. S. S., & Mahayadin, S. K. (2018, June). *Design and implementation of IoT-based automation system for smart home* [Symposium]. Networks, Computers and Communications (ISNCC), 2018 International Symposium, IEEE, pp. 1–6.

Kumar, S., & Lee, S. R. (2014, June). *Android based smart home system with control via Bluetooth and internet connectivity* [Symposium]. Consumer Electronics (ISCE 2014), 18th IEEE International Symposium, pp. 1–2.

Muhammad, T. R., Eman, M. A., Fariha, D., & Muhammad, A. M. (2017). Wireless android based home automation system. *Advances in Science, Technology and Engineering Systems Journal, Special Issue on Computer Systems, Information Technology, Electrical, and Electronics Engineering, 2*(1), 234–239.

Malik, N., & Bodwade, Y. (2017). Literature review on home automation system. *International Journal of Advanced Research in Computer and Communication Engineering (IJARCCE), 6*(3), 733–737.

Piyare, R., & Lee, S. R. (2013). Smart home-control and monitoring system using smart phone. *ICCA, ASTL, 24*, 83–86.

Piyare, R., & Tazil, M. (2011). *Bluetooth based home automation system using a cell phone* [Symposium]. Consumer Electronics (ISCE), IEEE 15th International Symposium, pp. 192–195.

Purohit, D., & Ghosh M. (2017, April). Challenges and Types of Home Automation Systems. *International Journal of Computer Science and Mobile Computing, 6*(4), 369–375.

Rajeev, P. (2013). Internet of Things: Ubiquitous Home Control and Monitoring System using Android based Smart Phone. *International Journal of Internet of Things, 1*(2), 5–11.

Singh, H., Pallagani, V., Khandelwal, V., & Venkanna, U. (2018, March). *IoT based smart home automation system using sensor node* [Conference session]. Recent Advances in Information Technology (RAIT), 4th International Conference, IEEE, pp. 1–5.

Subhamay, S., Mithun, C., & Anindita, B. (2014). Low cost embedded system/android based smart home automation system using wireless networking. *International Journal of Electronics and Communication Engineering, 7*(2), 175–186.

Tharishny, S., Selvan, S., & Nair, P. (2016). Android based smart house control via wireless communication. *International Journal of Scientific Engineering and Technology*, 5(5), 323–325.

United Nations Department of Economic and Social Affairs. (2020). *World Population Ageing 2019*. UN.

WHO Library (2011). *World Heath Report on World Report on disability*. Cataloguing-in-Publication Data, World Health Organization, 20 Avenue Appia, 1211 Geneva 27, Switzerland.

Yan, M., & Shi, H. (2013). Smart living using bluetooth-based android smartphones. *International Journal of Wireless & Mobile Networks (IJWMN)*, 5(5), 65–72.

# 8 Privacy-Preserved Access Control in E-Health Cloud-Based System

*Suman Madan*

## CONTENTS

8.1 Introduction .................................................................................................. 145
    8.1.1 Characteristics of Cloud Computing ............................................. 147
    8.1.2 Cloud Deployment Models ............................................................ 147
    8.1.3 Cloud Service Models .................................................................... 149
8.2 E-Health Cloud Security and Privacy Issues ............................................. 149
8.3 Recent Work in E-Health Privacy ............................................................... 151
8.4 Privacy Preservation E-Health Monitoring System ................................... 153
    8.4.1 Broadcast Group-Key Management Scheme ................................. 154
    8.4.2 Proposed Architecture ................................................................... 157
8.5 Experiment Results ..................................................................................... 157
8.6 Conclusion .................................................................................................... 160
References ............................................................................................................. 160

## 8.1 INTRODUCTION

The progression of information and communication technologies has outmoded the old-style methods making way for the new in health-care management in the world, and the progress is nowhere more evident than in a partial relinquishment of paper-based clinical treatment to electronic version mostly in the developed countries around the world (Zhang & Liu, 2010). Thus, the necessity to organize and collate e-health data from various sources like hospitals, pathological laboratories, health insurance firms etc., has changed, and in turn has changed how clinical decision-making, health-care delivery, shoveling diseases, etc are managed and disbursed to patients. Figure 8.1 shows e-health data collection sources (Privacy Analytics Inc, 2017), and it shows that the healthcare domain continues to be one of the most susceptible to information breaches from freely disclosed data falling in the hands of data thiefs. Thus, cloud technology can bring to the healthcare domain its advantages in the form of managing e-health records through collaboration, communication, and coordination between different e-health data collection sources but without the disadvantages of siloed data sets and complex processes. Besides, cloud technology can be helpful in storing, managing, sharing, and protecting the e-health records for longer durations as

**FIGURE 8.1** E-Health Data Collection.

patients receive better care and lesser worry with up-to-the-minute e-health records fortified by constant communication between different e-health collection sources. However, inability to keep up with regulations and standards, interoperability problems, security, privacy, etc. pose major obstacles in cloud adoption in this domain.

Computer security is a developing field of study in the IT industry which emphasizes protection of computer systems and e-data against not only unlawful access but also common threats and hazards like phishing, DoS attacks, etc. Again, user privacy is the eye around which the requirements of computer security develop as it imposes some rules and standards that control how much information about people can be gathered, accessed, and communicated to a third party, which makes security and privacy not clearly distinguishable and, thus, they are considered as overlapping in many ways (Suman & Puneet, 2019a; Goswami & Madan, 2017). Literally, invaders can utilize information mining procedures and methods to discover sensitive information and thus publish it to the general population's reading leading to information breach. Although, the security measures implementation is still a complex procedure, the dangers are imminent since as the data theft modalities become more sophisticated those of data protection become more relaxed leading to frequent overthrow of security controls.

Traditional access control models have clients with preordained rights that can be done in two ways: first, the group administrator provides the rights; or, one client gets the rights from another client, which is called delegation. Zhang et al. (2003) clarifies the situations where delegation is essential. Several privacy-protecting access control models are proposed (Jafari et al., 2011; Moniruzzaman et al., 2010; Madan & Goswami, 2018; Puneet & Madan, 2018; Suman & Puneet, 2019b) in literature.

This chapter mainly focuses on proposing a privacy model which permits data provisioners to define privacy guidelines for various associations accessing their data and information. Access control is defined by using broadcast encryption where a sender is permitted to send ciphered text to some selected groups wherein every client from that group may decipher that ciphered text with their private key; but, no one external to the group can decipher the message. Also, this chapter's agenda contains different schemes of cloud computing that can be utilized in healthcare domain to be discussed, with various security and privacy issues that are considered as obstacles in extensive adoption of cloud in healthcare domain, while it also proposes new privacy-preserved access control scheme.

Cloud computing is the buzzword in the IT industry rather than a scientific term. According to (Mell & Grance, 2011; Suman & Goswami, 2018) "cloud computing is a model for enabling ubiquitous, convenient, on-demand network access to a shared pool of configurable computing resources like storage, servers, networks, applications that can be rapidly provisioned and released with minimal management effort or service provider interaction".

### 8.1.1 CHARACTERISTICS OF CLOUD COMPUTING

As per the official characterization, cloud computing has five main characteristics: on-demand self-service, shared resource pool, elasticity, broad-network access, and measured service (Mell & Grance, 2011).

- **On-demand self-service:** The cloud can be inattentively configured by the clientele, if required, and thus the interference from the service engineers is not required. Clientele can make their own scheduling and implement their required storage along with computing power.
- **Shared resource pool:** Clientele can share resources like storage, servers, networks, software, memory, and processing concurrently. Resources can be dynamically allocated according to the demand variations and the customer is totally ignorant about the physical locations of cloud services.
- **Elasticity:** Customers feel that resources are limitless since cloud is very flexible, adaptable, and configurable.
- **Broad-network access:** Broad-access to the network is permitted by the cloud utilizing Internet from any device.
- **Measured service:** Different metrics exist for measuring the cloud services and exhaustive usage reports may be generated for the customers and cloud service providers.

### 8.1.2 CLOUD DEPLOYMENT MODELS

Figure 8.2 shows various cloud deployment models. While large numbers of IT services are delivered by the public clouds via the internet at subscription-based low-priced models having low pricing and higher elasticity and scalability, the private cloud is committed to being used by only a single organization; thus, there is no sharing of resources which makes it more desirable for handling sensitive data. The hybrid cloud is a combination of private and public cloud services, thus improving cloud costs without sufficient leverage from advantages

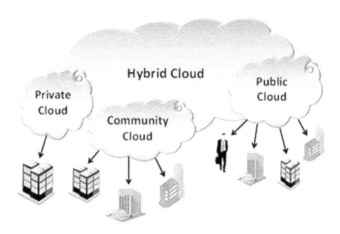

**FIGURE 8.2** Different Cloud Deployment Models.

that only private or only public cloud technologies can deliver. The community cloud looks a lot like private cloud, the only difference being users' groups. However, private cloud server is owned by only one customer while many other customers with analogous backgrounds can share the infrastructure and resources of a community cloud. Table 8.1 shows comparative study of cloud deployment models (Zhang & Liu, 2010).

**TABLE 8.1**
**Comparison of Cloud Deployment Models**

| Deployment Model/Parameters | Public | Private | Hybrid | Community |
|---|---|---|---|---|
| Owner | Cloud Provider | Enterprise | Enterprise | Community |
| Cost effective | Very cost effective | High Cost | Moderate | Moderate |
| Scalable | Very high | Limited | Very high | Limited |
| Reliable | Moderate | Very high | Moderate to high | Very high |
| Complexity | Less | Depends on Customization | Moderate | Moderate |
| Security | Less | High | Moderate secure | Not much secure |
| Performance | Low to moderate | Good | Good | Very Good |
| Availability | Available in global network | Available in Monitored and secured network | Available in both global and monitored networks | Available over intranet and VPN |

### 8.1.3 CLOUD SERVICE MODELS

Given the need to manage the type of service by the need of the service to be accessed as per the requirement of customer, the cloud provides three types of services: application services, platform services, and infrastructure services. Cloud application services are commonly known as Software-as-a-Service (SaaS), that directly runs over the web browser only without the need of downloading and installing applications. Cloud platform services, commonly known as Platform-as-a-Service (PaaS), provides a platform for building a software and is similar to SaaS. Cloud infrastructure services, commonly known as Infrastructure-as-a-Service (IaaS), is responsible of managing data of application, middleware, and runtime environments. Table 8.2 shows traditional cloud service models along with their key benefits and examples (Zhang & Liu, 2010).

## 8.2 E-HEALTH CLOUD SECURITY AND PRIVACY ISSUES

Currently, with the need to access clinical records everywhere all the time, the healthcare domain uses cloud computing that encourages amalgamation of all clinical records that can be easily shared. Though cloud computing provides many advantages, still it presents security and privacy issues for e-health records (Dong et al., 2011), and the security issues need to be managed properly by cloud service providers for enhancing the trust level between health-domain data providers and the patients (Abbas & Khan, 2014). For better security and privacy in healthcare domain, a few cloud application requirements are enumerated here:

1. **Confidentiality:** Confidentiality means that unauthorized entities cannot access the patient's e-health data so that the patient can trust the healthcare system for protecting the confidentiality of his e-health records. Since data can be compromised with data-control given to the cloud and hence data becoming available to a bigger number of parties, devices and/or applications, it becomes imperative for patient-doctor relationship to flourish that the data shared with the doctor be protected; otherwise, the patient may become choosier and more careful about the information being shared with the doctor in future. Various encryption and access-control methods may be used to achieve confidentiality.
2. **Integrity:** Integrity means that e-health record stored and given to any individual or organization should be accurate and consistent and have not been altered at all. Cloud utilization for an important application like e-Health requires reliability assurance for the provided services and of error-free records. In case of integrity failure, the healthcare system should terminate current processing and report error.
3. **Availability:** The e-health record should be available all the time especially in critical situations, which means high-availability frameworks ought to forestall service disruptions because of hardware failures, power disruption, system upgrade, denial-of-service attacks etc. At the same time, it should preserve the usability of e-health record even after enforcement of HIPAA security and privacy rules.

## TABLE 8.2
## Cloud Service Models

| Cloud Service | Definition and Characteristics | Examples | Advantages |
|---|---|---|---|
| Software-As-a-Service (SaaS) | • Refers to applications that are enabled for cloud and have stateless application architecture.<br>• Supports architecture that can run on multiple instances of itself, irrespective of location. | • Salesforce<br>• Google Docs<br>• Zoho<br>• MobileMe | • Reduction in cost of application software licensing, servers, and other infrastructure along with personnel required to host the application internally<br>• Enables software vendors to control and limit use, prohibits copying and distribution.<br>• Application delivery uses the one-to-many delivery approach |
| Platform-As-a-Service (PaaS) | • Variation of SaaS<br>• Enables developers to write applications to run on cloud.<br>• Platform provides several application services for quick deployment. | • Microsoft Azure<br>• Force.com<br>• Google App Engine<br>• Cloud Foundry | • It is greatly increasing the number of people who can develop, maintain, and deploy web applications.<br>• Offers general developers to build web applications without needing specialized expertise. |
| Infrastructure-As-a-Service (IaaS) | • Highly scaled redundant and shared computing infrastructure accessible using internet technologies<br>• Consists of storage, servers, databases, security and other peripherals | • Google cloud storage<br>• Amazon EC2, S3 etc | • Offers the ability to scale infrastructure requirements like memory, storage, computing resources etc. based on usage requirements<br>• Ability to Pay as you go.<br>• Access to best-of-breed technology solutions and superior IT talent for a fraction of the cost |

4. **Non-repudiation:** Repudiation warnings are needed to avoid worries from clients who refute their signature's genuineness after accessing e-health record. Thus, healthcare domain cloud applications should force digital signatures to set authenticity and nonrepudiation along with encryption.

5. **Authenticity:** Authenticity means veracity of origins, ascriptions, assurances, and intents. Thus, it is simply ensuring the genuineness of the access requesting entity. In e-health records, the information authentication may present problems like man-in-the-middle attacks and is frequently moderated with a blend of usernames and passwords. For healthcare domain, e-health record should be verified at every access, regardless whether access is by e-health record provider or by patients.
6. **Ownership and access-control:** Owner is defined as the originator or collector of the e-health record and ownership plays a significant role in protecting e-health records against unauthorized access. Ownership of data may be safeguarded using a blend of watermarking and encryption techniques which will give secured e-health record so as to transmit, access, or release with mutual approval of each and every stakeholder owing ownership of access to e-health records. Thus, patients have control for allowing or denying the sharing of e-health record.
7. **Anonymity and re-linkability:** Anonymity means that the patient can't be identified from e-health records received for purpose of researching and analytics. Thus, before releasing data to a third party, the sensitive attributes must be de-identified either by eradicating sensitive data-parts or applying anonymity techniques like generalization or suppression (Suman and Puneet, 2019b). Re-linkability means that different versions of e-health record released to various entities should not be later re-linked to release the identity of individual. This is a privacy breach.

## 8.3 RECENT WORK IN E-HEALTH PRIVACY

The e-Health's privacy and security hazards are an area of serious interest and intensive study with researchers. This section is focused on the survey of recent work carried out for secured e-Health system architecture. In literature, most encryption strategies incorporate security measures against oversight, and, guarantee information privacy (Su et al., 2012). In fine-grained information access control, among numerous techniques proposed are compulsory access-control, optional access-control, and job-based access-control. In every one of these models, personality is given as clients and assets have interesting names to indicate it. This method is productive in static frameworks in which the arrangement of clients and administrations are known ahead of time. In present day framework, clients are dynamic and assets are specially appointed. Information on clients are related to his characteristics or those of the assets and can be single-esteemed or multi-esteemed. Thus, character-based access control that is unbending is supplanted by quality-based access control which can be either the characteristics of the client or traits of the assets. These dynamic credentials can be compared or cross-checked against fixed qualities or against one another and called as connection-based access control.

Sahai and Waters (2005) proposed and investigated Attribute based encryption (ABE), a public key encryption scheme, in which the cipher text and secret key of clients rely only upon attributes. The administration of the framework was accountable for making encoding and decoding keys for information owners and clients. However, the attribute list based on which encoding key is made must be predefined. In the event that another client who goes into the framework has new attributes, administrator must do attribute redefinition, keys regeneration, and re-encoding. The cipher text decoding is conceivable if cipher attributes synchronize with client's keys attribute. Pirretti et al. (2010) proposed an attribute-based framework for conveyed condition and presumed that it is a proficient, compelling, and secure method of overseeing information in inexactly coupled condition.

Hamid et al. (2017) addresses secrecy of healthcare multimedia information of patient existing in cloud by suggesting a triparty agreement constructed by bilinear-pairing cryptography and using one-round verified-key protocol. The suggested technique generates a session key for secure communication among the members; then, using a technique having fog computing facility access control is implemented and the sensitive health-domain information is accessed and stored securely. However, the technique brings about a computational overhead expenditure in loss of substantial security.

In frameworks that have explicit groups of clients, the cynosure of attention and exploration becomes group-key management (Challal & Seba, 2006; Zou et al., 2008). Primarily, a trusted key server was accountable for sharing keys among clients relying upon secrets (Chu et al., 2002; Harney & Muckenhirn, 1997). In this technique, forward and backward secret conservation turns into a troublesome process. To defeat the challenges of the past techniques, Hierarchical key management plans were proposed (Sherman & McGrew, 2003; Wong & Lam, 2000). The benefit from the exercise is the size of key-size was diminished with every client retaining redundant keys which may be planned hierarchically.

Marwan et al. (2017) proposed a new technique based on Shamir's Secret Share Scheme (SSS) and multi-cloud concept that boosted the cloud storage reliability so as to fulfil security requirements to evade unauthorized access, loss of information, and privacy leak. Their proposed technique splits the sensitive information into numerous little parts with the goal that any part shouldn't reveal any sensitive data about medical records. In addition to multi-cloud architecture, information is spread over a diverse array of cloud-storage systems. For these setups, cloud clientele makes use of SSS technique for encrypting their information and thus ensuring secrecy and privacy. Though healthcare information is divided into different parts so that information confidentiality is guaranteed, still the ideal number of parts for the sustained trade-off between security and efficacy remains unresolved.

Yongdong et al. (2013) presented attribute-based access strategy that stirred numerous troublesome computationally exhaustive tasks to cloud servers and in turn allowing movable gadgets like mobile which have restricted resources with no trade-off on security perspectives to gain leverage. Smithamol and Rajeswari (2017) proposed a model for addressing the information confidentiality and access-privacy which uses well-ordered set for developing the access structure that is group-based and Ciphertext-Policy Attribute-Based Encryption (CP-ABE) for providing fine-grained e-health records access-control. The approach minimized the overall encryption time and the computational overhead and also the performance examination reflected the proposed model's efficiency making it apt for real-world use. Edemacu et al. (2019) presented an overview of collective e-Health and provided

**TABLE 8.3**
**Comparative Analysis of ABE, Single-Layered, and Double-Layered Encryption**

| Encryption Property | Attribute-Based | Single-Layered | Double-Layered |
| --- | --- | --- | --- |
| Cryptographic technique | Asymmetric | Symmetric | Symmetric |
| Secured attribute-based group communication | √ | √ | √ |
| Access control delegation | × | × | √ |
| Efficient reversal | × | √ | √ |

comparisons of attribute-based encryption schemes privacy issues. Azeez and Van der Vyver (2019) proposed secure and dependable architecture for e-health that guaranteed reliable, efficient, and structured access-framework for the health data.

The comparative analysis of few properties of attribute-based encryption (ABE) vs. single and double layered encryption is given in Table 8.3.

## 8.4 PRIVACY PRESERVATION E-HEALTH MONITORING SYSTEM

A general observation on patients is the impracticability of the expectation that every patient every time will bring their clinical report and all clinical-related archives with them without fail on every hospital visit. A cloud-based e-health application system with privacy-preservation to secure patient's sensitive information is expressed in Figure 8.3. All clinical reports and all clinical-related information are encrypted and then stored in the cloud which in turn allows it to be accessed from anywhere in world.

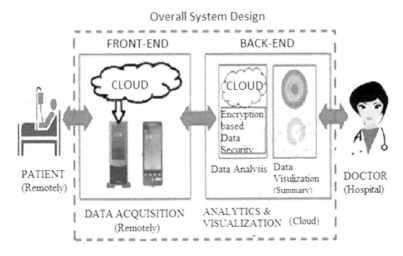

**FIGURE 8.3** E-Health Monitoring System.

A few proposed techniques rely upon broadcast key management plans. Shang et al. (2010) have used an encryption method with a single layer of encryption in which the information owner should be consistently active, which is resolved by a twofold layer encryption. Nabeel et al. (2013) proposed one two-fold layer encryption in which coarse-grained encryption was performed by the information owner to guarantee the confidentiality of the information existing in the cloud. At that point, the cloud does the fine-grained encryption of the encrypted data given by the information owner relying upon access control policy by the information owner. In this twofold layer encryption, the broadcast encryption structure utilizes static form in which the unified group administrator is included once during the setup stage (Thuraisingham, 2011).

In the proposed framework which is dynamic in nature, client can enter into or exit from the application whenever he wishes, preserving privacy and decentralizing the framework. The proposed privacy model is utilized along with access control model to reserve privacy access rights. Thus, in this dynamic twofold layered decentralized broadcast encryption structure, there is no particular centralized controlling administrator for group-key administration process. Every bit of information the client accesses has privacy policies in their information objects that will improve security of the access control model.

### 8.4.1 Broadcast Group-Key Management Scheme

Group-Key management is a method for securely dispensing a message to a client group with classification as the main key. Here, individuals in a group have a symmetric group-key "K" and whenever they are sharing any message, it must be encoded with "K" and then sent to the all individuals of group. As "K" is known to all individuals in the group, they can decode and get the message. At any point when a new individual enters or leaves a group, a group key is recreated and circulated among group individuals. The new individual in the group can't gain access to prior communicated messages, and is called backward secrecy; while any individual who has left the group can't learn anything from future communications in the group, and is called forward secrecy. Thus, backward and forward secrecy are the explanation behind re-keying when the client dynamics change. However, group-key management has certain issues as follows:

1. It advocates group-individual identification and confirmation between the pair (Group Controller-Key Server) and clients.
2. Major concern here is access control to approve and check the new client joining activity, Key recreation, and circulation.
3. Since, it requires forward and backward secrecy to avert inauthentic access, thus keys ought to be consistently refreshed.
4. Additionally, storage overhead produce a major issue in key management since all the keys utilized at each level are stored, and, more the key-generation number, more the storage overhead. Thus, number of keys needs to be minimized as the smallest number of keys required to work for this system is already very huge.

Consequently, this methodology isn't appealing if there are successive exits and joins and, in broadcast group-key management structure for dispensing keys to clients, every client gets at least one secret that is joined with some variable to get the group key. Whenever there is change in group dynamics, there will be change only in public data. The encoded information is transported to the cloud and every client is given the keys just for the set(s) of information that it can access as indicated by the guidelines. This methodology ensures information secrecy and authorizes fine-grained access control guidelines. Key management is matter of concern here, as each client must be given the accurate keys regarding access control guidelines the client satisfies. The solution is to provide a hybrid solution by encoding keys with attribute-based encryption or/and proxy re-encoding which itself encloses a part of the shortcoming.

Attribute-based group-key management utilizes the following algorithms:

- **Initial_Setup (S):** This algorithm initializes all parameters including set of secrets "U", secret space "US" and key space "Ks" using the security parameter "S".
- **Secret_Generator (user, att):** This algorithm chooses a random string "s" from "U" at random from "US", adds "s" to "U" and outputs "s". For every user, a secret key is assigned which is used for generating group key.
- **Key_Generator (U, policy):** This algorithm picks group key "g" from key space "Ks" and generates tuple "GT" that is calculated from "U".
- **Key_derivative(s, GT):** This algorithm uses "U" and "GT" to get group key.
- **Update (U):** This algorithm executes Key_Generator algorithm when there is entry or exit in group.

At the point when an emergency clinic intends to use the cloud for keeping up-to-date the sensitive e-health records of patients, the privacy must be safeguarded in the cloud. Figure 8.4 shows the cloud architecture for e-health maintenance system in hospitals. The partners in medical clinic are specialist, drug specialist, assistant,

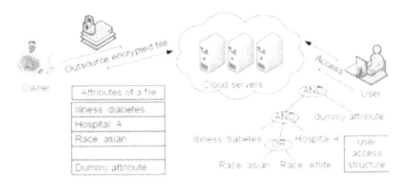

**FIGURE 8.4** Cloud Architecture for E-Health Maintenance System in Hospitals.

clerk, nurture, information passage administrators, and so on. Every partner needs a distinctive degree of access of detail, and thus, the wellbeing record support need fine-grained access control. Attributes like occupation, job title, and area-zone are used as behavior attributes that are collected suitably so that privacy of patients is not breached.

In the proposed framework, for scrambling the e-health record, a twofold encryption is done to diminish the information proprietors' burden and isolate the access control to the cloud. The information is deteriorated and put away across various mists and recovered utilizing keys given by clients. The proposed framework utilizes symmetric key cryptography. Given clients are not given genuine key by group key management and, with dynamic access control components, only the just approved client with legitimate key can decode the information. The framework has four stakeholders Data Owner, User, Token Provider, and Cloud, as shown in Figure 8.5.

The double encryption method for attaining fine-grained access-control ensures six phases for the whole process as follows:

1. Tokens are issued to users by token provider depending on attribute-identity. Access-control policy is disintegrated between the data owner and cloud.
2. The identity tokens of users are registered with the owner and the cloud.
3. Secrets are shared among the user, data-owner, and cloud.
4. The data-owner randomly encodes and uploads keeping one set of tokens with itself and another set to cloud preventing cloud alone to decode the data.
5. Re-encrypt and impose policy.
6. The encoded data is downloaded from cloud and by using derived keys, decoding is done twice.

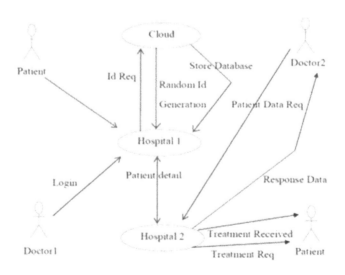

**FIGURE 8.5** Flow of Information in Hospital Cloud.

## 8.4.2 Proposed Architecture

In the proposed framework, for scrambling the e-health record, a twofold encryption is done to diminish the information proprietors' burden and isolate the access control to the cloud. The information is deteriorated and put away across various mists and recovered utilizing keys given by clients. The proposed framework utilizes symmetric key cryptography. Given the clients are not given genuine key by group key management and, with dynamic access control components, only approved client with legitimate key can decode the information. The framework has four stakeholders: Data Owner, User, Token Provider, and Cloud, as shown in Figure 8.5.

The double encryption method for attaining fine-grained access-control ensures six phases for the whole process as follows:

1. Tokens are issued to users by token provider depending on attribute-identity. Access-control policy is disintegrated between the data owner and cloud.
2. The identity tokens of users are registered with the owner and the cloud.
3. Secrets are shared among the user, data-owner, and cloud.
4. The data-owner randomly encodes and uploads keeping one set of tokens with itself and another set to cloud preventing cloud alone to decode the data.
5. Re-encrypt and impose policy.
6. The encoded data is downloaded from cloud and by using derived keys, decoding is done twice.

In this proposed system, the system architecture is as shown in Figure 8.6.

   i. Authorize: PHR owner initially authorizes a Central authority (CA).
   ii. Key Generation: Central Authority is responsible for generating Master key (MK) and Secret Key (SK).
   iii. User-Key issue: CA issues MK and SK to user.
   iv. ABE: PHR owner will Encrypt the file using ABE.
   v. Outsourced Encrypted PHR: The encrypted PHR will be outsourced to cloud server.
   vi. Obtain attribute value: Meanwhile users provide attribute value to owner.
   vii. Providing write Key: User gets the write key from PHR owner.
   viii. Search Encrypted PHR: The user can read or write file by providing their respective access policies and secret keys.
   ix. Bloom filter: The encrypted file search is done using the Bloom filter.
   x. Result: The encrypted PHR's is given to user.

## 8.5 EXPERIMENT RESULTS

Privacy saving access control for distributed storage in cloud is executed with a private cloud arrangement. Visual studio 2006 and SQL server 2005 are utilized for building ASPX pages and can utilize the usefulness of the .NET Framework. Reenactment results that are acquired utilizing cloud CPABE toolbox for key age is

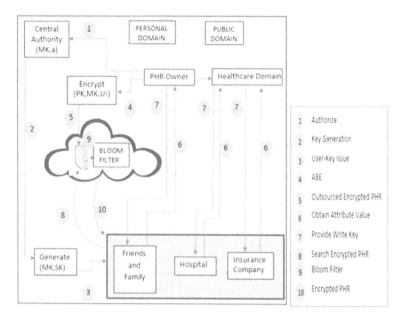

**FIGURE 8.6** Proposed Architecture.

appeared in Figure 8.7. Figure 8.7 shows that when number of attributes are less, key generation takes less time. However, with increase in attributes, key generation takes on an exponential growth. Figures 8.8 and 8.9 show the ideal opportunity for creating encryption and decoding keys. Figure 8.8 shows encryption time started with slow-growth, followed with moderate growth and then steady growth along with the number of leaf nodes in the policy. However, in Figure 8.9, with number of leaf nodes, the decryption time follows the shape of sine curve but with a displacement of 90 degrees to the left. Figure 8.10 shows the average time taken to produce single layer and twofold layer encryption for differing group sizes. For experiments, the number of attribute-conditions considered for experiments is 1000

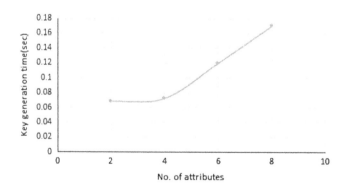

**FIGURE 8.7** Key Generation Time.

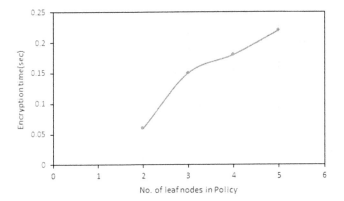

**FIGURE 8.8** Encryption Time.

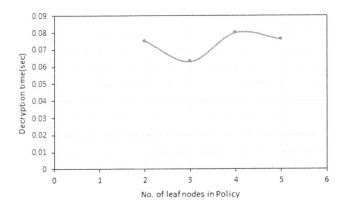

**FIGURE 8.9** Decryption Time.

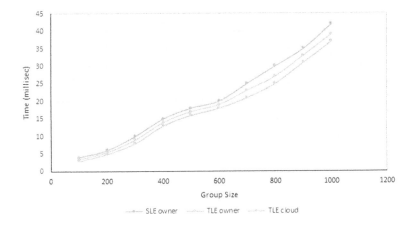

**FIGURE 8.10** Average Time to Generate Keys for Single and Two Layers.

and per policy, maximal attribute conditions is 5. So, Figure 8.10 shows that the Single layer execution (SLE) time is higher than twofold layer execution (TLE). Thus, all the figures indicate that obviously normal time taken to create keys in twofold layer encryption is not exactly single layer encryption.

## 8.6 CONCLUSION

Recent patterns in cloud innovation show the need for and related administrations are keen on pushing capacities to outsider suppliers for accomplishing adaptability and versatility. However, ongoing studies have indicated that a fundamental factor for associations to move to the cloud is information privacy and security concerns. The strengths of cloud computing make it a very promising option while overlooking its threats. However, shifting e-health records on the cloud is a strategic as well as complex decision, and the e-health providers should tackle the security challenges before making it available on cloud. This chapter focuses on accomplishing decentralized two-fold encryption, versatility, and information classification for information access control in cloud. The proposed framework, likewise, empowers the information proprietor to designate the majority of calculation-intensive errands to cloud servers without uncovering information substance or client access benefit data, and furthermore, client responsibility. Future work will focus on lessening the computational expense by taking on incomplete connections.

## REFERENCES

Abbas, A., & Khan, S. U. (2014). A review on the state-of-the-art privacy-preserving approaches in the e-health clouds. *IEEE Journal of Biomedical and Health Informatics, 18*(4), 1431–1441. 10.1109/jbhi.2014.2300846.

Azeez, N. A., & Van der Vyver, C. (2019). Security and privacy issues in e-health cloud-based system: A comprehensive content analysis. *Egyptian Informatics Journal, 20*(2), 97–108. https://doi.org/10.1016/j.eij.2018.12.001.

Challal, Y., & Seba, H. (2006). Group key management protocols: A novel taxonomy. *International Journal of Information Technology, 2*(2), 105–118.

Chu, H., Qiao, L., Nahrstedt, K., Wang, H., & Jain, R. (2002). A secure multicast protocol with copyright protection. *SIGCOMM Computer Communication Rev, 32*(2), 42–60.

Dong, N., Jonker, H., & Pang, J. (2011). Challenges in eHealth: From enabling to enforcing privacy. In Z. Liu & A. Wassyng (Eds), *Foundations of health informatics engineering and systems. FHIES 2011. Lecture notes in computer science* (pp. 195–206). Springer.

Edemacu, K., Park, H. K., Jang, B., & Kim, J. W. (2019). Privacy provision in collaborative ehealth with attribute-based encryption: Survey, challenges and future directions. *IEEE Access, 7*, 89614–89636. 10.1109/ACCESS.2019.2925390.

Goswami, P., & Madan, S. (2017). A survey on big data & privacy preserving publishing techniques. *ACST, 10*(3), 395–408.

Hamid, H. A. Al, Rahman, S. M. M., Hossain, M. S., Almogren, A., & Alamri, A. (2017). A security model for preserving the privacy of medical big data in a healthcare cloud using a fog computing facility with pairing-based cryptography. *IEEE Access, 5*, 22313–22328.

Harney, H., & Muckenhirn, C. (1997). *Group key management protocol (GKMP) specification* [Technical Report], Network Working Group, United States.

Jafari, M., Fong, P. W. L., Safavi-Naini, R., Barker, K., & Sheppard, N. P. (2011). *Towards defining semantic foundations for purpose- based privacy policies* [Conference session]. Data and Application Security and Privacy (CODASPY), First ACM Conference, San Antonio, Texas, USA, pp. 213–224.

Madan, S., & Goswami, P. (2018). *A privacy preserving scheme for big data publishing in the cloud using k-anonymization and hybridized optimization algorithm* [Conference session]. Circuits and Systems in Digital Enterprise Technology, International Conference, pp. 1–7. 10.1109/ICCSDET.2018.8821140

Marwan, M., Kartit, A., & Ouahmane, H. (2017). *Protecting medical data in cloud storage using fault-tolerance mechanism* [Conference session]. Smart Digital Environment, International Conference, pp. 214–219.

Mell, P., & Grance, T. (2011). *The NIST Definition of Cloud Computing [Recommendations of the National Institute of Standards and Technology-Special Publication 800-145] Gaithersburg*. NIST. http://csrc.nist.gov/publications/nistpubs/800-145/SP800-145.pdf.

Moniruzzaman, M., Ferdous, M. S., & Hossain, R. (2010. *A study of privacy policy enforcement in access control models* [Conference session]. Computer and Information Technology (ICCIT), International Conference, Bangladesh, pp. 352 – 357. 10.1109/ICCITECHN.2010.5723883.

Nabeel, M., Shang, N., & Bertino, E. (2013). Privacy preserving policy-based content sharing in public clouds. *IEEE Transactions on Knowledge and Data Engineering*, 25(11), 2602–2614.

Pirretti, M., Traynor, P., McDaniel, P., & Waters, B. (2010). Secure attribute based systems. *Journal of Computer Security*, 18(5), 799–837.

Privacy Analytics Inc. (PIC- 2017). *Patient-Level Data*. Privacy Analytics Inc.

Puneet, G., & Madan, S. (2018). Hybrid privacy preservation model for big data publishing on cloud. *International Journal of Advanced Intelligence Paradigms*. 10.1504/IJAIP.2 018.10025582

Sahai, A., & Waters, B (2005). Fuzzy identity-based encryption. *Advances in Cryptology V EUROCRYPT*, 3494 of LNCS, 457–473.

Shang, N., Nabeel, M., Paci, F., & Bertino, E.(2010). *A privacy-preserving approach to policy-based content dissemination, ICDE '10* [Conference session]. Data Engineering, IEEE 26th International Conference.

Sherman, A. T., McGrew, D. A. (2003, May). Key establishment in large dynamic groups using one-way function trees. *IEEE Transactions on Software Engineering*, 29(5), 444–458. doi: 10.1109/TSE.2003.1199073

Smithamol, M. B., & Rajeswari, S. (2017). Hybrid solution for privacy-preserving access control for healthcare data. *Advances in Electrical and Computer Engineering*, 17(2), 31–38.

Su, J. S., Cao, D., Wang, X. F., Su, Y. P., & Hu, Q. L. (2012). Attribute-based encryption schemes. *Journal of Software*, 6, 1299–1315.

Suman, M., & Puneet, G. (2019a). *A novel technique for privacy preservation using K-anonymization and nature inspired optimization algorithms* [Conference session]. Sustainable Computing in Science, Technology & Management, IEEE International Conference. http://dx.doi.org/10.2139/ssrn.3357276

Suman, M., & Puneet, G. (2019b). K-DDD measure and MapReduce based anonymity model for secured privacy preserving big data publishing. *International Journal of Uncertainty, Fuzziness and Knowledge-Based Systems*, 27(2), 177–199.

Thuraisingham, B. (2011). *Towards privacy preserving access control in the cloud*. 7th International Conference on Collaborative Computing: Networking, Applications and Work-sharing, ser. Collaborate Com'11, pp. 172–180.

Wong, C., & Lam, S. (2000). *Keystone: A group key management service* [Conference session]. Telecommunications, International Conference, ICT 2000.

Yongdong, W., Zhuo, W., & Deng, R. H. (2013). Attribute-based access to scalable media in cloud-assisted content sharing networks. *IEEE Transactions on Multimedia, 15*(4).

Zhang, R., & Liu, L. (2010). *Security models and requirements for healthcare application clouds*. 3rd IEEE International Conference on Cloud Computing (CLOUD), Miami, USA, pp. 268–275.

Zhang, X., Oh, S., & Sandhu, R. (2003). *PBDM: A flexible delegation model n RBAC*. 8th ACM symposium on Access control models and technologies (SACMAT), New York, NY, USA, pp. 149–157.

Zou, X., Dai, Y., & Bertino, E. (2008). *A practical and flexible key management mechanism for trusted collaborative computing*. INFOCOM 2008 – The 27th Conference on Computer Communications. IEEE, pp. 538–546.

# 9 Eye Gaze Mouse Empowers People with Disabilities

*Sonia Rathee, Amita Yadav, Harvinder Rathee, and Navdeep Bohra*

## CONTENTS

9.1 Introduction .................................................................................................. 163
9.2 Background .................................................................................................. 164
9.3 Eye Gaze Mouse ........................................................................................... 168
    9.3.1 Limitations of Previous Methods .................................................... 169
        9.3.1.1 Mouse Movement Using Pupil Detection ........................ 169
        9.3.1.2 Additional Headgear ......................................................... 169
    9.3.2 Working of Eye Gaze Mouse ........................................................... 169
        9.3.2.1 Video Processing .............................................................. 169
        9.3.2.2 Image Conversion to Grayscale ....................................... 170
9.4 Face Detection Using Haar Cascade Classifier ........................................... 171
    9.4.1 Facial Landmarks Localization ....................................................... 172
    9.4.2 Application Flow Detection ............................................................ 172
        9.4.2.1 Eye-tracking to Give Direction for Mouse ...................... 173
    9.4.3 Eyeball Tracking Left and Right as an Extra Input ........................ 173
    9.4.4 Blink Detection ................................................................................ 173
        9.4.4.1 Eye Features Detection .................................................... 173
9.5 Results .......................................................................................................... 175
9.6 Future Research Directions ......................................................................... 176
9.7 Conclusion ................................................................................................... 177
References ............................................................................................................ 177

## 9.1 INTRODUCTION

Internet browsing, online shopping, socializing, and entertainment: these are a few of the various activities that make up our day and that we engage in using a computer. Computers have come a long way from mathematical problem solving and word processing to become a necessity for getting us through every aspect of our daily activities.

In a simple computer system, the interaction between user and computer takes place using common devices, additional peripheral devices such as mouse and keyboard for inputs, and monitors and loudspeakers for output and display of

DOI: 10.1201/9781003154686-9

information. But the input devices are useless in case of people who are motor-impaired, who cannot move their hands or other parts of their body. For example, speaking for people with amyotrophic lateral sclerosis (ALS), these patients have lost motor neurons, which reduces their ability to use the voluntary muscles, and are thus paralyzed. This tends to show up as disabilities like patients who are unable to speak and to whom communication is very hard. Furthermore, this condition may also involve dysphonia or aphonia. Another such scenario are patients with cerebral palsy having restricted movement of muscle and whose coordination between body and brain is disrupted showing up in disabilities like patients who are unable to speak and are generally bound to a wheelchair. In today's world, computers are the best tools for communication and which can be modified to help these people in communicating with others. To invent for these people we need data on how these people communicate and we can gain information from electrical signals from brain muscles of the disabled, or by detecting partially active limb micro-movements.

The major challenge before us would be to facilitate the interaction between the user and the computer in complying with the user requirement that they can neither give voice commands nor move their hands. So, we can develop a system where we track the movement of the user's eyes to interact with the computer and to move the mouse around, helping to communicate with the computer and/or with others.

This chapter will explain an eye gaze mouse enabling these motor-impaired patients to use the mouse of a computer by means of moving their eye or head. The program uses the movements of the user's head to map these with the direction of the movement of the mouse, and uses left and right winks for the respective clicks. The speed of the mouse also varies as we move away from the center anchor point so that, the farther we go from the anchor point, the faster the mouse moves. A special scroll mode is also available where the user can scroll easily through a page without the need for clicking the down button on the scrollbar. In the following section, related work has been discussed. Afterwards, Eye Gaze system and the limitations of previous work have been explained. It is then followed by results of proposed system, future enhancements, and lastly, the conclusion.

## 9.2 BACKGROUND

Orman et al. (2011) described that in the broad field of computer vision the most challenging problem for the system developer is making sense of the face and eye detection. They study various literature on the face and eye detection, and gaze estimation, and found that with the latest trends in the industry the need for eye and face detection is increasing. The major challenges arise due to lighting, glasses, facial hair, and orientation. Yet, many advancements have been possible.

The principal component analysis was used by Raudonis et al. (2009) to reduce dimensionality problems by finding the six principal components of the eye image. Using additional hardware and Artificial Neural Network for classifying pupil position, they had been able to track the eye position. But the slow performance of the system makes it unfit for problem-solving in real-time. It is untested and limitations may include lighting, shadows, distance, and many more. Low computational hardware is not preferred.

Mehrubeoglu et al. (2011) proposed a system to detect and track the eyes by using template matching. The region of interest (ROI) is established which contains only the eye, and smart cameras are used to track the eye's position. They were able to establish that this method is fast as well as acceptable. However, not having carried out the tests of performance in a variety of conditions reduces the reliability of the algorithm, while affecting its dependability as it's unable to classify whether the gaze is up, down, left, or right.

A high-performance algorithm developed by Fu and Yang (2011) created templates for each eye curated from the video frame for evaluation, and then, uses a normalised 2D cross-correlation identification of face to map with the template. To identify the eye gaze, Hough circle and edge detection are performed. The proposed method can be used to control an application but has not been tested in different conditions.

Kuo et al. (2009) used a particle filter to find a sequence of hidden parameters that can be used for the tracking of an eye. A histogram which indicates the number of pixels of an image that share the same grey level was used as the feature of the particle filter. To optimise the algorithm, low-level features were used albeit it was accurate. Due to theirs not using a known database, real-time performance is not known and real-world application may have reduced accuracy.

Soukupová and Cech (2016) proposed an algorithm which can detect the blink of the eyes in the video sequence. The proposed algorithm requires a standard camera and estimates the facial landmark positions, calculating the eyes' aspect ratio (EAR) which computes the ratio of horizontal to vertical points and their difference in the eyes. The blink in the eyes is detected either by a support vector machine classifier or by hidden Markov model which roughly calculates the state of the eyes compiled by a state machine which can recognise the blinks in accordance to whether the eyes are open or close.

A camera mouse was made by Tu et al. (2005) using a modular framework visual face 3D modelling tracking method which speeded up manifold the human-computer interaction by the means of a hands free control for the system. Using various facial movements, multiple activities can be performed on the system. Once calibrated, the visual face tracking system can get these motion criteria in real-time and head movement can be used to steer the mouse cursor. This technique can act as a substitute for the input device for users with hand and speech dysfunction, and can be implemented in vision-based game and interface.

Betke et al. (2002) made a camera mouse for those people with severe disabilities. Body features were tracked to control the mouse movements by using a visual tracking algorithm and mapping the location of the features. Features such as nose tip, or a finger, can be used.

Fu and Huang (2007) developed a system called "hMouse" to handle hand-free perceptual user interfaces. The system is composed of a head tracker, an estimator which calculates motion of the head, and a virtual mouse control module. It shows authentic results by computing the person's horizontal and vertical motion for mouse control. It also uses head roll, tilting, and yaw for the same. The position where the cursor is can be located, explored, and calibrated by computing the overall situation of the window. Exploratory outcomes manifest to the need for the proposed system as prevailing in various situations like client hopping, occlusion, degree pivot, outrageous development, and multiuser impediment.

A technique to track the alteration of the movements of the head in the vector space was proposed by Siriteerakul et al. (2011) using texture detection in a low-resolution video. To approximate the direction of the user's head, a local binary pattern to the contrast between current and previous video frame repressing head rotation in an acquainted angle was suggested.

Song et al. (2011) developed the head and mouth tracking technique that uses image processing, a method that by means of ad boost to detect face first and, then, movements of the head analyzes the face position. Center of the face was used as cursor point which was mapped to the mouse pointer. The proposed methodology is quick and finds its application for people with locomotor disability.

In a study conducted by Ball et al. (2004), it was described that the impressions of effective communication for ALS patients were almost the same for the speakers and their frequent listeners across 10 different social situations. On reviewing the ALS speakers and their listeners, the range of communication effectiveness depended upon the seriousness of the catastrophe of specific social situations, which means that they can communicate and understand what the other person is saying.

Zhao et al. (2012) proposed a method to identify the movements of the head based on image processing which also used the Lucas-Kanade algorithm. To identify the face and nostrils, respectively, their positions are identified and the Lucas-Kanade algorithm is used to observe the optical flow of the nostrils to accurately identify head's movements. Nevertheless, the location of the face in the video may be an estimation and, thus, the feature point coordinates may not accurately reflect the head's movements. Therefore, they used the interframe difference to represent the head's movement in the coordinate of feature points. The proposed approach is quick and finds its application in the detection of head's movement. However, it does not address the challenges that are faced in a real-life application.

Xu et al. (2012) proposed a system for restoring 3-D head pose from a video sequence using a 3-D cross model for continuous tracking of the motions of the head. To approximate the eyes, the model is projected onto an initial prototype and used as a reference. The optical flow process can then be used to detect full motion of the head in video frames. It has used a non-head-mounted camera.

Cuong and Hoang (2010) have suggested an approach for the identification of eye-gaze signals from a web camera, where it is a real-time application. The measured data was sufficient to describe the eyes' movements since the web camera is fixed to the eyes. First, a dynamic threshold has been binarized to the image. Then, the eye image geometry features were extracted from the binary image. In extracting eye image geometry, first, the location of the two corners of the eyes were measured using an estimation method based on the geometry structure of the eyes. Afterwards, the iris center was identified by matching an iris boundary model with the contours of the image. Finally, the location where the eyes are viewing the display is determined using the relative position information between the center of the iris and the corners of the eyes based on the relationship between vision coordinate and monitor coordinate. This program only needed an inexpensive web camera and a personal computer.

MacLellan (2018) has proposed that people with disabilities, who lack dependable motor control for manipulation of a computer mouse (standard), required an alternative method to access a computer and further to communicate their needs properly.

Królak and Strumiłło (2012) proposed a vision-based human-computer interface. The interface detected eye-blinks which were voluntary and interpreted them as commands for controlling the computer. They assigned image processing methods such as, for detecting the face, they used Haar-like features, and template matching based eye tracking followed by eye-blink detection. Interface performance has been tested by forty-nine users (of which few patients had physical disabilities).

Robertson et al. (2004) proposed a vision-based virtual mouse interface that makes use of a robotic head, visual tracking of the head, and hand positions of the user, and also recognizes hand signs to control an intelligent kiosk. The user interface assists smooth control of the mouse pointer and buttons using hand signs and movements. The algorithm and architecture of the robot and real-time vision are reviewed.

Lupu et al. (2012) stated that most people with locomotor disabilities can understand the eye-processed image and can use their eyes for communication. The proposed solution is a reliable and robust system based on the eye-tracking mouse. The movement of the eyes is identified by a head-mounted device and, using the detected movement, the mouse cursor is moved on the screen. A click event denotes a pictogram selection and is performed if the patient does not blink and look at the screen for a certain time.

Rotariu et al. (2008) developed an electro informatics system that allows severely handicapped persons to perform special communication (through a video interface and radio links). The TELPROT system is chiefly made from two main functional components: the patient equipment and the care-taker equipment. Both sub-systems are composed of a control unit and a communication unit. The patient and nurse can communicate through radio waves, within a dedicated LAN.

San Agustin et al. (2010) proposed a webcam-based gaze tracking system. The evaluation of the performance in a task, where the eyes were used to type out the required text was done using two different typing applications. Participants in the above-mentioned evaluation could type between the speed of 3.56 and 6.78 words per minute. It mostly depended on the typing system used. A prior study to analyse the serviceability and feasibility of the system was carried out with a user who had severe motor impairments. As a result of using the gaze tracking system, the user was successfully able to type on a wall-projected interface with aid of his eyes' movements.

Lupu et al. (2013) proposed a novice technology to help patients with neuro-locomotor disabilities to communicate using embedded systems and eye-tracking approach. Firstly, the movement of the eyes is detected by a specifically assigned device and, to select a particular option, the patient needs to do a voluntary blink which is mapped to the keyword selection. The image processing technique based on the binarization algorithm is used for implementation.

Bozomitu et al. (2011) discussed the setting up of hardware component of an exciting new technology used for communication with people having major neuro-locomotor disorders using ocular electromyogram. Signals provided by five electromyogram sensors are amplified and processed to move a pointer on display per the person's gaze. Hence, the system helps to understand that the patient needs assistance.

Hori et al. (2004) presented an interface for communication which is controlled by eye movements, and, for selecting a particular option, they used voluntary eye blink. It was developed to help disabled people who have motor paralysis and hence could not speak. They experimented on a virtual keyboard to check the feasibility of the proposed methodology. Horizontal and vertical electrooculogram were measured, and four-directional cursor movements were realized by logically combining the detected two-channel signals.

Betke et al. (2002) presented the "Camera Mouse" system. Inspired to help people with severe disabilities, mainly who can't move, the system tracks the user's movements. For example, the tip of the user's nose or finger is tracked with a video camera and traces it into that of a mouse pointer on the screen. Twelve people with severe traumatic brain injury or cerebral palsy tried the system, nine of whom showed positive results. They experimented by spelling out messages and surfing the Internet.

Lupu et al. (2013) suggested a new technology used for connecting with people having major neuro-locomotor disability by using ocular electromyogram. The technology works using the Electromyogram which seizes the patient's gaze and selection of ideogram is done by voluntary blinking. The patient's request is sent to the nurse/caretaker using a wired/wireless transmission system.

Lupu et al. (2012) presented a new technology to communicate with people having neuro-locomotor disability by determining gaze direction. The proposed technology ensures communication with the patient in the given manner: (1) ideograms are displayed on the screen in four sections; (2) The patient continuously gazes on the required section; (3) an infra-red mini/micro-video camera placed on the patient's head sends the image of the eye to a computer; (4) the gaze is identified by image analysis; (5) voluntary blinking shows the selection of a certain word.

## 9.3 EYE GAZE MOUSE

The eye gaze mouse enables motor-impaired patients to use the mouse of a computer by means of moving their eyes or head. The program uses the movements of the user's head to map these to the direction of movement of the mouse, and uses left and right winks for the respective clicks. The speed of the mouse also varies as we move away from the center anchor point so that the farther we move away, the faster moves the mouse. A special scroll mode is also available where the user can scroll easily through a page without the need for clicking the down button on the scrollbar. The working of the eye gaze mouse is explained below.

### 9.3.1 LIMITATIONS OF PREVIOUS METHODS

The previous methods to implement this kind of applications were either very much complicated, inefficient, or required too much hardware. There were two kinds of existing methods for developing an eye gaze mouse:

#### 9.3.1.1 Mouse Movement Using Pupil Detection

In this method, a webcam mounted on a computer screen or the inbuilt webcams are used to detect the movement of the eyes and map these to the movements of the mouse. But this method is inefficient when it comes to upwards or downwards movement as no white region of the eye can be detected properly and we cannot determine the movement of the eyes.

#### 9.3.1.2 Additional Headgear

In this method, an additional headgear is mounted on the user to track the movements of their eyes up close to avoid facing issues ensuing from the first method. But this additional hardware can put a strain on the user's eye, can block the vision, and bring up the cast of the product as well.

Our method aims to eliminate these two problems by incorporating the head's movements as well as those of the first method for ease in operation.

### 9.3.2 WORKING OF EYE GAZE MOUSE

#### 9.3.2.1 Video Processing

As we are working on a video, the first step is to collect frames from the video. Further, each frame is individually processed given we are live streaming the video. We use the OpenCV library to perform the task. A VideoCapture object is made to record the video from the webcam attached to the system from which we separate each frame and use it to map the mouse's movements.

```
import cv2
# capture the video
cap = cv2.VideoCapture("file path")
while(True):
# Capture frame-by-frame
ret, frame = cap.read()
# Display the frame
cv2.imshow('frame',frame)
# press q to exit
if cv2.waitKey(1) & 0xFF == ord('q'):
break
# When everything is done, release the capture
cap.release()
cv2.destroyAllWindows()
```

The VideoCapture() function is used to record the video from the specified camera as its parameter. It is broken down into frames by cap.read(), and returns whether the image is captured properly or not. cv2.imshow() is used to display the frame on a window. cv2.waitKey() delays the process for n milliseconds specified in its parameters.

cap.release() is used to release the hardware from the program; else, it will be locked and other programs or applications won't be able to use it.

### 9.3.2.2 Image Conversion to Grayscale

Each frame of the video capture is converted to grayscale using cv2.COLOR_BGR2GRAY for processing (Figure 9.1).

A greyscale or a grayscale is the one where the value of a pixel in the image is a single which reflects the amount of light. The pixels carry light intensity information which ranges between black, white, and grey monochrome representing weakest, strongest, and moderate intensities, respectively.

1. In image processing, there are applications where colored images are needed, for instance, to detect edges (step change in pixel) or to identify objects of a known hue. In other applications, where colors are not needed for identification, they are treated as noise.
2. Grayscale images help to reduce the complexity of code since additional color information is mostly not of much need.
3. Multichannel processing is preferred rather than performing full-color imaging which can deviate from the important insights.
4. In grayscale images, it is easier to conceptualize using the watershed algorithm as compared to other color spaces as they have additional dimensions; whereas, grayscale has only two spatial dimensions and one brightness dimension.
5. The human brain can perceive and identify color with ease but processing of the colored images can be long and tedious.
6. With the help of parallel programming and modern computers, it is easier to process megapixel images by breaking it down to pixels, but tasks such as object

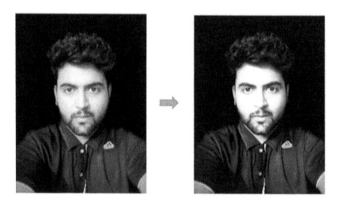

**FIGURE 9.1** Before and after Image Conversion to Grayscale.

recognition, or facial recognition, are very time-consuming. This is because three-channel color images are harder to process than are grayscale images which have only two channels and, thus, can improve the process substantially.

## 9.4 FACE DETECTION USING HAAR CASCADE CLASSIFIER

Haar Cascade is a concept proposed by Jones in 2001 used for object detection. By means of a machine learning algorithm using a cascade function, which is trained from the dataset of images, we know that it consists of both images containing target objects and images not containing a target object.

Following steps are followed:

1. Haar feature segregation and selection.
2. Forming integral images.
3. Ada-boost Training.
4. Cascade classification.

It is able to detect body parts and faces in the image and video but, in theory, can be trained to identify almost any object. For detecting a face, we need a dataset of positive images containing the face, and negative images cannot contain the face, which are used to train the classifier. The first step is extracting features from the image for which we use a grayscale image, which is made of black and white pixels. Every group of adjacent pixels falls into one of the following types (Figure 9.2).

Light intensities of adjacent regions at a specific location are summed up and the difference between these sums is calculated. But this results in a huge number of features, out of which many are irrelevant, so, to select the best features and train, the classifier Adaboost is used. To carry out detection, a window of target size is placed over the frame of the image or the video, and each window is divided into subparts. Haar features for each subpart are evaluated. The calculated differences

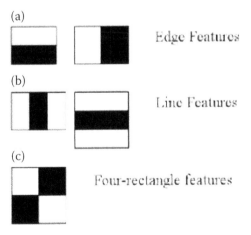

**FIGURE 9.2** HAAR Cascade Features.

are compared against the threshold to separate objects from noise. Individually, a Haar feature is very poor in carrying out classification, thus, a huge number of them are grouped to describe an object, resulting in a strong classifier. Cascade classifier is a group of levels and at each, a collection of weak learners is present. Weak learners are simple classifiers and are also called decision stumps. Boosting is used to train each level, where boosting is a technique in which the weighted mean of a weak learner is taken to increase the accuracy of the classifier. Each level classifier tags the subpart of the window for their respective location as positive or negative, where positive denotes the subpart as containing an object; whereas, negative denotes the subpart does not contain any object. If a negative tag is given, then further classification is not required and processing for that particular subpart is stopped. Object detection is recorded for a current subpart of the window when the final level classifies it as positive. The levels are trained such that negative regions get rejected as soon as possible. The underlying principle is that negative regions are frequent and positive regions are rare, thus a positive region requires further processing to verify that they exist. Three situations are possible: true positive is correctly classified, a negative gets wrongly classified as positive, or a positive is wrongfully classified as positives. If the occurrence of latter two is high, then the classifier is not an accurate one; thus, the first should have high frequency and others should have a very low frequency. Still, there is a high occurrence of false positive, which has to be rectified at higher levels. As the number of levels grows, accuracy also grows; and, the rate of false-positive decreases. But there is an issue with false-negative because, in this case, processing stops; thus, by stopping, it prevents the chance of correction.

In positive images, dataset regions of interest are labelled and the images without these labels are automatically considered as negative images. The number of levels, feature type, and other parameters are used to tune the detector for higher levels of accuracy.

### 9.4.1 Facial Landmarks Localization

Detecting the facial feature is a subtask that has to be performed to localize and predict the region of the face in each frame. The classifier works in the bottom-up fashion finding these features and, if such features can be characterized in proper orientation, the classifier gives approximate coordinates of the predicted region.

Facial features that can be used to characterize each face are Eyes, Eyebrows, Nose, Mouth, Jawline.

### 9.4.2 Application Flow Detection

Based on the model and application, flow detection can be any one of the following

1. Eye-tracking to give direction for mouse.
2. Eyeball tracking left and right as an extra.
3. Blink tracking as an input.
4. Left Blink and Right Blink for left-click and right-click.

### 9.4.2.1 Eye-tracking to Give Direction for Mouse

Every time mouse mode or scroll mode is activated, two variables are defined using:

anchor_pointx=int((landmarks.part(36).x+landmarks.part(39).x)/2)

anchor_pointy=int((landmarks.part(37).y+landmarks.part(41).y)/2)

These act as XY coordinates for the anchor point from where a vector to the eye is drawn and, using this direction and magnitude, the movement for the mouse is calculated. Euclidean Distance from the anchor point to the eye is calculated and scaled down to integer speed. This process is repeated for each frame giving the mouse dynamic speed and smooth movement.

### 9.4.3 EYEBALL TRACKING LEFT AND RIGHT AS AN EXTRA INPUT

Grayscale images help to segregate darker regions from lighter ones, and this chapter suggests utilizing this property to segregate the pupil which is a darker portion of the eye from the sclera, the white portion of the eye. The region of the eyes can be divided into two parts: left and right, where presence of white portion on either of the sides can help predict the orientation of the eye. For doing this, preprocessing is done to only consider the biggest contour and treat rest as noise.

Thus we use the principle that if the white portion is more towards the left then the eye is looking right; else, if the white potion is present more on the right, then the eye is looking towards left.

Whenever we are looking left or right, one half of the eye has a lot more white portion than the other half; we calculated the white portion in both halves, and used this to calculate Eye Gaze Ratio. Threshold for this ratio is set to 1.7, where less than 1.7 is left, and more than 1.7 is right.

We use this as an extra input to switch between scroll mode and mouse mode. Looking to the left for a certain number of frame switches to mouse mode; and, looking to the right switches to scroll mode. The threshold for frames can be adjusted in the menu. A helper frame is also available which shows different colors to assist the user in verifying whether he is looking in the correct direction. On successful mode selection, a sound is also generated according to the mode to acknowledge mode selection.

### 9.4.4 BLINK DETECTION

Soukupová and Čech, introduced the EAR which is known as eye aspect ratio. The following is the method followed by traditional image processing:

#### 9.4.4.1 Eye Features Detection
*Processing Image to Find the White Part of the Eye*
To categorize the blink, this white area disappears for a certain period.

$$\text{EAR} = \frac{\|p_2 - p_6\| + \|p_3 - p_5\|}{2\|p_1 - p_4\|}$$

**FIGURE 9.3** EAR Formula.

The eye aspect ratio which we used is a very simple solution that is based on the ratio of distances between feature points of the eyes. This approach for eye blink detection is simple and purely calculation-based and doesn't require any unnecessary hardware. This approach provides accurate and dependable results in many different scenarios. Each eye is represented by 6 (x, y)-Cartesian coordinates. The distance between the horizontal and vertical points have a relation which can be exploited to predict eye closure. Based on the method proposed by Soukupová and Čech in their 2016 paper, Real-Time Eye Blink Detection using Facial Landmarks is put to effect. The equation below helps to establish a relation called the eye aspect ratio (EAR) which can be used to predict mathematically if the eye has blinked, instead of taking recourse to physical detection methods (Figure 9.3):

Where p1, p2, p3, p4, p5, p6 are points used to denote landmark location of face, this equation's numerator calculates the distance between the vertical eye landmarks, while the denominator calculates the distance between the horizontal eye landmarks.

Left Wink and Right Wink for Left and Right Click

```
EAR are calculated for both eye and the following algo-
rithm is for left blink and right blink distinguished:
if  left_eye_ratio<EYE_AR_THRESHOLD   and   right_eye_
ratio < EYE_AR_THRESHOLD:
COUNTER_BLINK = COUNTER_BLINK+1
# else, the EAR is not below the set blink
# threshold
else:
if COUNTER_BLINK >= CONSEC_FRAMES_THRESHOLD:
TOTAL_BLINK += 1
mode detect=not mode_detect
COUNTER_BLINK = 0
if left_eye_ratio < EYE_AR_THRESHOLD and right_eye_
ratio > EYE_AR_THRESHOLD:
COUNTER_LEFT = COUNTER_LEFT+ 1
else:
if COUNTER_LEFT >=2:
TOTAL_LEFT += 1
print("Left eye winked")
COUNTER_LEFT = 0
mouse.click(Button.right,1)
```

```
if right_eye_ratio < EYE_AR_THRESHOLD and left_eye_r-
atio > EYE_AR_THRESHOLD:
COUNTER_RIGHT = COUNTER_RIGHT+ 1
else:
if (COUNTER_RIGHT >= 2):
TOTAL_RIGHT += 1
print("Right eye winked")
COUNTER_RIGHT = 0
mouse.click(Button.left,1)
```

Aspect ratio of the eye remains the same when the eye hasn't blinked, but as soon as the eyes are closed, the ratio shows a considerable drop, and by tracking this drop, the system can detect whether the user has blinked or not.

## 9.5 RESULTS

In this chapter, we researched various ways of camera mouse techniques. We identified the problem domain and found an alternative solution to it by setting up additional hardware. It can be concluded that our method can accurately track eye position for images with dominant pupils, and the user can also adjust the parameters for the system to best suit to their eye characteristics and lighting environment. Adjusting the threshold value can increase accuracy by making it easier or difficult to accept valid input on the basis of parameters set according to each user's personnel preference. We proposed to use face/head motion to control the movement of the mouse cursor. Thus, our system proves to be robust and capable of handling a wide range of scenarios in real-time application. We do pre-image processing and convolution of pixels to achieve best possible contrast for eye localisation, but in case of very poor lighting and low contrast, situation difficulty can arise to estimate the center of the eye. In this case, the user must change lighting or threshold values to test for best possible parameters. The estimated eye centers are shown using red spots in each frame.

The quantitative results of the proposed method defined the using of the standards of the normalised error by help of the equation

$$e \leq 1/d \max(el, er) \tag{9.1}$$

where el, er is the Euclidean distance used to evaluate space between eye centers, and d is the distance between the correct eye centers. Our approach yields an accuracy of 90% for eye tracking, 80% for an eye blink, 75% for left and right wink, and 80% for eyeball movement in left and right directions which can become even higher in situations of closed eyes and low contrast is not considered.

Different Poses: Face images can be vary due to relative camera-face pose, like; front, 45-degree, profile, and up-down. Also, some features of the face (like eyes, mouth…) may be partially or wholly occluded.

Structural components: There are a lot of obstructive elements like facial hair, scars, and spectacles, that can induce uncertainty for classifier; though the classifier can detect faces in a realistic situation, still the presence of these items in extreme degree can induce variable results.

Face expression: Changing facial expression can alter the face design. The classifier can take countermeasures to this problem to a certain extent but it is an obstruction for realistic performance.

Frame orientation: Camera axis relative to face can change the orientation of the frame, which was captured for processing.

Environment: When the frame is captured, some factors like lighting (from internal or external light sources in the room) and camera configuration affect the image of a face and its appearance.

## 9.6 FUTURE RESEARCH DIRECTIONS

This chapter involves building a camera mouse which works as a mouse replacement system. The following are the applications which can be extensions of this work:

1. The work is suitable for pc/desktop. These features can be implemented on mobile phones to facilitate hands-free access to it.
2. Camera Mouse is a way to increase the employability of disabled people. This will also help in increasing efficiency.
3. Other additional hardware can be combined with the application to detect symptoms like drowsiness of eye, erratic pupil movement, redness due to infection, and other symptoms that must be observed to ensure patient well being.
4. To improve the number of operations performed, we can add facial gesture recognition like smirk, yawn etc.
5. There has been a lot of development in the game industry on the concept of using physical movement as an extra input to mimic similar behavior inside the game world. Virtual reality is a promising field in which the user enters a computer-generated world in which he/she can perform a range of tasks from watching a movie, to reading and to playing games. Integrating an eye-tracking system with this virtual reality headset can be useful to provide an extra sense of natural direction which can be used to personalize this experience. For example, when browsing the library of movies, pupil direction can be used to traverse the library; or, while playing first-person games, the eyes can be used for aiming at a target. There are a huge number of applications of eye-tracking in this field.
6. In addition to eye and head movement, we can inculcate a speech module.
7. Eye-tracking can be utilized in commercial tracking to figure out which commercial or product customer gazes on more frequently. This data can be used to do a targeted advertisement and even discount schemes on selected merchandises to increase revenue.
8. Eye gaze tracking can also be implemented for the car entertainment system so that the user can interact with it without removing attention from wheels.

## 9.7 CONCLUSION

In this chapter, the authors discussed an Eye Gaze Mouse for the motor-impaired which can perform the basic functions of a mouse such as a click, scroll, or jump across the screen based on the user's eye movements. This work will act as a further impetus for those who are thinking to make life easier for them who are paralyzed neck down or don't have hands to hold the mouse while using a computer.

The eye gaze mouse enables these motor-impaired patients to interact with the computer by using their gaze as an input medium and mapped to directions of the mouse pointer's movements. The program uses the movements of the user's head to map it to the direction of the mouse's movements, and uses left and right winks for the respective clicks. The speed of the mouse also varies as we move away from the center anchor point so that the farther we move away, the faster the mouse moves. A special scroll mode is also available where the user can scroll easily through a page without the need for clicking the down button on the scrollbar.

With this proposed work, a patient anywhere in the world can communicate better using the eye gaze mouse. It also allows patients to use web browsers and stroll through websites, which in today's world is one of the most important applications to communicate with the world.

Apart from these, this work will also allow the using of various applications which are not yet upgraded to use voice commands or are yet to be equipped to help these patients in reality. The users can also use the findings here to mobilize older applications which have not been updated. The users can also use word processors or chat applications in conjunction with this work by using a virtual keyboard.

## REFERENCES

Ball, L. J., Beukelman, D. R., & Pattee, G. L. (2004). Communication effectiveness of individuals with amyotrophic lateral sclerosis. *Journal of Communication Disorders*, *37*(3), 197–215.

Betke, M., Gips, J., & Fleming, P. (2002). The camera mouse: Visual tracking of body features to provide computer access for people with severe disabilities. *IEEE Transactions on Neural Systems and Rehabilitation Engineering*, *10*(1), 1–10.

Bozomitu, R. G., Barabaşa, C., Cehan, V., & Lupu, R. G. (2011, October). *The hardware component of the technology used to communicate with people with major neuro-locomotor disability using ocular electromyogram* [Symposium]. Design and Technology in Electronic Packaging (SIITME), IEEE 17th International Symposium, IEEE, pp. 193–196.

Cuong, N. H., & Hoang, H. T. (2010, December). *Eye-gaze detection with a single webcam based on geometry features extraction* [Conference session]. Control Automation Robotics & Vision, 11th International Conference, IEEE, pp. 2507–2512.

Fu, B., & Yang, R. (2011, October). *Display Control Based on Eye Gaze Estimation* [Congress]. Image and Signal Processing, 4th International Congress, IEEE, Vol.1, pp. 399–403.

Fu, Y., & Huang, T. S. (2007, February). *hMouse: Head tracking driven virtual computer mouse* [Workshop]. Applications of Computer Vision (WACV'07), IEEE, pp. 30-30.

Hori, J., Sakano, K., & Saitoh, Y. (2004, September). *Development of communication supporting device controlled by eye movements and voluntary eye blink* [Conference session]. IEEE Engineering in Medicine and Biology Society, 26th Annual International Conference, IEEE, Vol. 2, pp. 4302–4305.

Królak, A., & Strumiłło, P. (2012). Eye-blink detection system for human-computer interaction. *Universal Access in the Information Society, 11*(4), 409–419.

Kuo, Y. L., Lee, J. S., & Kao, S. T. (2009, September). *Eye-tracking in the visible environment* [Conference session]. Intelligent Information Hiding and Multimedia Signal Processing, Fifth International Conference, IEEE, pp. 114–117.

Lupu, R. G., Ungureanu, F., & Bozomitu, R. G. (2012, August). *Mobile embedded system for human-computer communication in assistive technology* [Conference session]. Intelligent Computer Communication and Processing, 8th International Conference, IEEE, pp. 209–212.

Lupu, R. G., Ungureanu, F., & Siriteanu, V. (2013, November). *Eye-tracking mouse for human-computer interaction* [Conference session]. E-Health and Bioengineering Conference (EHB), IEEE, pp. 1–4.

MacLellan, L. E. (2018). *Evaluating Camera Mouse as a computer access system for augmentative and alternative communication in cerebral palsy: A case study* [Doctoral dissertation], Boston University.

Mehrubeoglu, M., Pham, L. M., Le, H. T., Muddu, R., & Ryu, D. (2011, October). *Real-time eye tracking using a smart camera* [Workshop]. IEEE Applied Imagery Pattern Recognition Workshop (AIPR), IEEE, pp. 1–7.

Orman, Z., Battal, A., & Kemer, E. (2011). A study on the face, eye detection and gaze estimation. *IJCSES, 2*(3), 29–46.

Raudonis, V., Simutis, R., & Narvydas, G. (2009, November). *Discrete eye-tracking for medical applications* [Symposium]. Applied Sciences in Biomedical and Communication Technologies, 2nd International Symposium, IEEE, pp. 1–6.

Robertson, P., Laddaga, R., & Van Kleek, M. (2004, January). *Virtual mouse vision-based interface* [Conference session]. Intelligent User Interfaces, 9th International Conference, pp. 177–183.

Rotariu, C., Costin, H., Cehan, V., & Morancea, O. (2008, August). *A communication system with severe neuro-locomotor handicapped persons* [Conference session]. Biomedical Electronics and Biomedical Informatics, pp. 145–149.

San Agustin, J., Skovsgaard, H., Mollenbach, E., Barret, M., Tall, M., Hansen, D. W., & Hansen, J. P. (2010, March). *Evaluation of a low-cost open-source gaze tracker* [Symposium]. Eye-Tracking Research & Applications, pp. 77–80.

Siriteerakul, T., Sato, Y., & Boonjing, V. (2011, December). *Estimating change in head pose from low-resolution video using LBP-based tracking* [Symposium]. Intelligent Signal Processing and Communications Systems (ISPACS), IEEE, pp. 1–6.

Song, Y., Luo, Y., & Lin, J. (2011, November). *Detection of movements of head and mouth to provide computer access for the disabled* [Conference session]. Technologies and Applications of Artificial Intelligence, IEEE, pp. 223–226.

Soukupová, T., & Cech, J. (2016, May). *Eyeblink detection using facial landmarks* [Workshop]. 21st Computer Vision Winter Workshop, Rimske Toplice, Slovenia.

Tu, J., Huang, T., & Tao, H. (2005, May). *Face as a mouse through visual face tracking* [Conference session]. Computer and Robot Vision (CRV'05), 2nd Canadian Conference, IEEE, pp. 339–346.

Xu, Y., Zeng, J., & Sun, Y. (2012, August). *Head pose recovery using a 3D cross model* [Conference session]. Intelligent Human-Machine Systems and Cybernetics, 4th International Conference, IEEE, Vol. 2, pp. 63–66.

Zhao, Z., Wang, Y., & Fu, S. (2012, August). *Head movement recognition based on the Lucas-Kanade algorithm* [Conference session]. Computer Science and Service System, IEEE, pp. 2303–2306.

# 10 Strategies for Resource Allocation in Cloud Computing Environment

*Nikky Ahuja, Priyesh Kanungo, and Sumant Katiyal*

## CONTENTS

10.1 Introduction ........................................................................................................ 179
    10.1.1 Resource Allocation ............................................................................ 180
    10.1.2 Resource Allocation System ............................................................... 180
    10.1.3 Significance of Resource Allocation System ..................................... 181
    10.1.4 Challenges of Resource Allocation .................................................... 181
10.2 Literature Review ............................................................................................... 182
10.3 Strategies ............................................................................................................ 184
    10.3.1 Existing Strategies ............................................................................... 185
    10.3.2 Proposed Strategies for Admission Control and Resource Allocation ............................................................................................ 188
10.4 Analysis and Result ............................................................................................ 195
10.5 Conclusion and Future Scope ............................................................................ 199
References ..................................................................................................................... 199

## 10.1 INTRODUCTION

Application development in a commercial and scientific environment has witnessed a remarkable change due to the growth of networking technology and infrastructure. Cloud computing technique is a revolutionary change in the direction of providing infrastructural facilities to the customer for transfer of information over the internet with a little or no effect on the system's performance (Zhang et al., 2010), effectively, a revolution in the way we think of using computers and computer networks for our purpose. Given cloud offers on-demand resources and storage capacity making cloud an attractive computing platform, it's the cloud's dynamic behavior that enables customers to change their business requirements anytime and still have at their disposal virtual machines, bandwidth of the network, storage space or memory available, processors' speed, etc. which are the cloud resources. In the initial stages, customers were supposed to pay rent for the entire time slot hired for the task, whether they have used the resources or not. However, with present-day technology, the focus is on the strategy to pay as much as are resources used for solving the problem of high cost of resource renting. This pricing flexibility has been created by the renting of virtual machines arranging requests in parallel, resulting in guaranteed budget and deadlines. The virtual machines are made available to serve user request that helps in providing choices to customers so that they can select the number and the type of VMs according to their demand. Also, if a rented VM is idle for a certain period of time, it can be used for any other task or request.

### 10.1.1 Resource Allocation

The process of selecting and making available the suitable resources for the user to satisfy their demand is known as resource allocation. It is an extremely intricate and challenging task because it requires balancing the optimum allocation of resources satisfying the quality constraints and cost reduction requests with profit maximization. A number of research efforts have been conducted for optimum utilization of resources in Cloud Computing Environment (Backialakshmi & Sathya Sofia, 2014). These resources are utilized via virtual environment, and the virtualization technique helps service provider to manage multiple user requests at the same time, given balancing the workload between virtual machines or VMs is essential for improving cloud performance and for improving the utilization of resource for task completion in a competent manner.

### 10.1.2 Resource Allocation System

The cloud service provider aims at the optimum utilization of available resources while there is maximum scheduling of tasks/ applications. The concern that raises its head in this matching of cloud resources with user applications is to select the best suited physical and virtual resources that can meet the requirements of tasks provided by the users. It is the Resource Allocation System that acts as a mechanism to sort the problems of cloud provider while assuring the fulfillment of user resource requirement, even as it checks for the current status of cloud's resources for their better deployment that results in the minimization of the systems' operational cost (Vinothina et al., 2012).

Cloud gives its users the option to request two different types of resources: physical or virtual. Since users have varying resource requirements like a physical resource requirement might request CPU, memory, storage, and network-related resources having their own bandwidth and delay specifications, and with these resources commonly placed in datacenters, the users can share these datacenters with multiple clients' calls handled by all the data centers, effectively, rolling into place dynamic resource allocation as per user requirement. From the perspective of virtual resources, users find this resource pool as unlimited and this is made possible with the help of the Resource Allocation System, which is an optimum Resource Allocation System when it fulfills these uncertain and random requests in a flexible and transparent way. The flexibility requirement must permit the dynamic utilization of physical resources to avoid the problems of over- and under-provisioning of cloud's resources. Thus, it becomes important for a service provider, before resources are allocated for any new user request, to consider the way in which resources are modeled. With there being various differentiated levels of service abstractions for developers in Cloud Computing Environment, and various differentiated parameters that could be altered to make an efficient allocation, an optimal modeling and resource description will consider these parameters to facilitate Resource allocation strategy to function properly. It is evident that the input of a resource allocation system are cloud resources (both virtual and physical), resource modeling, and application user requirements, and the output is an optimum resource allocation.

### 10.1.3 SIGNIFICANCE OF RESOURCE ALLOCATION SYSTEM

Cloud computing has various advantages for business enterprises with a need to shed an overweight physical infrastructure, or reinforce the old infrastructure, like low cost, resource provisioning and re-provisioning, remote accessibility, etc, because instead of hiring or purchasing physical infrastructure when they are in an expansionary mode, say, they use virtual resources which results in lowering the cost by saving on the deployment cost. Cost apart, it is also flexible to the demands of business which means it expands with more activity going on; or to its size, which means it can increase infrastructure with expansion going on. It also allows users to access the cloud resources and services anytime from any geographical location across the world. For earning maximum benefits from the cloud, resources must be deployed optimally as per the requirement of the applications running over cloud (Chopra et al., 2015).

In the Cloud Computing Environment, the process of allocating available resources to the essential applications on a cloud over the network is known as resource allocation (Komal & Saroha, 2017). It is essential to manage resource allocation properly, since any mismanagement may result in the scarcity of required services. The technique of resource provisioning resolves this problem by permitting the cloud service providers autonomy to manage resources for each individual module.

### 10.1.4 CHALLENGES OF RESOURCE ALLOCATION

Resource allocation in cloud faces following major challenges in Cloud Computing Environment as shown in Figure 10.1 (Gonçalves et al., 2011). These include:

a. **Resource Modeling and Description:** Modeling describes the schematic way in which a cloud system works and what are the indivisible constituents of its workbase, or customers, resources, the Resource Allocation System, etc. Given the modeling is important for all the functions in the cloud to execute without hassles, there are various service delivery models on the cloud that developers represent in algorithms i.e. optimization, control, and management algorithms rely on these models selected by the operators. A Resource Allocation System takes on its first challenge which is the establishment of an optimal resource model and its description. For example, if any request consists of all the physical specifications of each and every machine, then the cloud will have to fulfill all the minor details which

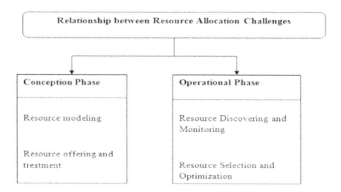

**FIGURE 10.1** Relationship Between Resource Allocation Challenges.

another state of affairs might make irrelevant. This makes the matching of the resource requirements with cloud resources difficult.
  b. **Resource Offering and Treatment:** The cloud offers and manages the allocated resources via Resource Allocation System which acts as an interface between the cloud and the customers. Resource Allocation System manages the requirements of applications at higher level and cloud resources at lower level. However, matching and handling these application requirements with available scarce resources while fulfilling the SLA constraints is a challenge for Resource Allocation System.
  c. **Resource Discovery and Monitoring**: Resource discovering means finding the set of most suited resources to match the requirements of incoming requests; and, monitoring relates to keeping track of resources in use and those recourse that are available for use. Monitoring is a continuous process and finding the suitable available resource is a challenge for any Resource Allocation System.
  d. **Resource Selection and Optimization**: After finding the resources available in cloud from all the available possibilities to match with the user requirement, the resources using the infrastructure in the most efficient manner is selected, which is a challenge for a Resource Allocation System.

These challenges are categorized into the conception phase and operational phase. To begin with, a service provider faces the problems of resource modeling and resource offering, but as soon as a new request arrives, the next two challenges of resource discovering and selection must be tackled by the cloud provider. Cloud, here, checks for the availability of requested resources and Resource allocation system allocates the appropriate resource to the request.

## 10.2 LITERATURE REVIEW

Cloud computing plays an important role in business, health, or education. Maintaining the confidentiality of data, identifying key business performance indicators, and security of results and data assimilation (Vouk, 2008) have helped business in more ways than one. For one, cloud computing has become the crucial part of an organization facilitating its management and control of data over the internet whose users can access the resources and services provided by the clouds from any location in the world at anytime as per their convenience in the same manner that we access our e-mail anytime and from any geographical area (Buyya et al., 2010). Second, cloud computing facilitates users to store data on VMs available in the cloud, thereby reducing the need for the installation of applications or storage of files and data on the user's computers. The cloud service providers manage the task of storage, scheduling, backup development etc, for the user. Thirdly, this also makes the data virtually available at any environment irrespective of the location of applications being processed. Document portability apart, and finally, the cloud also helps in improving the quality of applications, providing platform independence, and simplifies operation via virtual abstraction (Vaquero et al., 2009).

Since a cloud puts many users with different service quality requirement on the same table simultaneously, Xu et al. have focused on devising a scheduling strategy that could manage multi workflows with vivid QoS requirements. In their research, authors have proposed a Multiple QoS constrained scheduling strategy of Multi-Workflows (MQMW) to schedule randomly initiated multiple workflows considering their robust QoS requirements (Xu et al., 2009). According to Hamdy et al., resource allocation strategies help the minimization of struggle for limited resources, resource fragmentation, and over provisioning and under provisioning of resources. The parameters affecting applied resource allocation strategies include virtual machines, scheduling methods, optimization methods inspired from nature (i.e. solutions based on animals natural behavior modeling), SLA, priority based methods, auction mechanism, and gossip protocol based method (focusing on dynamic utilization of global cloud utility function). The authors have discussed some of the methods for resource allocation like hardware resource dependency and management of virtualized resource, and incorporated a decentralized user by the inclusion of another layer called domain in the middle of resource and user. They have further stated that the selection of allocation strategy has its direct impact on the utilization, throughput, latency, and response time of resources (Hamdy et al., 2017).

To address the challenges associated with resource allocation and optimization in cloud, Vinothina et al. (2012) have presented various resource allocation strategies which could benefit both cloud users and researchers to develop some secured and smarter resource allocation algorithms and frameworks to reinforce the paradigm of cloud. Dynamic cloud environment makes it impossible for cloud providers to accurately predict the nature and demands of the user as well as the application. Limited resources, resource heterogeneity, restrictions associated with locality, necessities of environment, and resource robustness become important for the cloud service provider to devise an efficient resource allocation system that suits the needs of dynamic cloud environment. To devise a resource allocation strategy, the cloud service provider should factor in services, cloud infrastructure, and applications (Vinothina et al., 2012) into the strategy.

Mahendran et al. have presented some of the resource management techniques, strategies, and optimal resource allocation matrices that are helpful for measuring the allocation of a resource. This assessment addresses some major challenges in resource management and load balancing which includes system and application security, identity management, availability of resources, and privacy management. Given resources are allocated on the basis of a service level agreement to define the cost and penalties of service infringement, the system is designed in a way to provide fast and transparent application execution to the user. If required, applications maybe migrated from one machine to another for improving service and availability of the system. For service level agreement to take effect, the strategies for load balancing become important, which are devised on the basis of different parameters like performance, fault tolerance, response time, migration time, resource utilization and scalability, system throughput, energy consumption, and overhead associated with the movement of task and data communication. Authors are of the opinion that an effective resource allocation strategy is a must for attaining the objective of user satisfaction and also for the maximization of profit by a cloud service provider (Mahendran et al., 2013).

Wu et al. have presented an innovative and cost effective admission control and job scheduling algorithm for accepting a new user request in cloud without leaving an impact on it, mapping multiple requests with different QoS requirements with available VMs, and selection of prospective resources as per request demands. Strategies suggested for optimum resource allocation include initiation of new VM strategy, wait strategy, insert strategy, and finally, penalty delay strategy. Performance result as compared with reference algorithms i.e. StaticGreedy and MinResTime obtained by varying QoS and provider parameters generates better results compositely. Final robustness analysis shows that degradation in the performance of VM results in reduction of average total profit and doubles average response time. Thus, authors have suggested considering a slack time during scheduling in cases where a risk exists for a SaaS provider to enforce SLA violation with resource and consideration of penalty compensation clause (Wu et al., 2012).

Choi and Lim have evaluated resource provisioning strategies for SLA based cloud computing architecture. They have proposed a combinatorial auction system for the allocation of resources keeping in view the SLA constraints of the contract. Suggesting that the profit can be increased by reducing SLA violations and penalty cost associated with it, the authors have compared their proposed system with conventional methods with their test results showing that the proposed system is more profitable and shows higher rates of successfully accomplished jobs (Choi & Lim, 2016) compared to the legacy systems.

## 10.3 STRATEGIES

A methodology for assuring the optimum allocation of cloud's resources whether virtual and physical to cloud users is known as Resource Allocation Strategy or RAS (Vinothina et al., 2012). An optimum RAS works towards the integration of the activities of cloud provider for employing and assigning the limited cloud resources to meet the application demands. RAS helps in reducing the struggle for scarce resources in cloud, their fragmentation, and over- or under-provisioning which requires information about category and quantity of resources required by each and every request for the accomplishment of tasks. An optimal resource allocation strategy also takes sequence and resource allocation time as an input. To make an optimum Resource Allocation Strategy a service provider must avoid the following constraints.

   a. **Resource Contention:** A condition that occurs when two applications demand the same resource simultaneously and try to access at the same time.
   b. **Scarcity of Resources:** A situation that arises when resources are in limited quantity.
   c. **Resource Fragmentation:** A condition when sufficient resources are available but are impossible to allocate to the desired application i.e. when the available resources get isolated from the application.
   d. **Over-provisioning:** When any service in cloud gets resources allocated in a higher amount than the desired one there is over-provisioning of the resource.
   e. **Under-provisioning:** When any service in cloud gets resources allocated in a lower amount than the desired one there is under-provisioning of resource.

### 10.3.1 EXISTING STRATEGIES

Given provider-estimated time for resource use may result in the over-provisioning of resources, there is the opportunity to prevent it by the cloud user's giving an estimation of resource required to get the job accomplished. On the other hand, service provider's estimation of time of resource use may result in an under-provisioning of resources too. To overcome this indecisiveness, a cloud system needs input from both the service provider and the user for devising suitable resource allocation strategy. From the perspective of the user, major inputs for a RAS are the terms specified in SLA and the requirements of the application. From the providers perspective, the input required are resource availability, resource offerings, and present status of the resource for managing and allocating resources to the applications on the host. The output of every sound RAS must fulfill the parameters such as latency, response time, and throughput. Although resources supplied by the cloud are reliable, their allocation and management in a dynamic cloud environment is a crucial task.

Figure 10.2 will show that Wu et al. have proposed four different strategies for choosing user request to be accepted for minimizing the impact on performance and for avoiding penalty cost due to SLA violation thus increasing service providers' profit. These strategies are:

   a. **Initiate new VM**: A strategy that investigates all VMs in the system and compares the request deadline with estimated finished time. In the next step, it calculates the return on investment to identify whether accepting the request would be profitable or not.
   b. **Wait Strategy**: Strategy that checks for the flexible time of the current request. If the current request is ready to wait for all prior accepted requests and to produce a satisfactory return, it is accepted else it is rejected.

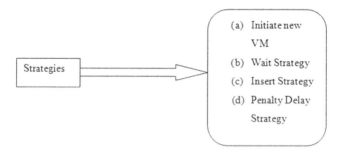

**FIGURE 10.2**  Resource Allocation Strategies.

c. **Insert**: Strategy that checks whether there exists accepted request which can wait for the new request to be finished, then checks for the new request if it can be finished before stated deadline. Thereafter, return is calculated, and if found feasible, the new request is given priority over accepted one.

d. **Penalty Delay Strategy**: Strategy which checks if the budget is satisfactory enough to balance the loss due to delayed performance of request that has waited for all the accepted request's performance. If loss could be balanced, the new request is accepted; otherwise, not.

Mahendran et al. have stated five strategies as shown in Figure 10.3 for resource management in CCE that aims at resource allocation and reallocation for load balancing. The stated strategies are:

a. **Identity management:** This strategy states that cloud enterprises must have their own identity management systems to help them in monitoring and controlling information and cloud resources' access.

b. **Physical and personnel security:** Service providers assure the security of physical machines, their access, and client's data stored on these machines.

c. **Availability:** An assurance by service providers that user's data and applications will be available as and when needed.

d. **Application security:** Every outsourced or packaged application code is implemented, tested, and accepted via proper techniques to guarantee the security of user application in Cloud Computing Environment.

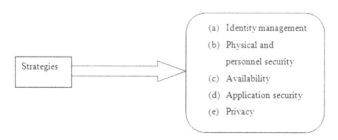

**FIGURE 10.3**  Strategies for Resource Allocation.

# Cloud Strategies of Resource Allocation

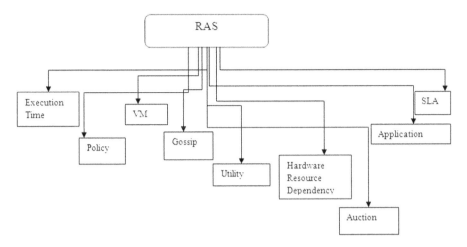

**FIGURE 10.4** Various Resource Allocation Strategies.

e. **Privacy:** A service provider in cloud ensures the security and confidentiality of all the critical data of user. Their job is to ascertain the authorized access to this data.

Vinothina et al. have presented different strategies as shown in Figure 10.4 for resource allocation in cloud computing environment based on the nature of application, cloud infrastructure, and services that require cloud's resources. The strategies include:

a. **Execution time:** A strategy that helps in overcoming the constraint of resource contention and enhanced utilization of resource by considering actual execution time for any request and preemptive scheduling technique.
b. **Policy**: A strategy that helps in facing the challenges of centralized user and resource management by introducing an additional layer known as domain between both the users and the virtualized resources i.e. decentralization of resource and users.
c. **Virtual machines**: A system which comprises of a network of virtual machines that can easily move across dynamic and multi-domain cloud infrastructure, and automatically scale its resources. This strategy allows users to choose their VMs as per need.
d. **Gossip**: A strategy that states the benefit of resource sharing among enterprises using co-operative resource (VM) management technique. Resource sharing helps in reducing the cost for enterprises, and thus, increases their profit. It also helps in the meeting the challenge of allocation of scarce resources. This strategy considers both public and private clouds, and searches for resource availability in the remote node for every change in user demand.

e. **Utility Function**: This strategy works towards the prioritization of tasks based on achieving objectives stated in QoS requirement of users. Here, VMs can perform live mitigation to optimize cost, and aims towards application satisfaction and enhanced profits by considering response time in a multi-tier system.
f. **Hardware Resource Dependency**: A strategy that proposes the use of Multiple Job Optimization (MJO) scheduler, and focuses on the input-output resources & CPU capacity to manage and improve hardware utilization.
g. **Auction**: A strategy that is based on a sealed-bid auction and focuses on the simplification of service provider's resource decision with the presentation of allocation guidelines for resource allocation to meet ordering problems. Service provider determines cost on the basis of highest bid and allocates resources accordingly.
h. **Application**: A strategy that considers nature of application to allocate resources and estimate request execution time. For task scheduling and resource allocation, the techniques of FIFO, Naïve, Optimized, and services group optimization become important.
i. **SLA:** A strategy that aims at profit maximization of SaaS providers by adhering to the terms of SLA and QoS requirements.

## 10.3.2 Proposed Strategies for Admission Control and Resource Allocation

1. Focusing on the quality requirements of clients as defined in SLA will result in no or minimum SLA violation. This will reduce the number of errors, and hence, the penalty delay cost incurred due to the violation.
2. Before accepting any contract, a check for the contract's feasibility in terms of SLA is made that will guarantee reduction in SLA violation and help in controlling cost of resources.
3. Scheduling and rescheduling of requests must be performed to get the task finished within deadline. The tasks must be queued as per priority, and the one with lower priority should be made to wait. Also, prime focus must be on balancing of workload to get the task finished on time.
4. Following up on above strategies is the one with surprise element of loyalty point scheme for customers entering into multiple contracts with service provider or the placing of consecutive requests helping service provider earn customer loyalty.

SLA is a vital element for the success of above stated strategies that will help in reducing penalty cost for any work delayed due to violation of SLA terms (Ahuja et al., 2019). Figure 10.5 shows the proposed Resource Allocation Strategies for optimum utilization of cloud resources.

Following are the strategies for the improvement of system performance as per QoS demand and profit maximization as suggested by present research work:

# Cloud Strategies of Resource Allocation

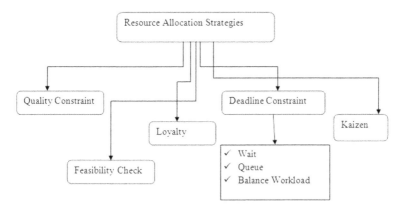

**FIGURE 10.5** Proposed Resource Allocation Strategies.

---

**Strategy 1:**

**Meeting Quality Constraints as a Strategy (ProfQS):** A knowledge based cost effective algorithm has been used to devise effective strategies for optimum scheduling and allocation of infrastructural resources. This enables cost controlling without compromising with the quality of service being provided. Focus is on reducing penalty cost by not violating SLA terms following the contract on time and within stipulated budget. A check is applied on revenues generated at each step to guarantee minimum return from each renting, where minimum expected return from a request is calculated using following formula:

$$expInvRet_{ij}^{new} = p_1 + \frac{tCost_{ij}^{new}}{T_{ij}^{new}} \quad \forall\ i \in I, j \in J \quad (10.1)$$

Here $p_1$ = expected investment return ratio varies as per service provider's need.

---

Also, SLA is established between IaaS and SaaS provider to reduce risk due to non-compliance of quality requirements of the user. This risk sharing guarantees minimum level of quality by IaaS provider, which also helps in building long-term relationship between customer and service provider. This also improves the customer satisfaction as numbers of errors are reduced.

## Strategy 2:

**Meeting Deadlines as a Strategy (ProfDL):** With this strategy, efforts are made to schedule and re-schedule tasks on the basis of priority and profit, so as to meet the deadline on time and reducing the number of penalties that can be imposed on service provider for extending service length. In this algorithm, a job with lower priority waits in a queue as soon as deadline is met. A check on workload regarding processing of tasks as per schedule and that tasks do not get interrupted in middle, which may also result in extension of contract time, is implemented. For achieving deadline, following sub-strategies can be followed:

- **Wait Strategy**: As stated earlier, wait list is prepared per low priority tasks queued on the basis of deadline. As, soon as a new request arrives, algorithm looks whether it is possible for the request to wait for all the prior scheduled request to get completed without violating the SLA terms. If yes, then request is accepted; else, the request is rejected. The wait time of a new request is calculated with the help of Eq. 10.2.
- **Queuing as a Strategy**: Queuing of a new request is done on the basis of profit along with a check on the contract length and penalty for extension in contract. VM is then allocated on the basis of queuing to maximize profit and reduce penalty.
- **Balanced Workload as a Strategy**: Balance of workload is also necessary to get the task finished on time, as for load balancing knowledge, resource is used to measure the load that a request can build and its effects on concurrent task execution.

## Strategy 3:

**Feasibility Check as a Strategy (ProfFS):** On the basis of meta-information submitted by customer, a check is made for finding a best-suited server meeting customer request. For this, qualitative assessment of available resources along with server availability is checked. Then SLA terms of both the provider and the user side are matched to check whether it is feasible to enter into a contract. A check is also made for estimated task execution time and cost associate with it. If contract is feasible, then the VM is checked for availability and storage. After, finding the match, the VM is deployed to the request; else, if contract isn't found feasible, the request is rejected and new search is made. The system tries to reduce the number of rejections; however, a check is always made for minimizing SLA violation and earning a guaranteed level of profit from each request made.

# Cloud Strategies of Resource Allocation

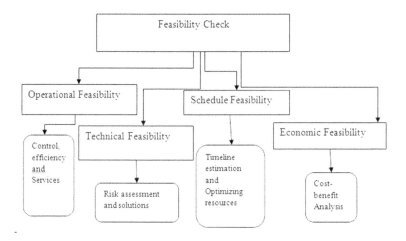

**FIGURE 10.6** Feasibility Check.

A feasibility study as shown in Figure 10.6 could be performed in following ways:

- **Operational Feasibility**: This study is conducted to find the efficiency of services being provided to end user. If the services deployed are not useful the user will not be benefited.
- **Technical Feasibility**: Technical feasibility is conducted for finding the resources having the requirement of theirs to be allocated for the service and risk associated with it (Figure 10.6). The request with maximum probability of rejection or ending up as failed may be due to lack of suitable resources or non-compliance of SLA terms. This is a threat to the system's performance, and thus, an attempt is made to find an appropriate solution for it.
- **Schedule Feasibility**: The study performed to check if the deadline is feasible for the project; otherwise, it may result in penalty cost. For each new request after finding a suitable VM, a check is made whether it can wait for all its previously scheduled requests to execute completely. In that case, the request is placed in wait list; else, if the request can't wait, it is replaced with a previously queued request of less priority in terms of budget and return with loyalty points. The waiting time for a new request is calculated using the formula (10.2) as given here:

$$WT_{ijl}^{new} = DL^{new} - \sum_{(k=1)}^{K} ProcT_{ijl}^{k} - subT^{new} \quad \forall\ i \in I, j \in J, l \in A_j \quad (10.2)$$

And the waiting time of an already scheduled request is calculated using formula 10.3:

### TABLE 10.1
### Parameters and Constraints for Schedule Feasibility

| Notation | Description |
|---|---|
| $WT_{ijl}^{new}$ | waiting time of new request $c^{new}$ to process |
| $DL^{new}$ | Deadline of new request $c^{new}$ |
| K | total number of accepted request |
| L | all VMs |
| J | all resource provider |
| l | type of VM |
| $A_j$ | all types of VM provided by resource provider j |
| $subT^{new}$ | submission time of $c^{new}$ |
| $WT_{ijl}^{K}$ | waiting time for an already accepted request $c^k$ (that can wait for new request $c^{new}$ to complete) |
| $DL^k$ | Deadline of request $c^k$ |
| $ProcT_{ijl}^{new}$ | processing time of new request $c^{new}$ |
| $subT^{new}$ | submission time of $c^{new}$ |

$$WT_{ijl}^{K} = DL^k - \sum_{\substack{n=1 \\ n \neq k}}^{K} ProcT_{ijl}^{n} - ProcT_{ijl}^{new} - subT^{new} \quad \forall\ i \in I, j \in J, k \in K,$$

$$l \in A_j \tag{10.3}$$

Table 10.1 lists the parameters and constraints for calculating schedule feasibility

- **Economic Feasibility**: It checks for the cost of procuring the resources and revenues that will be earned from a particular assignment. If total returns generated from a request is higher than or equal to minimum expected request return of SaaS provider, then request is accepted. A service provider sets a minimum profit margin as his earnings and checks for the request's feasibility as per limit. The formula 10.4 for checking economic feasibility is as given here:

$$\text{If } (prof_{ijl}^{new} >= cost_{ijl}^{new} * p1) \text{ and } \left( expInvRet_{ij}^{new} = p1 + \frac{tCost_{ij}^{new}}{T_{ij}^{new}} \right)$$

$$\forall\ i \in I, j \in J, l \in A_j \tag{10.4}$$

*(Here $p_1$= varies as per providers need and is expected investment return ratio)*
  Then accept
  Else

  Check customer data

  If (Customer is regular then)

**CASE 1  CHECK SYSTEM'S CAPACITY UTILIZED**

If Cap. Utilization >= 50% but <=75%
Check $prof_{ijl}^{new} >= cost_{ijl}^{new} * p_2$ + loyalty points
and $expInvRet_{ij}^{new} = Loyaltyret_{ijl}^{new}$   $\forall\ i \in I, j \in J, l \in A_j$
(Here, $p_2$ = expected investment return ratio less than $p1$)
Popup message for loyalty points redemption & acceptance of request
If (Customer agrees)
    Set loyalty points = 0 and accept request
Else
    Reject

**CASE 2  IF CAPACITY UTILIZATION IS <= 50% AND $prof_{ijl}^{new} >= cost_{ijl}^{new}$ ($P_2$/100)**

$expInvRet_{ij}^{new} = Loyaltyret_{ijl}^{new}$   $\forall\ i \in I, j \in J, l \in A_j$
Then accept
Else
    Check for loyalty points
Popup message for loyalty point redemption and acceptance of request
If (Customer agrees)
    Redeem loyalty points and accept request
Else
    Reject request
    Then accept

## Strategy 4:

**Loyalty as a Reward Strategy**: Loyalty of a customer is the measure of probability to do repetitive business with the provider. A satisfied customer with positive experience enhances the value of service and results in customer loyalty. A loyal customer is rarely going to get affected by the availability or pricing of resources. They are even ready to pay more for the services they love or for the services they are familiar with. Apart from this, a loyal customer doesn't search for another service provider, for they understand if there is any issue in service delivery or if the system fails, and helps in improving service quality by providing proper feedback.

Customer loyalty has various benefits like boosting of profits (a mere 5% retention rate may boost profits by 25% to 95%), effective planning, and retention of existing customer is less expensive than getting new (new customers are five times more costly than loyal ones), conversion rates are higher (around 60% to 70% for loyal customer), and they spend more than first-time customers. Thus, present strategy aims at gaining customer loyalty by providing them loyalty points as a reward. In this strategy, customers get loyalty points for every request accepted and contract completed. There is a limit of thousand points after which a customer can redeem their points to pay for the service rented. This will act as an add-on for building customer relationship. The formula 10.5 for calculating loyalty points and return after loyalty point's redemption is as follows:

Loyalty points = net profit $* (L/100)$ where L = % of net profit earned provider wants to give users as a reward for their loyalty.

$$\text{Loyaltyret}_{ijl}^{new} = \frac{\text{Prof}_{ijl}^{new} + \text{loyaltypoints}_{ijl}}{T_{ijl}} \quad \forall\, i \in I, j \in J, l \in A_j \quad (10.5)$$

---

**Strategy 5:**

**Kaizen as a Strategy for Continuous Improvement**: For bringing small but positive changes in the provider's services, the concept of Kaizen can be used. It will help to improve quality of the allocated service and to control cost of the resources. Also, Kaizen will play a significant role in reducing the number of errors in resource allocation, task scheduling, or their rescheduling. This will finally lead to an increase in overall profit of the service provider. Figure 10.7 will show the different phases involved in scheduling of applications and resources in the cloud. Different phases are:

- **Knowledge Phase**: Meta-data regarding customers, their resource requirements, quality constraints along with data about system's present status and performance is recorded.
- **Control Phase**: Data is collected for monitoring and controlling of system's performance. This gives a clear idea of any possible problems with root cause. Data thus collected is used for further analysis in next phase.
- **Analysis Phase**: Analyzes data obtained from control phase to plan and devise solutions for enhancing system's performance.
- **Planning Phase**: Planning improves quality of service on the basis of information received from the data analysis. Different solutions are framed and a suitable idea is worked out.
- **Testing Phase**: The idea is tested to see the results and the solution is implemented in case of positive results.

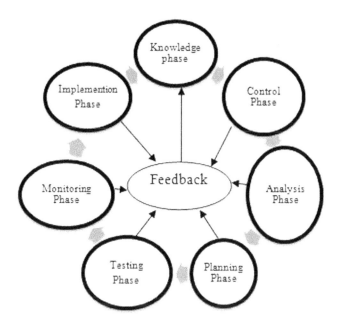

**FIGURE 10.7** Kaizen Cycle for Continuous Improvement.

- **Monitoring Phase**: System is continuously monitored for the evaluation of results and measuring of the system's performance. A solution is selected on the basis of screening.
- **Implementation Phase**: This phase involves implementation of selected solution for the enhancement of the system's performance. The process is repeated regularly for continuous improvement of the system and feedback is sent back to previous phases for monitoring and control of performance. This also helps in keeping a check on the quality and reducing violation of SLA norms.

SLA plays a vital role in every phase of each of the strategies as SLA violation not only results in penalty, but if it occurs repeatedly, then it hampers the relation between user and service provider. Costs and returns are calculated in every strategy to determine the feasibility of the request and effectiveness of the strategy.

Table 10.2 shows comparison of proposed Resource Allocation Strategies.

## 10.4 ANALYSIS AND RESULT

A simulation model of proposed system was developed using CloudSim as a tool with NetBeans for the evaluation of performance. For the testing of algorithms, Java force was used as an application programming software. The proposed mathematical model was tested on a simulator with the help of synthetic workload. Data was generated randomly with the help of a suitable statistical distribution.

## TABLE 10.2
### Comparison of Resource Allocation Strategies

| Constraint | ProfQS | ProfFS | ProfDL |
| --- | --- | --- | --- |
| No. of Request Accepted | Less requests as compared to ProfFS Strategy | Higher no. of user requests due to loyalty points | Lowest number of requests arrived |
| Number of initiated VMs | Comparatively have highest number of VM initiated | Lowest number of VM initiated | As compared to ProfFS Strategy have higher number of VM initiated |
| Cost incurred | Highest of all | Least of all | Greater than ProfFS but less than ProfQS |
| Customer Satisfaction Level (CSL) | Second highest level of CSL | Highest level of satisfaction | Lowest of all |
| Response Time | Lowest | Highest | Second Highest |
| Overall Profit | Second highest | Highest | Lowest |

Figure 10.8 shows the impact of variation in incoming user requests (from 600 to 1000) on the number of accepted requests. It can be observed from Figure 10.8 that ProfFS accepts 0.6% and 6.6% more requests than ProfQS and ProfDL do, when number of users is equal to 1000 users. It accepts 0.4% and 5% more requests than ProfQS and ProfDL do, when number of users are 600 users. Thus, it is evident that for a varying number of user requests, admission control strategy following feasibility check accepts more user requests than do the other two strategies.

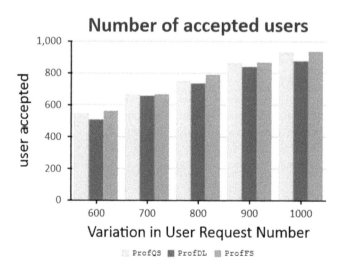

**FIGURE 10.8** Accepted Users v/s Number of Request.

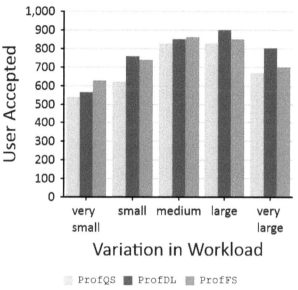

**FIGURE 10.9** Accepted Users v/s Variation of Workoad.

Figure 10.9 shows the impact of varying workloads on the acceptance of incoming user requests, while keeping other parameters like budget and deadline constant. It can be seen that admission control ProfDL strategy accepts 15% more requests than ProfFS does, and 20% more requests than ProfQS strategy does when workload is very large. Hence a SaaS provider must exploit the concept of ProfDL when workload is very high to increase the number of user requests in acceptance.

Figure 10.10 studies the impact of variation in deadline over number of user requests accepted. It can be observed that ProfFS accepts 6% more user requests than ProfQS does, and 5% more user request than ProfDL does. The reason for the small difference in number of requests acceptance is in the fact that whereas all the three strategies exploit the penalty cost to maximize output, ProfFS attempts to reschedule requests so that maximum number of user requests are accepted.

Figure 10.11 evaluates the impact of changing budget (from very small to very large) over user request acceptance number. It can be observed from the figure that with an increase in user budget, profFS accepts 42% more requests than ProfQS does, and 38% more user requests than ProfDL does. Therefore, it can be used by the SaaS provider to encash the benefits of heavy budget requests. The reason behind this is in ProfFS strategy exploiting the concept of requests rescheduling that helps in the acceptance of some high priority and large budget requests.

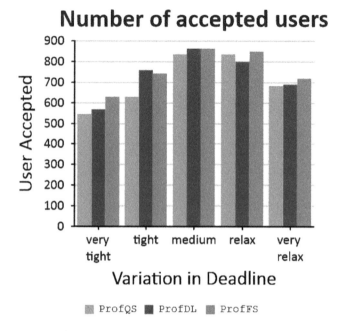

**FIGURE 10.10** Accepted Users v/s Variation in Deadline.

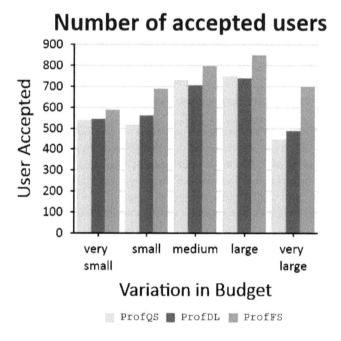

**FIGURE 10.11** Accepted Users v/s Variation in Budget.

## 10.5 CONCLUSION AND FUTURE SCOPE

Resource Allocation Strategies presented in the chapter are useful for SaaS providers who obtain third party resources on lease or utilize their in-house hosted resources. Resources in the cloud are shared among different SaaS providers. The dynamics of request arrival and software usage might produce an impact on the performance of hosted software service. The proposed strategies help in the identification of requests which are more profitable and feasible and, therefore, must be accepted. These strategies also focus on minimizing the SLA violations to reduce the cost of services and enhance the revenues generated. The proposed RAS strategies are helpful in achieving various business objectives of the enterprise, whether in the terms of profit maximization, cost minimization, achieving customer loyalty, or the expansion of their market share.

Although, the study has conducted an in depth evaluation of the research topic, there remain some open issues which may act as a starting point for future studies. These are:

- Service providers can formulate vivid robust pricing models for their services so as to achieve the objectives of the maximization of profit, earning customer loyalty, and expansion of market share. They can provide customers with the facility of self-service, as in case of buying a laptop or personal computer, so that customers can configure the software packages as per their demands in a more dynamic and profitable way. It would help SaaS providers to devise strategies for SLA based resource management and meet their business objectives. For the formulation of effective and efficient resource management strategies, it is really important to understand the components offered by this software(s); the variation in resource consumption with changing requirements of these components; and finally, the method to formulate optimum pricing policy to address the challenges and issues associated with these variations.
- For attaining the objective of cost minimization, SaaS providers can utilize resources with varying pricing policies which could help them fulfill customer demands. One can formulate innovative resource management strategies for handling the variations incipient in dynamic pricing policies by getting a thorough understanding of the impact of the utilizing vivid pricing models on SLA based resource management system.

## REFERENCES

Ahuja, N., Kanungo, P., & Katiyal, S. (2019). SLA based rescheduling of task for optimum allocation of resources in cloud computing environment. *National Journal of System and Information Technology (NJSIT)*. SRIMCA, ISSN: 0974-3308, *12*(1), 65–78.

Backialakshmi, M., & Sathya Sofia, A. (2014). Survey on scheduling algorithms in cloud computing. *International Journal of Engineering Research and General Science*. ISSN 2091-2730, 2(6), 12–22.

Buyya, R., Broberg, J., & Goscinski, A. M. (Eds.). (2010). *Cloud computing: Principles and paradigms* (Vol. 87). John Wiley & Sons.

Choi, Y., & Lim, Y. (2016). Optimization approach for resource allocation on cloud computing for IoT. *International Journal of Distributed Sensor Networks, 12*(3), 3479247.

Chopra, A., Rani, M., & Mehta, T. (2015). Time based resource allocation in cloud computing. *International Journal of Computer Science and technology, 6*(3), 42–44.

Gonçalves, G. E., Endo, P. T., Cordeiro, T. D., Palhares, A. V. A., Sadok, D., Kelner, J., … & Mangs, J. (2011). Resource allocation in clouds: concepts, tools and research challenges. XXIX SBRC-Gramado-RS. Inbook: Minicursos do SBRC 2011 (pp. 197–240, Chapter 5). SBC.

Hamdy, N., Aboutabl, A. E., ElHaggar, N., & Mostafa, M. M. (2017). Resource allocation strategies in cloud computing: Overview. *International Journal of Computer Applications* (075-8887), *177*(4), 18–22.

Komal, & Saroha, V. (2017). Resource allocation in cloud environment-A review. *International Journal for Research in Applied Science & Engineering Technology (IJRASET)*. ISSN: 2321-9653, *5*(5), 245–249.

Mahendran, D., Gopi, M., Priyadharshini, S., & Karthick, R. (2013). A study on cloud resource management techniques. *International Journal of Innovative Research in Computer and Communication Engineering, 1*(1), 58–61.

Vaquero, L. M., Rodero-Merino, L., Caceres, J., & Lindner, M. (2009). A break in the clouds: Towards a cloud definition. *ACM SIGCOMM Computer Communication Review, 39*(1), 50–55.

Vinothina, V., Sridaran, R., & Ganapathi, P. (2012). A survey on resource allocation strategies in cloud computing. *International Journal of Advanced Computer Science and Applications, 3*(6), 97–104.

Vouk, M. A. (2008). Cloud computing – issues, research and implementations. *Journal of Computing and Information Technology, 4*, 235–246. University of Zagreb, Croatia.

Wu, L., Garg, S. K., & Buyya, R. (2012). SLA-based admission control for a software-as-a-service provider in cloud computing environments. *Journal of Computer and System Sciences, 78*(5), 1280–1299, Elsevier.

Xu, M., Cui, L., Wang, H., & Bi, Y. (2009). *A multiple QoS constrained scheduling strategy of multiple workflows for cloud computing* [Symposium]. Parallel and Distributed Processing with Applications, IEEE, Chengdu, pp. 629–634.

Zhang, Q., Cheng, L., & Boutaba, R. (2010). Cloud computing: State-of-the-art and research challenges. *Journal of Internet Service Applications, 1*, 7–18, Springer.

# 11 Optimization Mechanism for Energy Management in Wireless Sensor Networks (WSN) Assisted IoT

*Urmila Shrawankar and Kapil Hande*

## CONTENTS

| | | |
|---|---|---|
| 11.1 | Introduction | 201 |
| 11.2 | Literature Review | 202 |
| 11.3 | Methodology | 202 |
| 11.4 | Algorithm of Proposed Method | 205 |
| 11.5 | Experimental Results | 206 |
| 11.6 | Conclusion | 210 |
| References | | 211 |

## 11.1 INTRODUCTION

Wireless Sensor Network (WSN) is a group of organized machines for cooperative networking. WSN features the low power sensor node which periodically senses the data of different sink nodes. Wireless Sensor Network provides the environment monitoring facility as an efficiency feature in many applications including civil and military (Liu et al., 2008). The increases in the WSN's lifetime is a result of evenly distributed load on all discovered paths, though the distribution of load on all paths is usually not regular due to different network topology, hop factor, network connectivity, and nature of the WSN application.

In this chapter, we propose clustering of nodes and distribution of traffic on paths using protocol for load balancing in WSN. The proposed protocol effectively smooths the traffic evenly on all discovered routes. The proposed protocol combines the ideas of traffic load balancing and distributed network.

This chapter is organized in the following sections: the literature review for this paper is carried out in section 11.2; the methodology for registration of applications is explained in section 11.3; in which during transmission of data traffic how data is distributed in network and random and explain the process of election of cluster head; the proposed methodology depicted in the algorithm is given in section 11.4;

the simulation result of proposed work is elaborated in section 11.5; and the conclusions arrived by this study are elucidated in section 11.6.

## 11.2 LITERATURE REVIEW

This section gives the details of research on different optimization mechanisms for energy management in wireless sensor networks that have attracted attention of the world till date. The routing protocol is presented for the wireless sensor network using multiple cluster heads. (Tang et al., 2011). The improved version of LEACH is presented for wireless sensor network in (Xu et al., 2012). The target tracking in Wireless Sensor Networks is implemented using energy aware fault tolerant clustering scheme (Bhatti et al., 2010). The multipath routing protocol is used in wireless sensor networks using traffic splitting protocol (Ebada & Mouftah, 2011). The energy efficiency in wireless sensor networks is achieved using contours broadcast channel in realistic power models (Akbari et al., 2012).

## 11.3 METHODOLOGY

In this section, the methodology of the proposed load distribution protocol, called clustering and traffic distribution, to balance the load in the network and increase the network lifetime is described. There are three steps in the proposed algorithm: initialization of the network, clustering, and traffic allocation. The methodology proposes two types of nodes: sensor nodes for forwarding data, and cluster nodes for data transmission, that is it can make decisions on how to send data from source to destination. When data requiring energy N has to be sent, the traffic protocol swings into action by packing all nodes with energy equal or less than N to the cluster head, which has the highest energy N or more than N. During clustering phase, we adopt the same concept of rounding with E-LEACH. The traffic allocation method distributes the data packets in a split manner according to capacity of routes, that is, if a route has capacity higher than N, where N > n, with n being the energy of data packets after splitting it into smaller bits of energy n, then that route gets selected. Figure 11.1 shows the basic idea behind the working of the mechanism. As Figure 11.1 Figure 11.1 shows, S is the source node and D is the end node. The Source node forwards the data packets $D_{p1}, D_{p2}, D_{p3}$ towards the destination node D according to the capacity of the route. For example, if data packet size is 5mb, then capacity of route to carry the data packet is equal to 5mb or greater than 5mb. Sensor nodes collect the data packets arriving from the source node and forward them to the nearest cluster head. A cluster head sends the data packets to the cluster head which is nearest to the end node. With intelligent data transmission, energy is saved.

The chapter stated that the N nodes are distributed with following assumptions:

    i. Fixed Base station (BS) located away from the square area.
    ii. Cluster size is random.
    iii. The level of energy sent to the BS.

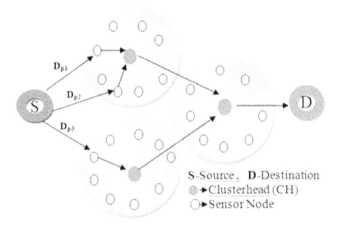

**FIGURE 11.1** Basic Working Idea.

Although clustering protocols like LEACH, LEACH-V, and ELEACH will all extend network lifetime, each is different according as the plane which uses multi-hop routing or dynamic routing. However, there remains the issue of energy overload from energy consumption. Cluster heads (CH) are selected without considering the energy at destination node. The node with less energy but with same priority causes the system energy to increase.

The Sensor nodes calculate $\varepsilon_j$, where $\varepsilon_j$ is energy of data packet, as the energy that remains with it and then forwards it to the Base Station. The node whose value of $\varepsilon_j$ is high, that is at least that of energy required to transmit the whole data packet, is selected as CH of the network, and BS broadcasts the information to the member nodes. For CH selection to happen of any node in next round, it depends on the selection made by basestation and use of random (N) between 0 and 1. When the threshold number **T(n)** becomes more than the current cluster head, there is re-clustering and the node with energy higher than threshold will become cluster head is next rounds (Kwon & Shroff, 2012).

$E_{initial}$ is initial energy of destination node and $E_{current}$ is current energy of station at the r round, P is probability of node being cluster head using G which is group of station with cluster head in 1/P round (Figure 11.2).

Traffic allocation procedure deals randomly among all discovered paths. Traffic allocation starts from source station till the data is transmitted to the destination node. Traffic allocation starts after finding the multipath routes using on-demand routing strategies. The path is selected which has the highest energy, provided there are extra stations on the same route. i.e. reinforced path. Traffic allocation consists of two parts: first, route weight is collected which is to distribute the load evenly among the all discovered paths.

   i. Route Weight Assignment:

The proposed method calculates the weight for each route and assigns weight to each. The weight reflects the route capacity to deliver the data packet from source to destination.

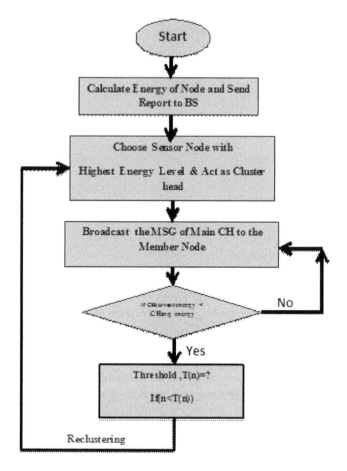

**FIGURE 11.2** Flowchart for Cluster Head Selection.

Path$_j$ is weight of route j,
$\varepsilon_j$ is remaining energy at node,
$\lambda_j$ is load for the route j,
$h_j$ is the number of hops for route j,
$\beta$ is the network hop factor, it is defined as the impact of route's number of hops, or the sum of weights for route j.

For calculating the weight of a route, we have to calculate the network factor which is considered for weight route calculation.

The network factors are:

a. Network Size factor:
It is the average number of hops from all nodes to destination (Deng et al., 2011).
$S_f$ is the network size factor

**L** is lowest layer order
**L$_i$** is collection of station layer i
**N** is the total number of nodes of network
b. Network routing factor:
Network routing factor calculates the average number of routes discovered by the node.
$\gamma_f$ is the routing factor
**T** is total number of routes discovered by all nodes
**N** is the total number of nodes in network
**t$_i$** is total number of routes discovered by node i
c. Network connectivity Factor:
Network connectivity factor calculates the average number of connected nodes to any node (Sergiou & Vassiliou, 2012).
**C$_f$** is the network connectivity factor
c$_i$ is number of connected nodes to node i
d. Network Hop Factor ($\beta$):
$\beta$ defines impact of number of hops on the weight assigned to route. As $\beta$ gets smaller, the weight assigned to route becomes bigger, which means protocol selects the longer route. If $\beta$ gets bigger, the weight assigned to routes gets smaller.
$(0 < \beta > 1)$ (7)

ii. Traffic Distribution:

Traffic distribution starts after route assignment method ends. The traffic is distributed among all discovered routes according to weight of route. For example, if weight of route j is 5mb then j route allows for transmission of only 5 mb or less than 5mb data packets (Leung & Li, 2006).

Wj(k$_i$) is the expected work load i.e. theoretical work load calculated on the basis of network factor.
Objective is to get the optimized route to node j
k$_i$ is the number of messages forwarded by the node i
The goal of traffic allocation method is

- To minimize the workload on all routes.
- To minimize energy consumption and increase network lifetime.
- To minimize the end to end delay.
- To achieve the maximum packet delivery ratio.

Figure 11.3 depicts a flowchart for traffic allocation.

## 11.4 ALGORITHM OF PROPOSED METHOD

Step 1 Initialize the network
Step 2 Calculate the energy factor of each node

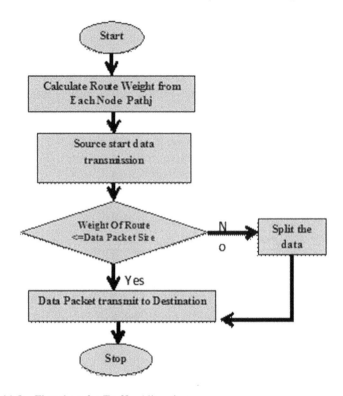

**FIGURE 11.3** Flowchart for Traffic Allocation.

Step 3 The node having highest energy factor considered as cluster head (CH) at initially
   **3.1** select the number (N) between 0 and 1,
   if (N < T(n))
   **then**, node become CH
Step 4 Start packet data transmission from start to end node
Step 5 Select the node which consists of highest number of routing factor to transfer the packet data
Step 6 Calculate the weight of routes
   **6.1** if weight of route is less than the packet data, go to the **step 8; else** go to step 7
Step 7 Spilt packet data according to route capacity
Step 8 Transfer the data packet to destination (Repeat step 3 to 8 till transmission completed.)

## 11.5 EXPERIMENTAL RESULTS

This chapter presents a detailed implementation of the WSN simulation module using Microsoft Visual Studio.NET 2008. Visual Basic programming language is derived from the language called BASIC. The instructions in this language are used to demonstrate the topology in WSN and the node motion mode. It also creates

# Optimization Mechanism for Energy in IoT

statistical data track file. To verify, the proposed algorithm is summarized with earlier protocol E-LEACH and TSP.

A fifty sensor station is arranged in the field of 200 sqm × 200 sqm. As shown in Figure 11.4 one source node and one destination node along with base station are present. The continued bit rate (CBR) traffic is used. All the stations are mobile in logical working space of Network Simulation. Our simulation setting and parameters are summarized in Table 11.1.

The evaluation is done using various parameters such as the clustering and load balancing shapes. In this work, the network load, network lifetime, and amount of data messages received by base station are used to differentiate the

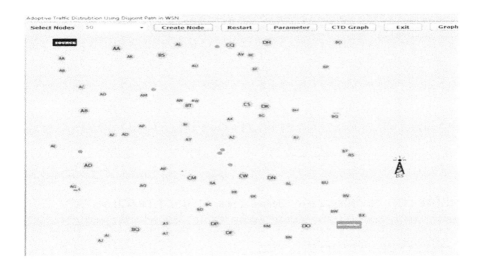

**FIGURE 11.4** The Distribution of WSN Nodes.

**TABLE 11.1**
**Simulation Parameters**

| | |
|---|---|
| No. of Nodes | 50 or 100 |
| Area Size | 200 * 200sq.m |
| Queue Type | Drop Tail |
| Queue Size | 100 |
| Mac | 802.11 |
| Routing protocol | AODV |
| Range of Transmission | 15 m |
| Time of Simulation | 50s |
| Traffic Type | CBR |
| Data Rate | 100kbps |
| Initial Energy | 5 J |

performance protocol with traffic splitting and E-LEACH. If node's energy is equal to zero, we define the node as a dead node.

In Figure 11.5, the number of dead nodes reflect the balance of energy consumption in WSN. If overall percentage of dead nodes is minimum, then automatically it will increase the network lifetime. Figure 11.5 is the simulation

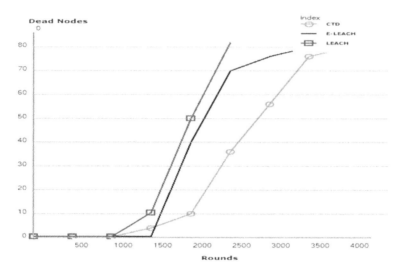

**FIGURE 11.5** The Difference of Lifetime Among LEACH, E-LEACH and CTD.

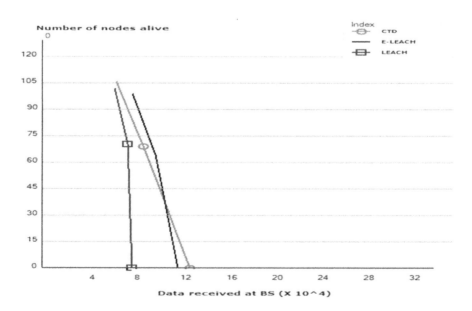

**FIGURE 11.6** The Value of Data Collected by the BS with Number of Stations Alive.

result of network lifetime for proposed protocol and ELEACH protocol. The system lifetime increase by 30% of proposed protocol than ELEACH.

In Figure 11.6 X-coordinate represents the Data received at BS whereas Y-coordinate represents Number of nodes alive. The proposed method delivered two times more messages than LEACH and ELEACH did.

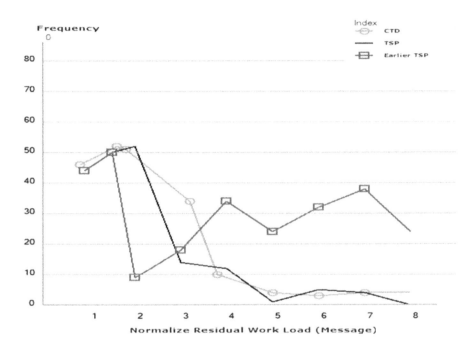

**FIGURE 11.7** Work Load Comparison Between Earlier TSP, TSP and CTD.

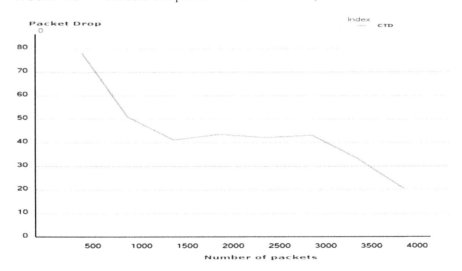

**FIGURE 11.8** Packet Drop Ratio.

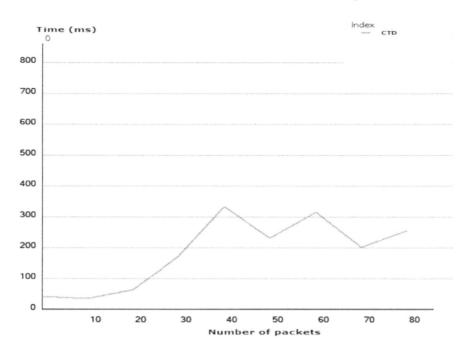

**FIGURE 11.9** End to End Delay.

Figure 11.7 shows the work load comparison between earlier traffic splitting protocol, traffic splitting protocol, and the proposed method. X-coordinate represents normalized residual work load, whereas Y-coordinate represents frequency. The normalized residual work load is the difference between actual work load and expected work load. Figure 11.7 shows that load decreases by 20% of proposed method than improves the traffic splitting protocol.

The ration of data forwarded to destination is collected by CBR. Figure 11.8 shows that data packets sent by source node on x-axis, and delivered fraction, on y-axis. The result of only proposed protocol is depicted.

Figure 11.9 highlights the relative performance of the proposed method.

## 11.6 CONCLUSION

The proposed protocol permits a sensor node to select itself as a cluster-head on its residual energy, and distributes the data traffic evenly among the network by using multiple paths. The result obtained in simulation with proposed protocol provides the improvement in both the ELEACH and TSP protocols in terms of increase in network lifetime, decrease in energy consumption, and decrease in load balancing, respectively. Energy utilization by proposed protocol is 20% less as compared to E-LEACH, and increases the lifetime of WSN. We tested the two additional parameters of the proposed work: first, packet ratio; and second, delay in end to end data transmission.

# REFERENCES

Akbari, A., Héliot, F., Imran, M. A., & Tafazolli, R. (2012). Energy efficiency contours for broadcast channels using realistic power models. *IEEE Transactions on Wireless Communications*, *11*(11), 4017–4025.

Bhatti, S., Xu, J., & Memon, M. (2010, September). *Energy-aware fault-tolerant clustering scheme for target tracking wireless sensor networks* [Symposium]. Wireless Communication Systems, 7th International Symposium,IEEE, pp. 531–535.

Deng, S., Li, J., & Shen, L. (2011). Mobility-based clustering protocol for wireless sensor networks with mobile nodes. *IET Wireless Sensor Systems*, *1*(1), 39–47.

Ebada, M., & Mouftah, H. T. (2011, December). *Traffic splitting protocol for multipath routing in wireless sensor networks* [Workshop]. 2011 IEEE GLOBECOM Workshops (GC Wkshps), IEEE, pp. 503–507.

Kwon, S., & Shroff, N. B. (2012). Energy-efficient unified routing algorithm for multi-hop wireless networks. *IEEE Transactions on Wireless Communications*, *11*(11), 3890–3899.

Leung, K. C., & Li, V. O. (2006). Generalized load sharing for packet-switching networks. I. Theory and packet-based algorithm. *IEEE Transactions on Parallel and Distributed Systems*, *17*(7), 694–702.

Liu, L., Wang, Z., & Zhou, M. (2008, August). *Cooperative multipath routing and relay based on noncoherent detection in wireless sensor networks* [Conference session]. Automation Science and Engineering, IEEE International Conference, IEEE, pp. 128–132.

Sergiou, C., & Vassiliou, V. (2012, May). *Source-based routing trees for efficient congestion control in wireless sensor networks* [Conference session]. Distributed Computing in Sensor Systems, IEEE 8th International Conference, IEEE, pp. 378–383.

Tang, D., Liu, X., Jiao, Y., & Yue, Q. (2011, September). *A load balanced multiple cluster-heads routing protocol for wireless sensor networks* [Conference session]. Communication Technology, 13th International Conference, IEEE, pp. 656–660.

Xu, J., Jin, N., Lou, X., Peng, T., Zhou, Q., & Chen, Y. (2012, May). *Improvement of LEACH protocol for WSN* [Conference session]. Fuzzy Systems and Knowledge Discovery, 9th International Conference, IEEE, pp. 2174–2177.

# 12 Intelligent Systems for IoT and Services Computing

*Dr. Preeti Arora, Dr. Laxman Singh, Saksham Gera, and Dr. Vinod M Kapse*

## CONTENTS

| | | |
|---|---|---|
| 12.1 | Literature Survey | 214 |
| 12.2 | Introduction | 215 |
| 12.3 | Proposed Method | 216 |
| 12.4 | Three-Layer IoT Architecture and Its Components | 218 |
| | 12.4.1 Hardware Requirements | 218 |
| | 12.4.2 Software Requirements | 218 |
| 12.5 | Scope of the Sensor Networked Devices | 219 |
| | 12.5.1 Scope of the Wireless Things | 219 |
| | 12.5.2 IoT and Its Market Segments | 219 |
| | 12.5.3 Experimental Analysis of the scanning Method | 219 |
| 12.6 | The IoT Contains an Enormous Variety of Connected Objects | 220 |
| | 12.6.1 Tiny Stuff: Smart Dust Enormous Stuff: An Entire City | 220 |
| | 12.6.2 Digital Locks | 220 |
| | 12.6.3 Smart Buildings | 220 |
| | 12.6.4 Data Collection by Existing Process | 220 |
| 12.7 | Common Problems with Data | 221 |
| 12.8 | Working Methodology | 221 |
| 12.9 | A Bunch of Convolutional Neural Networks | 222 |
| 12.10 | Selection of Loss Function (Training) | 222 |
| 12.11 | IoT & Machine Learning Another Application for Urban Intelligence | 224 |
| | 12.11.1 Bringing Data-Driven Management to Complex, Fast-Paced Environments | 224 |
| 12.12 | Challenges | 224 |
| | 12.12.1 Solution | 224 |
| 12.13 | Key Benefits | 226 |
| 12.14 | Sample Use Cases for the Restaurant Industry | 226 |
| | 12.14.1 Food Quality Control | 226 |
| | 12.14.2 Automated Restaurant Management | 227 |
| | 12.14.3 Data Sharing to Drive Commercial Purchases | 227 |
| | 12.14.4 Distributed Marketplace for Commodities | 227 |
| | 12.14.5 Inventory Management | 227 |

DOI: 10.1201/9781003154686-12

| | | |
|---|---|---|
| | 12.14.6 | Queue Management ................................................... 227 |
| | 12.14.7 | Smart Waste ............................................................... 227 |
| | 12.14.8 | More Use Cases .......................................................... 228 |
| 12.15 | How It Works in Brief ................................................................. 228 |
| | 12.15.1 | RFID Tags ................................................................... 228 |
| | 12.15.2 | The Foundation for IoT .............................................. 228 |
| 12.16 | IoT Smart Public Transportation System ................................... 229 |
| 12.17 | IoT Based Smart Electricity Distribution System for Industries and Domestic Usages ......................................................... 229 |
| 12.18 | Remote Health Monitoring System for Fetal and Mother Health Monitoring ....................................................................... 230 |
| 12.19 | Smart, Safe and Clean Streets .................................................... 230 |
| 12.20 | IoT Based Healthcare Solution for Tracking Human Spine Movement 230 |
| 12.21 | IoT Based Smart Parking System ............................................... 231 |
| 12.22 | IoT Based Assisting System for Miners to Prevent Accidents .... 231 |
| 12.23 | IoT Based Disease Prevention System for Smart Healthcare ..... 231 |
| 12.24 | Intelligent Toll Collection System for Highways ...................... 232 |
| 12.25 | Forecasting Potential Health Threats ........................................ 232 |
| 12.26 | Conclusion .................................................................................. 232 |
| 12.27 | Limitations/Future Work ............................................................ 233 |
| References ............................................................................................... 233 |

## 12.1 LITERATURE SURVEY

Learning in rats can be better understands that when rats find food product which good smell and condition to eat, they are in habit to take the small portion of the meal/food product and later rest of the eating depends on the flavor and behavior of the rat. In contrast if the rat finds any unfavorable condition of the food item in that case, they do not eat the subsequent portion of the food (Choi, Chung, & Young 2019). It is clear indication all the selection activity is done by the learning skill. The animal basically used his previous experience and be an expertise to make a selection for next food items for the sake of safety concern. Animal faced a bad experience in past and termed as bad food than he uses that selection skill in future. It is an example of learning.

Now we can talk about the learning of machines if we want to train a machine for spam email detection a naïve solution would be appropriate for selection of non-spam emails the way rat choose the food items and avoid the food pieces which is not good for their survival. The machine will simple save/record the emails that are selected/named as spam emails by human user. Every incoming email search the previous data of spam emails if there is a match of any previous email it will be discarded and thrown into spam folder afterwards it will be deleted from inbox folder. That approach is useful at few cases but not very commonly used for unseen mails (Chandrasekar & Sangeetha, 2014). Machine Learning is "Explicit programming can be learnt by ability of Computers" termed by Arthur Samuel in 1959 an American scientist.

Basically, machine learning uses algorithms trained in advance that analysis of data is done at receiver machine within range provided. To get the optimized results new algorithm data is fed to machine to achieve the better performance and to make the machine more intelligent.

When we perform the complex task manually the actual need of machine comes in to reality. Two major concerns problem complexity and adaption of machine basically meant for complex problems (Choi et al., 2019). There are various tasks performed by humans commonly on day-to-day basis it is not frequently asked how do they do it is not satisfactory answer for a program. Most common examples of speech recognition, image processing, driving learning skill. The said skills if performed by trained machines that satisfactory results can be met very easily (Zuo, 2010). Tasks beyond Human Capabilities: Machine Learning easily perform analysis of the complex data as astronomical data, turning medical archives in to knowledge. New horizons are opens up for ever increasing processing speed of computers Combination of programs learn in the promising domain with the unlimited memory capacity of complex data sets (Zeiler & Fergus, 2013).

## 12.2 INTRODUCTION

Smart management and payment billing systems is basically use in shopping complexes in metro cities and popular malls. Existence of barcode reader is bit older and it has been using from last many years only thing major role of its to keep track of products and to make the transaction easier for sale purchase record etc. It is not only the solution because due to time taken for barcode reading no body wants to keep waiting and to maintain the queue for billing etc. specially in busy or festive days. Benefits of the suggested billing system like as it proves a better experience of shopping as comparison to older methods, man power reduction due to elimination of checking process by counters at all. Such type of systems would be beneficial for the custmers and the supplier (Chaudhari, Gore, Kale, & Patil 2016). Rapid increase in the production of semiconductors due to decrease in the price and to make the selling buying system more attractive. Radio Frequency identification tags are proven useful for markets and shopping sites. There are lots of methods proposed in this field. High Speed billing system is proposed by Ananth Bharti in which RF linked with the server directly for billing purposes with the help of detector which is placed with in the cart (A. A. Anil, 2018). Smart Trolleys have been invented by (Rajeshkumar et al., 2016) while shopping bill calculation is done by RFID reader with MCU of each product (Rezazadeh, Sandrasegaran, & Kong 2018). Zigbee has also been proposed a smart cart for shopping by P. Chandrasekar, T. Sangeetha in which product and bill calculation is done by Product Identification Device (PID) of the RFID reader card. Zigbee used this concept in development of central billing system for bills transmission (Chaudhari, Gore, Kale, & Patil 2016). Many researchers have already been proposed systems using the MCU along with RFID tags for better trolley communication (Rajeshkumar et al., 2016).

## 12.3 PROPOSED METHOD

Now a day's results are producing manually due to lack of automated Goods Billing Systems which can be beneficial for e Commerce industry and upload results and further information on online systems which results in to error prone operations and helpful in Selection of required Goods with in minimum period of time. There is no other way to check overall graph for all the available Goods. The above challenges lead to insecure result data, time consuming and missing business Intelligence. Even trolleys are not reliable for effective means of communication for wireless medium. Most often trolley is not used by many individuals for lesser item purchase list. In such scenario, such systems would not work. That is the major limitation of the such systems (J. Rezazadeh et al., 2018).

In this chapter we proposed a new concept which will be very fruitful for central systems. Most of the android application bar code scanner is used to for maintenance of bills transaction and product bills updating processes.

Microcontroller PIC16F877A 8-bit is used to control the entire system (Chandrasekar & Sangeetha, 2014) Traditional scanning from android phone now has been transformed by bar code scanner reader (Chaudhari, Gore, Kale, & Patil 2016). A common platform is used for communication as Bluetooth between MCU and bar code scanner. PLX-DAQ for parallel systems is the means for database maintenance on laptop or personal computer to make a communication by serial port (Rohith & Madhusudan, 2015). Record tracking like individual product cost, total cost of items, customer name, date and time for product purchase, payment method and complete transaction of sales is maintained by this system. RFID is used for the payment system not for the card users. Product, bill or payment-based information is displayed by LCD alphanumeric method. The complete procedure of the proposed system is illustrated in Figure 12.1. This approach could be useful for real world applications. After entering in shopping site, individual has to connect his smart phone with connecting device Bluetooth (Chaudhari, Gore, Kale, & Patil 2016). Afterwards he/she has to scan his card. MCU is responsible to forward the feedback and welcome note after verifying his identity. Trolley can be put off manually. Every picked item needs to be scanned by using their barcodes. There is no such limit has been set for scanning the no of

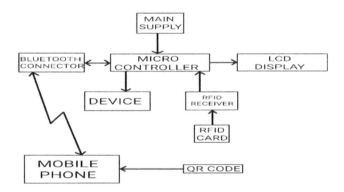

**FIGURE 12.1** Payment System Graph.

items for purchase list he/she can scan as many as items according to their needs and interest. At last, individual will receive a completion message by receiving a notification. Total amount shall be updated by central database and local transaction system as well. Information will also be sent to purchaser as well (Figure 12.2).

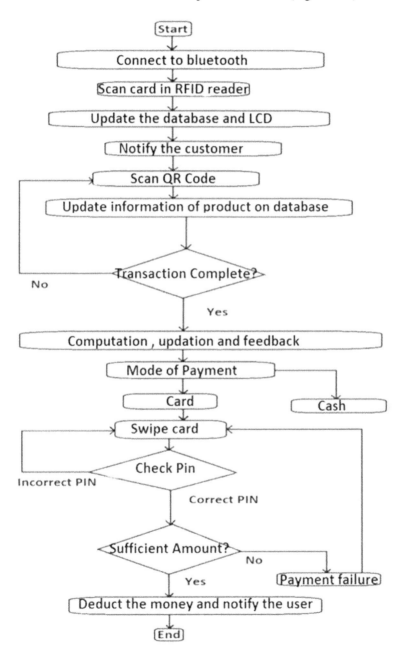

**FIGURE 12.2** Data Flow Chart of PIC16F877A Microcontroller.

PIC16F877A microcontroller belongs to PIC16 family and is 8-bit microcontroller developed by Microchip Technology (Chandrasekar & Sangeetha, 2014).

## 12.4 THREE-LAYER IOT ARCHITECTURE AND ITS COMPONENTS

Thinger.io is widely popular platform for online IoT. Internet of things is mainly used to allow for integration of many more devices to provide incorporation with cloud named as THINGER.IO. THINGER.IO cloud shows the real time data for processing. Ubuntu operating system is basically used for better usage of Thinger.io cloud generally used UP squared board. Real Time Light Sensor Value is read by the system and results are plotted in Time Series Graphically. Dark light sensed by the sensor in that case notification will be received electronically. For high performance and low power consumption UP Squared board is the deal for IoT applications.

Intel processor (N3350) or Pentium® processor (N4200) generally used UP Squared board (Figure 12.3).

### 12.4.1 Hardware Requirements

Sensor Tool KIT, Pi Board, UP Squared Board, HDMI Interface with Monitor Screen, Input Devices like Mouse and Keyboard having USB ports, A Visual Graphics Adapter or HDMI cable, For maintaining connection network connection along Internet access, UP Squared WiFi Enabled Kit.

### 12.4.2 Software Requirements

For the UP Square IoT Development Kit, an Ubuntu 16.04 Server Image is installed in the UP Squared board. To upgrade or refresh, follow UP Squared IoT Development Kit Ubuntu 16.04 Server Image. Gaining access to the UP Squared GPIO pins, verification and updation of Kernel is required to do in order to follow the instruction manually.

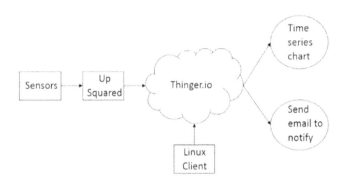

**FIGURE 12.3** Three-Layer IoT Architecture and Its Components.

# Intelligent Systems for IoT and Services

## 12.5 SCOPE OF THE SENSOR NETWORKED DEVICES

### 12.5.1 Scope of the Wireless Things

There is a flood of Internet of Things usage applications. Numerous "smart devices from miniscule chips to mammoth machines are available frequently to communicate with each other due to wireless Technology. IoT is moving very rapidly. Numerous objects in 2006 and more than 200 billion would be completed by the end of this year. Every human being is surrounded by IoT objects in their day to day lifes.

### 12.5.2 IoT and Its Market Segments

Big Industries, Business and health sector units are using IoT Devices widely not commonly used at respective homes and small mobile phones. This is widely use due to smart objects give these major industries the vital data they need to track inventory, manage machines, increase efficiency, save costs, and even save lives. By 2025, the total global worth of IoT technology could be as much as USD 6.2 trillion—most of that value from devices in health care (USD 2.5 trillion) and manufacturing (USD 2.3 trillion).

### 12.5.3 Experimental Analysis of the scanning Method

In this study, the Proteus software is used for result analysis and for verification the data results of the proposed method (Y. J. Zuo 2010). Figure 12.4 explains the experimental results of the models with different communication. Serial communication with different models is achieved by terminals. Various products and membership cards are scanned by common barcode scanning method. Afterwards payment needs be done, whenever the scanning process is over. Payment mode

**FIGURE 12.4** UP Squared Board.

depends on the individual wish and comfort level like online/card/. For card payment systems RFID cards are responsible for generating special 12 bytes code while swapping at the card at the machine (Xiong, Barash, & Frey 2011). Wireless transducer, antenna and encapsulating material are the major components of the card reading process. Two modes of the cards are available like active or passive mode. Chip power is used by active mode while magnetic field is used by passive mode of card reading system (J. Rezazadeh et al., 2018). Microcontroller receives the special code generated by card at the time of swipe is basically used for identification of individual like who is card owner and pin possess by individual. Flow chart of the scanning process is displayed in earlier section.

## 12.6 THE IOT CONTAINS AN ENORMOUS VARIETY OF CONNECTED OBJECTS

### 12.6.1 Tiny Stuff: Smart Dust Enormous Stuff: An Entire City

To diagnose problems in human body communicating devices like tiny computers smaller than a gain of sand be sprayed or injected like an intravenous vaccine to the environment anywhere to measure the chemical composition in soil. Dublin already dispersed fixed and mobile sensors throughout city to create a real time picture of what is happening and it will help the city react quickly in times of crisis.

### 12.6.2 Digital Locks

Smartphones plays an important role to grant or restrict access to employees and guest's top management of industries can change the codes frequently. Locking and unlocking of doors can be achieved by smart phones easily.

### 12.6.3 Smart Buildings

Brand-new buildings let owners and occupants "monitor, manage, and maintain all aspects of the building that impact operations, energy, and comfort," according to the Smart Buildings Institute, which has certified buildings in Saudi Arabia and San Salvador.

### 12.6.4 Data Collection by Existing Process

E Commerce **industry**/business/organizations are having more than 50,000 photographs of variety of products/Goods like shoes, dresses, pants and outerwear. Every task is matching with Labels to achieve matching some special tasks are assigned with every picture.

The perfect tag of a picture is used by applying the test set for special purposes. In Machine learning vocabulary it is a multilabel classification problem. Selection of a perfect tag among all the available pictures is done by many tasks.

Intelligent Systems for IoT and Services    221

At the time of examining the total purposes each and every picture cannot be allocated all the images individually.

Day by day, the lifestyle of today's generation is changing which enforce to have a greater number of shopping malls, supermarkets and retail outlets. Every Shopping area like a mall are crowded with people specially on weekends and festive seasons. During the sale/special offers/discount situation is even more over crowded. So therefor It is essential to have efficient billing management system in those places so that people coming for purchasing can avoid long queue on billing counter, save time and shop comfortably. By the requisite of such payment methods, we hereby "Smart Payment System and Goods Billing Management" is proposed to achieve above said goal.

## 12.7 COMMON PROBLEMS WITH DATA

Distorted training is one of the major problems. Whenever a particular picture from the test was recognized in label format either "camo" or "camouflage" by the label handlers. API Libraries for data analytics tools and artificial intelligence are building at customization level. To achieve the high performance across multiple architectures, like as GPU's, CPU's, FPG's and other accelerating systems. Optimization of the deep learning systems and high-performance python libraries can be achieved by acceleration of end-to-end machine learning systems and data science pipe lining.

## 12.8 WORKING METHODOLOGY

There is no need to check in and check out while using such kind of Goods Management and payment methods. To avoid waiting time and to wait in a line it is most advanced shopping management system. By using this we have to use just Google app to enter in to store purchase the products no waiting line etc. Self-driving cars, computer vision, sensor fusion are the examples of the deep algorithms. technology made Goods and management system free shopping experience at the time of check out. Whenever products are picked up or returned to the shelves and to keep entry in cart detected automatically by using such Payment systems. After purchasing products, you may leave the online store immediately without waiting time. Receipt will be forwarded to you electronically and your purchasing account will be updated immediate after product purchase. Bar code scanner of smart phone used well connected with the android application and connected to wireless systems for record updating, products, billing, voucher information in detailed view. For this every process for control management used PIC16F877A 8-bit Mi-controller. Scanning purposes methods known as Arduino Application used for pairing with barcode scanner with communicating device such as mobile phones. MCU and bar code scanner uses Blue tooth model as interface for communication between the devices. By using serial port with MCU for database record transaction handling PLX-DAQ is widely using by parallax. Record keeping is done easily by such Goods management systems and other activities like as product price, gross price of product.

Track record like as product cost, gross value of the product, customer information details like as name of customer, date of purchase an item, payment details as mode of payment done at the time of purchase etc. Radio Frequency Identification cards commonly used for payment mode. For displaying meaning full text like product bill/transaction details Alphanumeric LCD are popularly used. Benefits of implementing this approach has been described in various ways for utility of real-world entities.

After entering the shopping areas customers' needs to open application "Scan to Arduino" and connect their smart phones via blue tooth with counter for payment transactions.

Block diagram of the proposed method is shown in Figure 12.2.

After words he/she has to scan the shopping/membership card from smart communicating device like phone etc. Welcome note and feedback form will be received to user by MCU only after the customer identification. It will be considering as token for next shopping. Customer are having authority to pick the purchase items independently they have to pick and place the selected items in their respective trolleys after their scan by barcode reading system. Limited access is not provided to customer but in fact they can select the items as many as they want based upon their needs and desires. Only thing they have to keep in mind that they have to scan each and every item positively. When shopping is done by customer than he/she requires to notify the system about purchasing has been done by some textual means. Product purchase information like price, total purchase items, gross/total cost will be updated in Database immediately. Information about total cost/amount spent will also be notifies back to the customer (Figure 12.5)

## 12.9 A BUNCH OF CONVOLUTIONAL NEURAL NETWORKS

Here according to our system, we can consider 20 convolutional neural networks in one go. Architectures used in various variants are as follows:

Dense Net, Res Net, Inception, VGG. Image Net dataset contains duly weighted convolutional networks and weights are initialized ones. Here we consider different models in terms of cropping, normalizing, resizing and switching of color channels for preprocessing of data. All the models are well augmented. PyTorch framework used for neural network implementation.

## 12.10 SELECTION OF LOSS FUNCTION (TRAINING)

Bigger concern is to select the training loss function is a bit complex task. So many unique combinations of labels are there/available in the training data choices so that accuracy in output can be maintained with minimum loss factor. Only few labels can be fitted to the particular task. Actual reality is only a few coordinates were nonzero among the available. So that it's a hit and trial method to select a training loss function. As this was a multi label classification problem, choosing the popular cross entropy loss function:

# Intelligent Systems for IoT and Services

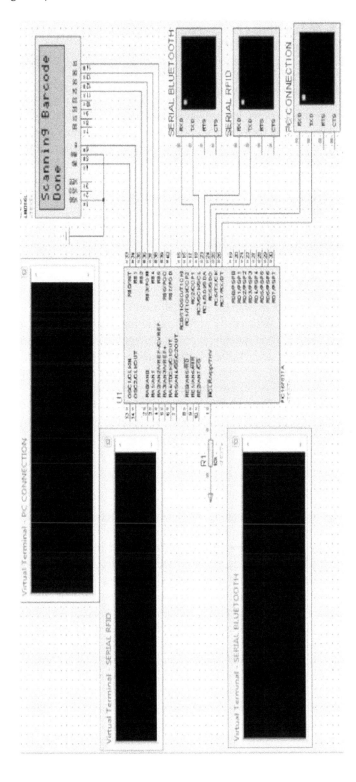

FIGURE 12.5  Bar Code Scanning Process.

## 12.11 IOT & MACHINE LEARNING ANOTHER APPLICATION FOR URBAN INTELLIGENCE

### 12.11.1 Bringing Data-Driven Management to Complex, Fast-Paced Environments

There are various industrious having large amount of data but due to non-availability of data mining tools and unawareness of techniques to infer the information by this data availability. Such type of industries like fast food chain is changing very rapidly and needs to calculate cost daily basis. ARDIC's IoT-Ignite united with Intel® technologies is a solution provider for data collection, filter, analysis of variables to provide a complete solution to automate the system. Availability of such type of solution a requisite now a days. Such type of industries is having good quality impact like customer satisfaction, working efficiency reduction in waste materials etc.

## 12.12 CHALLENGES

Over 40% of food produced globally is wasted every year. Fast food restaurants face particular challenges with costly waste, operational inefficiencies, and food safety. In order to deliver food quickly, it must be prepared in advance of customer orders, even though daily foot traffic is unpredictable. Consistency of product is essential, and this requires accurate measurement and temperature control, but managers often have little insight into the variables of equipment and staff activity throughout the day. Fast food operations need solutions that can scale across multiple locations and geographies, and holistic insight into growth challenges and opportunities than the drastic increase in growth rates of the roads and railway management systems.

### 12.12.1 Solution

ARDIC's IoT-Ignite platform running on Intel® architecture-based gateways is helping fast food companies better monitor food production and quality to improve customer experiences, increase operational efficiency, and minimize waste.

The innovative solution adapts a range of technologies to the specific requirements of the restaurant industry. Nonintrusive sensors and RFID tags collect and generate information relevant to different services. Smart scales and level counters add to the data pool. All data is gathered, filtered, and processed on a single Intel architecture-based gateway. Data parameters are determined in conjunction with the restaurant to ensure relevance.

Management is localized to each venue. Items are present in the cold room are likely to be expired then alert will be generated automatically and when waste accumulates after a certain limit and many other factors also considers like consumption trends, kitchen environment and situation as out of stock (Figure 12.6).

ARDIC is piloting the solution at an expanding number of a leading fast food franchise's restaurants in Turkey. RFID tags and readers, sensors, and smart

# Intelligent Systems for IoT and Services

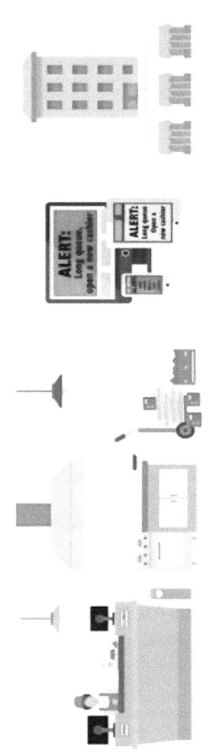

FIGURE 12.6 Bringing Data-Driven Management.

cameras transmit data to the gateway where it is mined for actionable edge intelligence and/or transmitted to the cloud for deeper analytics. Capabilities include counting customers, tracking the customer journey through the experience (e.g., wait times, peak cycles), tracking goods and inventory, and managing weight and environmental variables such as temperature and humidity. The solution includes smart scales for calculating food weight, a smart trash can that computes waste and unsold burgers, and a level counter that informs kitchen staff about changing occupancy levels, so they can adjust food production accordingly.

To maintain quality, the fast-food restaurant's staff must dispose of burgers not sold within 15 minutes. With the smart solution, food production is based on data, decreasing waste, and potentially saving millions of dollars globally each day. The open IoT-Ignite platform from ARDIC allows applications and features to be easily added or modified. The entire solution is highly automated, works with the existing franchise infrastructure, and does not disrupt ongoing operations.

With ARDIC, the fast-food franchise is gathering more data each day and mining it to realize its full value—with resulting increases in system efficiency. As the solution is deployed in more venues, cross-store data informs refinements and the opportunity for global deployment. As operations are standardized, they can be optimized and managed more efficiently.

## 12.13 KEY BENEFITS

ARDIC's IoT-Ignite and Intel architecture-based gateways support a wide range of benefits, including stock optimization, loss prevention, and lower OpEx. ARDIC works closely with its customers and their field operations to ensure analytics are pertinent, actionable, and meet business objectives. Maximum profit and efficiency is achieved by optimization. Comparison measures between such industries to tune optimum level of profits and to create a best practice for current and forthcoming situations. Adjustment of employees to identify the such industries which are not up to the performance mark. Increase in employee performance to do the daily counts and record keeping with minimum time consumption. Response at immediate effects after detection of abnormal activity like as higher rate of customer number with minimum sales. To gain a monopoly/good will in food industry it is a biggest advantage to maintain industry automated.

## 12.14 SAMPLE USE CASES FOR THE RESTAURANT INDUSTRY

### 12.14.1 Food Quality Control

Quality maintenance in public institutions and on store shelves is the key issue to food industry. Devery.io is one of the solutions, is a platform to provide digital identities for every available product. ID's are stored in the form of record on central format ledger which cannot be altered and history details can be retrieved later on whenever there is a requirement.

## 12.14.2 Automated Restaurant Management

Origin of the automated management is done for cost cutting to have lesser number of skilled cooks in the kitchen at various locations. Due to the semi-skilled cooks the quality cannot be maintained. More than 50% people thinks that the major cause of food poisoning is at restaurants rather than homes.

## 12.14.3 Data Sharing to Drive Commercial Purchases

Lack of information is the basic source of purchase of agricultural products and they are not willing to buy due to no information available regarding number of calories, storage condition and sugar content etc.

## 12.14.4 Distributed Marketplace for Commodities

Third parties are majorly concern with finalization of the deal between traditional market and commodity supplier of the various food manufacturing companies. Their role is to give priority according to farmer's needs. Governance checking, transparency, timeliness of deals makes the Goods management system faster for the farmers and suppliers.

## 12.14.5 Inventory Management

Usage of Reader Identification card Detection to track inventory or assets available at cloud Inventory tracking must be maintained by such industry. Placing orders at critical conditions by raising flags. Elimination of daily counts inventory entry. Tracking of reduction in loss. Interference of human up to minimum level with lesser errors. Accessing daily report and to know about business fluctuations.

## 12.14.6 Queue Management

Identification of customer behavior and to take prompt action for opening a new cash system to avoid to long waiting to get optimize results at back end. Queuing system is required to control people with minimal waiting time. To restrict the people entry during a specific visiting hour by queue management. Counting and tracking of customers journey like exit from the store is also done. Predefinition of logic by management of cashier and Generation of data for staff or labor availability is also done by the queue management.

## 12.14.7 Smart Waste

Track dumped precooked, cooked, and unsold food with near-real-time inventory tracking service and to monitor abnormal dumps. Generation of average dump levels for future evaluations. To check the available sold food items along with outgoing items between the stock room and waste bin.

### 12.14.8 More Use Cases

To achieve smart monetization companies can utilize the IoT Ignite to design, develop and deployment their own IoT services no double IoT ignite is an open platform. Today, IoT-Ignite enables services for a wide range of vertical market segments including retail, mobile services, agriculture, education, and energy. The opportunities to extend services to healthcare, manufacturing, and mobility are wide open.

The flexible, easy-to-deploy smart solution provides relevant data for optimization of connected industries.

## 12.15 HOW IT WORKS IN BRIEF

Processing and data filtering can be done by Intel architecture based smart gateway and to upload the data on IoT Ignite cloud. An RFID reader developed by Intel® Labs, smart scales, and smart cameras contribute meaningful data from throughout the restaurant venue.

### 12.15.1 RFID Tags

These tags are put on boxes of materials before leaving the distribution center. The boxes can be tracked in the distribution center, freezer room, and kitchen of the restaurant. Headquarters can see how many boxes are left in each branch and make sure that each type of goods is stored at its correct temperature. Cameras are used for queue management. Displays in the kitchen show a number from one to seven. If queues are long, this number is increased and staff in the kitchen can quickly respond.

To meet specific goals of deployment Wi Fi layer service is used and run algorithms build by ARCID.

All services available at ARCID platform. For Fast deployment of IoT solutions the IoT service providers uses the ARDIC IoT Ignite service. Enabling of networking, end to end security and interoperability can be achieved by Intel architecture-based gateways systems. Easy customization can be done by Android operating systems. Accessing od apps and data from anywhere or any communication device provided to customer. To provide a flexible, user friendly and it is basic necessity for deployment services for any type of market.

### 12.15.2 The Foundation for IoT

The ARDIC solution is just one example of how Intel works closely with the IoT ecosystem to help enable smart Internet of Things (IoT) solutions based on standardized, scalable, reliable Intel® architecture and software. These solutions range from sensors and gateways to server and cloud technologies to data analytics algorithms and applications. Intel provides essential end-to-end capabilities—performance, manageability, connectivity, analytics, and advanced security—to help accelerate innovation and increase revenue for enterprises, service providers, and the restaurant industry.

## 12.16 IOT SMART PUBLIC TRANSPORTATION SYSTEM

In India, a large number of people commute to their workplaces, home, market, hospitals, railway stations and much more, using public transport. Local trains, metros, trams, and buses are the common choices available to them. But unfortunately, not every city in our country is facilitated with all the above means of transportation. However, buses are ubiquitous. It could be either large buses, double decker buses or mini buses. So, people rely on them heavily. This also creates a heavy crowd in the buses at some specific peak hours and in some particular routes.

In our country, we have a large number of languages in use. And all of them have a different script of writing. For example, Gujarati, Kannada, Bangla, Hindi, Konkani and much more. These days people travel a lot across the country in various states either due to their work or for exploring the culture or as per their needs. Thus, people have to come across various languages and hence find it difficult to communicate. But when it comes to board a public transport bus, it becomes a bit difficult as the display board on the buses have the starting and ending point of the journey written in the local language and/or in the English language. A person who does not belong to the state finds it difficult to communicate with the people standing at the bus stop and enquire about his/her destination.

We find it daily and find it everywhere, that people are waiting at the bus stop for their respective bus. They get anxious, when will their respective bus come to their bus stop so that they can board it. They get even more anxious when it starts getting late in the night, particularly for women. It usually happens because of workplace/job commitments which everyone has to fulfill by the end of the day. And if they leave their office late at night, they get a little worried. They simply wait at the bus stop with a hope that they will get a bus sometime soon. They do not have an idea that, whether the last bus has left or not. They do not know how many buses are yet to come which can take them to their destination. They are also not sure that how much crowded the incoming bus is.

## 12.17 IOT BASED SMART ELECTRICITY DISTRIBUTION SYSTEM FOR INDUSTRIES AND DOMESTIC USAGES

Most of the Indian people are a part of organization or Employee so therefore they go to Office daily for job. MNC's or organizations provides facilities health car, salary account transaction etc. Indian Industries and organization are required to strictly follow the rules for electricity and frequent checking's are done to ensure that whether they are following the policies or not. In case Industries are not following the rules of electricity safety or installed cheap systems without quality check ISI mark. They hired untrained electricians due to less salary packages if their technician do not have proper training to maintain the electrical units then the chances of electrical short circuits are maximum extent. Technician by pass the line and provide supply to the industry without information at department level or to senior staff and repairs are done that may be highly dangerous to technician life. More than 10000 deaths occur due to short circuits/shocks. Due to risk factor at highest scale because technician could not follow the rules properly and repairs are done without information of senior staff. According to Indian government survey 1500 deaths in Bangalore for the last three years, Madhya Pradesh (MP) had 1664 deaths in 2014, Maharashtra 1373 deaths in 2014. It is requisite to deploy an optimal solution to prevent above stated problem.

## 12.18 REMOTE HEALTH MONITORING SYSTEM FOR FETAL AND MOTHER HEALTH MONITORING

Adverse outcomes of the pregnancy are fetal death after 20 weeks and more gestation. It's normal for babies to have quiet period in utero, and a temporary dip in activity could just mean that the baby is sleeping for low on energy. With this assumption mothers usually neglect the absence of fetal movement which may lead to fetal death or stillbirth due to this negligence. Develop an IoT based remote health monitoring system to avoid stillbirth and to save the life of the baby. Using this system pregnant women can know the fetal movement and take necessary measures.

## 12.19 SMART, SAFE AND CLEAN STREETS

Street light failure has become a problem in cities and rural areas. As the failure reports are sent very rare to government authorities, these failures are cleared only if a higher involvement takes place. Also, nowadays a major problem is drainage overflow. The entire street becomes wet, boggy and inconvenient to walk. The smell is also disgusting. Most places it is seen careless. Drainage Overflow causes inconvenient to walk in streets and even causes diseases. Some streets or roads have no movement of vehicles or persons, but the street lights are switched ON for the entire night. This is wastage of power. A person has to make a call to the corporation to clean the drainage. Develop an IoT solution to solve the problem mentioned above.

## 12.20 IOT BASED HEALTHCARE SOLUTION FOR TRACKING HUMAN SPINE MOVEMENT

In the medical field, especially around physiotherapy, a lot of treatment involves curing pain related to human body. Most of the time the patient is advised exercises and some special courses related to affected body part. Once a doctor gives a treatment to a patient it is only the doctor's judgement which tells the progress in the treatment. There is no way by which the doctor can track the improvement of the patient's body except by waiting for a long time to see if the patient is recovering or not. This process can be tedious for both the patient and the doctor as they must wait for long duration to determine if the recommended exercise is working or not.

Physiotherapists do not have any means to track their patients progress on a timely basis. As a result, it takes a long time for the doctor to realize if the prescribed exercises are working or not. This might result in delayed recovery in some cases and in cases where time is a critical issue it might cause challenges to both the doctor as well as the patient. Develop an IoT solution to solve the problem explained above.

Intelligent Systems for IoT and Services 231

## 12.21 IOT BASED SMART PARKING SYSTEM

In the existing parking systems, the driver needs to manually search for the parking area in the neighborhood thus resulting him in parking the car on the streets. This process takes time and effort, leading to traffic congestion on the roads and maybe the worst case of failing to find any parking space if the driver is driving in a new city with high vehicle density. To allow a smooth traffic, the system is needed such that it helps users to find parking space availability in the underground parking spots near the neighborhood, thus saving time and reducing traffic. The current intelligent parking systems do not provide the features for reserving parking spots & providing information about available parking spots nearby and do not solve the problem of vehicle refusal.

## 12.22 IOT BASED ASSISTING SYSTEM FOR MINERS TO PREVENT ACCIDENTS

Major safety concern in mining Industry is safety of miners the Owner has bear millions of rupees on miner's safety measures. It leads to huge loss to Mining Owner to maintain safety polices/rules. Workers of mining Industry plays a ruff life Mining Industry is highly hazardous. Famous problems occurred in mining Industry like as poisonous Gas leakage, floods and collapse etc. due to said problems several people killed every year. Only way out to monitor and capture real time data by monitoring properly above said problems of mining workers will overcome and provide working and hazard free environment to mining people. By placing intelligent sensors using IoT Technology.

Research data statistics says that mining industry working environment is combination of hazards and causes in the form of accidents. Gas explosion can be consequence of flood and collapse. Toxic contaminations could be released by fire. Usage of explosives might be a cause of earthquake that can destroy the mine working and trapping of miners as done by many miners like 33 numbers who stucked underground from August to October in the year 2010. Explosive related fatalities were caused by miners being too close to the blast followed by explosive fumes poisoning, misfires and premature blasts according to surveys conducted by National Institute for Occupational Safety and Health (NIOSH) in underground mines. Seismicity addition in the same due to induction of mines.

Miners may lead to harmful diseases due to Carbon dioxide, carbon monoxide, methane, ethane, Propene, etc. Such gasses present in the surroundings of mining area.

## 12.23 IOT BASED DISEASE PREVENTION SYSTEM FOR SMART HEALTHCARE

Imagine a world where the germs are caught before they affect humans. Diseases lead to the deterioration of the community health. There is a need to find solutions in the domain of health care to prevent the spread of diseases and provide better cure. Diseases are spread by bacteria, viruses and other pathogens which are mainly transmitted through air, water and food or vectors such as mosquitoes.

Hence keeping a track on the harmful biological contaminants present in these modes of disease transmission or on the vectors which spread diseases is necessary if we want to detect the presence of virus or bacteria before they affect humans. We can say that, today, there is a need for using the highly developed field of electronics for finding solutions to the sensitive world of health care.

Different diseases are spread through different vectors. For controlling the growth of vectors responsible for different diseases we would need different sensor and detection systems. We have proposed an idea to keep a track on growth of mosquito larvae which lead to diseases such as malaria. Hence, we would be concentrating on early detection of malaria risk in our proposed idea. More than 40% people of world's population lives in highly prone areas of malaria and one million people die from the Malaria annually. Malaria is a disease caused by a virus and spread through mosquitoes. Malarial mosquitoes develop in regions having stagnant water stores as these are favorable for their breeding. According to facts provided by WHO, early diagnosis and prompt treatment of malaria prevents death. Hence, if we could design a system that would detect the threat of malaria before it affects people, it would reduce deaths due to malaria by a high percentage. Develop an IoT solution to solve the problem discussed above.

## 12.24 INTELLIGENT TOLL COLLECTION SYSTEM FOR HIGHWAYS

In current toll collection systems, passengers have to stop at the toll terminals to confirm their identify and process payments. Identification of the vehicle owner is not known at the toll terminals. To avoid traffic congestion at the toll terminals, a new system is proposed that will automatically transfer information from the owner of the vehicle to the server interface situated within the toll terminal.

## 12.25 FORECASTING POTENTIAL HEALTH THREATS

In the field of medicine, it is very difficult for us to predict certain cases like heart attacks and allergic reactions, even when we have practically thousands of cases to learn from. This problem can potentially be solved by monitoring, a patient's, pulse, oxygen levels, temperature, blood pressure, stress level etc. and generating a real-time report of the patient and by comparing the data to the various medical cases, using machine learning algorithms to predict incidents. The algorithm will run all possible combinations based on the patient's history and the data gathered instantaneously and predict the outcomes. It will keep in account the time constraint and notify the operator, if any unfortunate events predicted.

## 12.26 CONCLUSION

In this chapter we have demonstrated a good payment method for item purchase system. This makes our process easy, efficient and reliable one. To overcome the long queue, we have proposed a bar code scanning method. By this way as a result in improvement of payment systems in terms of efficiency, complicity so that improvement will be shown in solving the real time problems.

Fast food restaurants offer a great example of the challenges of managing a fast-paced environment with numerous dynamic variables. With ARDIC and Intel, restaurants, along with many other industries, can improve decision making and operational efficiency based on accurate, near-real-time data generated by on-site, undisruptive technologies—from sensors and cameras to RFID tags and readers and smart gateways. ARDIC closely collaborates with its customers to ensure its solutions powered by Intel® technology meet evolving requirements and generate useful, actionable data.

## 12.27 LIMITATIONS/FUTURE WORK

Recent methods use the long queue method of the billing, hardware scanning of barcode. A major limitation of the current system is that if we connect it with Bluetooth devices in that case it will connect with limited number of devices only. Another disadvantage of it is that it carries up to smaller range area only. Up to short range it can work efficiently. In future it is a possibility to build such algorithms to resolve such issues to reduce the queue time as a result number of users would be used the scanning system concurrently at a time smoothly.

## REFERENCES

Anil, A. A. (2018). RFID based automatic shopping cart. *International Journal of Advanced Science and Research*, *1*, 39–45.

Chandrasekar, P., & Sangeetha, T. (2014, February). Smart shopping cart with automatic billing system through RFID and ZigBee. In *International Conference on Information Communication and Embedded Systems (ICICES2014)* (pp. 1–4). IEEE.

Chaudhari, M., Gore, A., Kale, R., & Patil, S. H. (2016, May). Intelligent shopping cart with goods management using sensors. *International Research Journal of Engineering and Technology (IRJET)*, *3*(5), 3243.

Choi, D., Chung, C. Y., & Young, J. (2019). Sustainable online shopping logistics for customer satisfaction and repeat purchasing behavior: Evidence from China. *Sustainability*, *11*(20), 5626.

Rajeshkumar, R., Mohanraj, R., & Varatharaj, M. (2016). Automatic barcode based bill calculation by using smart trolley. *International Journal of Engineering Science and Computing*, March 2019, 20264.

Rezazadeh, J., Sandrasegaran, K., & Kong, X. (2018, February). A location-based smart shopping system with IoT technology. In *2018 IEEE 4th World Forum on Internet of Things (WF-IoT)* (pp. 753–748). IEEE.

Rohith, S., & Madhusudan, C. (2015). Easy billing system at shopping mall using hitech trolly. *International Journal & Magazine of Engineering, Technology, Management and Research*, *2*(7), 1942.

Xiong, H. Y., Barash, Y., & Frey, B. J. (2011). Bayesian prediction of tissue-regulated splicing using RNA sequence and cellular context. *Bioinformatics*, *27*(18), 2554–2562.

Zeiler, M. D., & Fergus, R. (2013). Stochastic pooling for regularization of deep convolutional neural networks. arXiv:1301.3557v1 [cs.LG] 16 Jan 2013.

Zuo, Y. (2010). Survivable RFID systems: Issues, challenges, and techniques. *IEEE Transactions on Systems, Man, and Cybernetics, Part C (Applications and Reviews)*, *40*(4), 406–418.

# 13 Framework for the Adoption of Healthcare 4.0 – An ISM Approach

*Vinaytosh Mishra and*
*Sheikh Mohammed Shariful Islam*

## CONTENTS

| | | |
|---|---|---|
| 13.1 | Introduction | 235 |
| 13.2 | Literature Review | 236 |
| 13.3 | Research Methodology | 237 |
| 13.4 | Results and Discussions | 237 |
| 13.5 | ISM Model for Healthcare 4.0 Adoption | 243 |
| 13.6 | MICMAC Analysis | 243 |
| 13.7 | Policy Implication | 244 |
| 13.8 | Conclusion | 245 |
| 13.9 | Limitations and Future Direction | 245 |
| Acknowledgement | | 245 |
| References | | 245 |

## 13.1 INTRODUCTION

The recent advances in the production process and their automation have led to Industry 4.0 (Piccarozzi et al., 2018). The same in healthcare provisioning has brought in Healthcare 4.0 which broadly consists of incorporation of novel ICTs by a healthcare provider to manage more efficient and agile processes in healthcare delivery (Tortorella et al., 2019). The healthcare industry has been traditionally a digitalization deficient sector (Habran et al., 2018). Increased penetration of the internet, reduced cost of storage and computing, and falling cost of data usage is driving transformation in healthcare. Healthcare 4.0 is a term which draws its analogy from the term Industry 4.0 or Manufacturing 4.0 (Jayaraman et al., 2020). Before we discuss the factors affecting the adoption of Healthcare 4.0 in India, it is apt to examine the earlier version of technical advancements prior to Healthcare 4.0 and other versions of digitalization in healthcare.

Different researchers have categorized the various stages of healthcare transformation from technology deficient to Healthcare 4.0 differently. Chen et al. (2020) define healthcare 1.0 as the era post-industrial revolution, which was more focused on solving the problems related to public health issues like Sanitation, Vaccination, and Germ Theory. The beginning of the 20th century

**TABLE 13.1**
**Revolution in Indian Healthcare Industry**

| Stage | Years | Salient Point | Description |
| --- | --- | --- | --- |
| Healthcare 1.0 | 1970–1990 | Initial Stage | Lack of Resources, Public Health Innovations |
| Healthcare 2.0 | 1990–2006 | Health with Information Technology | Digital tracing introduced, MI Systems came into the picture |
| Healthcare 3.0 | 2006–2015 | Electronic Health Record (EHR) Systems | EHR system was started, Wearable and Implantable systems were used |
| Healthcare 4.0 | 2016 + | High-tech with High Touch | Cloud Computing, Machine Learning, Artificial Intelligence, and Real-time computing |

marked the commencement of the era of Healthcare 2.0, which continued till the 1980s. This era marked the beginning of mass production in healthcare, which included big hospitals, classification of healthcare into specialties, developments in pharmaceutical industries, and the discovery of antibiotics. The next advancement in healthcare was about the use of computers and imaging. The use of evidence-based medicine also started in this era. This era was termed as Healthcare 3.0. Healthcare 4.0 marked the beginning of Artificial Intelligence, Telemedicine, and Precision Medicine (PM). Kumari et al. (2018) have a different definition of various stages of the revolution in the Indian healthcare industry (Table 13.1).

Technological advancements like the penetration of the internet and smartphones, and the advent of various remote devices are driving the adoption of technology in healthcare (Papa et al., 2020). The objective of this research is twopronged: (1) To pinpoint the factors behind adoption of Healthcare 4.0 in India, and (2) To ascertain the relationships among these specific factors. The next section of the paper reviews the extant literature to identify different issues related to Healthcare 4.0 in India.

## 13.2 LITERATURE REVIEW

Emerging technologies can help transform the Healthcare sector involving all stakeholders (Aceto et al., 2020). Aceto et al. (2018) in their paper have emphasized the importance of the information technology infrastructure in the successful implementation of Healthcare 4.0. Tortorella et al. (2019), in their paper, observe that healthcare 4.0 has been more commonly found in the hospital's information flow and lacks a holistic approach. The implementation of Healthcare 4.0 requires high capital expenditure and a more skilled workforce (Tortorella et al., 2020). This is not possible without top management commitment and commitment of the government (Chanchaichujit et al., 2019; Thuemmler & Bai, 2017). The participation of top management is critical for the successful implementation of Healthcare 4.0 (Luthra & Mangla, 2018). Ocloo and Matthews (2016), in their paper, emphasize that healthcare needs to move from tokenism to empowerment of patients, if it wants to drive improvement at a large scale.

There is a need of adept and secure architecture to support the big data revolution in Healthcare (Abouelmehdi et al., 2018; Manogaran et al., 2017). Kumari et al. (2018) discuss the importance of infrastructure for the implementation of technologies like Fog Computing. If not suitably taken care of, digitalisation may lead to a healthcare data security issues related to Electronic Health Records (EHR) (Hathaliya et al., 2019; Hathaliya & Tanwar, 2020; Jayaraman et al., 2020).

## 13.3 RESEARCH METHODOLOGY

For identifying the Healthcare 4.0 factors, the study reviewed the articles published between the years 2005 to 2020. The database included in the search were PubMed, EBSCO, and Google Scholar. The keywords used for searching the literature were (1) Factor affecting Healthcare 4.0 (2) Factors affecting digitalization of healthcare, and (3) Industry 4.0 and Health. A total of six factors were identified based on SLR and focus group discussion. The approach for determining the factors affecting Healthcare 4.0 adoption is depicted in Figure 13.1.

The methodology used in this study is ISM Modelling. ISM begins with the identification of the factors, and then the study uses the group decision-making technique to establish the relationship between these factors as put down in Table 13.2.

Once transitivity is taken in to account, the Final Reachability Matrix (FRM) was tabulated. The partitioning of the elements and extraction of the structural model called ISM is derived from the FRM. Using a relationship from FRM, a digraph is drawn, and transitivity links are removed. Once the digraph is drawn, the next step is to replace nodes with relationship statements to convert it into ISM Model. Finally, the model is checked for conceptual consistency, and necessary changes are made, if required. To incorporate group decision making for deciding the contextual relationship between two factors, a focus group comprising six experts was formed. The details of the six experts approached for this study are listed in Table 13.3.

Figure 13.2 gives a bird's eye view of ISM Modelling.

## 13.4 RESULTS AND DISCUSSIONS

The review of extant literature helped us in finding the six factors for the study. The details of the finding of the SLR are listed in Table 13.4.

Once the factors were decided upon, Structural Self Interaction Matrix (SSIM) was developed. The following notations were used for preparing SSIM (Table 13.5)

V: Factor i influences factor j
A: Factor i is influenced by factor j
X: Factor i and j factor each other
O: Factor i and j do not factor each other

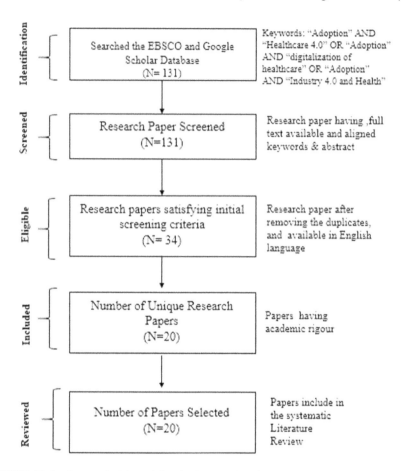

**FIGURE 13.1** Approach Adopted for the Systematic Literature Review.

### TABLE 13.2
### Rules of Transformation

| If the (i,j) entry in the SSIM is | Entry in the Initial Reachability Matrix | |
|---|---|---|
| | (i,j) | (j,i) |
| V | 1 | 0 |
| A | 0 | 1 |
| X | 1 | 1 |
| O | 0 | 0 |

## TABLE 13.3
## Details of the Focus Group

| SN | Expert | Specialty | Education and Experience |
|---|---|---|---|
| 1 | Expert 1 | Researcher | Ph.D., Healthcare Researcher |
| 2 | Expert 2 | Endocrinologist | MBBS, MD, Practicing doctor |
| 3 | Expert 3 | Cardiologist | MBBS, MD, Practicing doctor |
| 4 | Expert 4 | Radiologist | MBBS, MD, Practicing doctor |
| 5 | Expert 5 | Dentist | BDS, MDS, Practicing doctor |
| 6 | Expert 6 | Pediatric Surgeon | MBBS, MS, Practicing doctor |

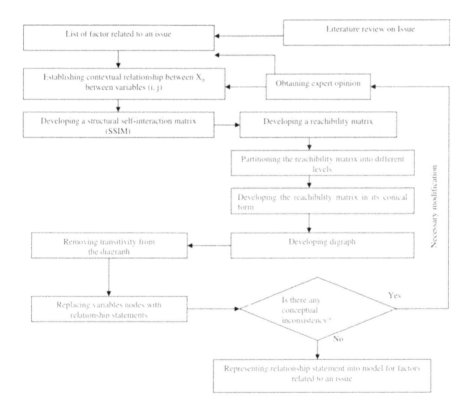

**FIGURE 13.2** Summary of ISM Modelling.

Using the transformation rule mentioned in Table 1 V, A, X, and O were replaced with 0 and 1 values. The resulting Initial Reachability Matrix (IRM) is given in Table 13.6.

The Final Reachability Matrix (FRM) was derived from the Initial Reachability Matrix after the application of the concept of transitivity. The resultant FRM is given in Table 13.7.

## TABLE 13.4
### Summary of Reviewed Literature

| Factors | Brief Description | Reference |
| --- | --- | --- |
| Government's Effort (**GE**) | Government's effort in developing ICT infrastructure and policies. | Agarwal et al. (2010), Anderson and Agarwal (2011), Angst and Agarwal (2009), Hillestad et al. (2005), Kumar et al. (2020) |
| Financial Investment (**FI**) | Financial investment is required as the adoption of healthcare requires advanced devices and IT infrastructure. | Hathaliya and Tanwar (2020), Luthra and Mangla (2018), Tanwar et al. (2020), Thuemmler and Bai (2017) |
| Empowered Customer (**EC**) | The customer is becoming aware of the new technologies, which is culminating in increased adoption. | Anshari (2019), Pousttchi and Dehnert (2018), Rantala and Karjaluoto (2016), Thorun et al. (2017) |
| Customer-Centric Care (**CC**) | Healthcare is becoming more customer-centric, and it necessitates the adoption of Healthcare 4.0. | Laurenza et al. (2018), Mettler (2017), Nokkala and Dahlberg (2018) |
| Data-Revolution (**DR**) | The availability of real-time data is driving Healthcare 4.0. | Ghassemi et al. (2015), Koster et al., (2016), Nokkala and Dahlberg (2018) |
| Data Privacy and Security (**DS**) | Data security and privacy are essential to allay the apprehensions of the patients adopting Healthcare 4.0. | Hathaliya et al. (2019), Hathaliya and Tanwar (2020), Jayaraman et al. (2020) |

## TABLE 13.5
### Structural Self-Interaction Matrix for Factors

| SN | Factor Affecting Adoption | DS | DR | CC | EC | FI | GE |
| --- | --- | --- | --- | --- | --- | --- | --- |
| 1 | Government's Effort (**GE**) | V | V | O | V | O | |
| 2 | Financial Investment (**FI**) | V | V | O | O | | |
| 3 | Empowered Customer (**EC**) | V | V | V | | | |
| 4 | Customer-Centric Care (**CC**) | V | V | | | | |
| 5 | Data Revolution (**DR**) | V | | | | | |
| 6 | Data Privacy and Security (**DS**) | | | | | | |

After the FRM is prepared, the reachability and antecedent set for each factor were found. The factor having the same reachability set and intersection set was eliminated at Level-I. Once the Level-I is decided, all factors used for the Level-I are removed. Now the same exercise is repeated for the other remaining factors.

# Framework for Adoption of Healthcare 4.0

**TABLE 13.6**
**IRM for the ISM Model**

| SN | Factor Affecting Adoption | DS | DR | CC | EC | FI | GE |
|---|---|---|---|---|---|---|---|
| 1 | Government's Effort (GE) | 1 | 1 | 0 | 1 | 0 | 1 |
| 2 | Financial Investment (FI) | 1 | 1 | 0 | 0 | 1 | 0 |
| 3 | Empowered Customer (EC) | 1 | 1 | 1 | 1 | 0 | 0 |
| 4 | Customer-Centric Care (CC) | 1 | 1 | 1 | 0 | 0 | 0 |
| 5 | Data Revolution (DR) | 1 | 1 | 0 | 0 | 0 | 0 |
| 6 | Data Privacy and Security (DS) | 1 | 0 | 0 | 0 | 0 | 0 |

**TABLE 13.7**
**FRM for the ISM Model**

| SN | Factor Affecting Adoption | DS | DR | CC | EC | FI | GE | Driving Power |
|---|---|---|---|---|---|---|---|---|
| 1 | Government's Effort (GE) | 1 | 1 | 1* | 1 | 0 | 1 | 4 |
| 2 | Financial Investment (FI) | 1 | 1 | 0 | 0 | 1 | 0 | 3 |
| 3 | Empowered Customer (EC) | 1 | 1 | 1 | 1 | 0 | 0 | 4 |
| 4 | Customer-Centric Care (CC) | 1 | 1 | 1 | 0 | 0 | 0 | 3 |
| 5 | Data Revolution (DR) | 1 | 1 | 0 | 0 | 0 | 0 | 2 |
| 6 | Data Privacy and Security (DS) | 1 | 0 | 0 | 0 | 0 | 0 | 1 |
| | **Dependence Power** | 6 | 5 | 3 | 2 | 1 | 1 | |

**TABLE 13.8**
**Iteration-1 for LP**

| SN | Factor Affecting Adoption | Reachability Set | Antecedent Set | Intersection Set | Level |
|---|---|---|---|---|---|
| 1 | Government's Effort (GE) | 1,3,4,5,6 | 1 | 1 | |
| 2 | Financial Investment (FI) | 2,5,6 | 2 | 2 | |
| 3 | Empowered Customer (EC) | 3,4,5,6 | 1,3 | 3 | |
| 4 | Customer-Centric Care (CC) | 4,5,6 | 1,3,4 | 4 | |
| 5 | Data Revolution (DR) | 5,6 | 1,2,3,4,5 | 5 | |
| 6 | Data Privacy and Security (DS) | 6 | 1,2,3,4,5,6 | 6 | Level –I |

The different iterations of level portioning are shown in Table 13.8 to Table 13.11. The level partitioning (LP) for the Iteration-1 is given in Table 13.8.

Once the Level-I is found, all factors used to find it were removed from all sets. This elimination resulted in the same reachability and intersection set for Factor-5.

## TABLE 13.9
## Iteration-2 for LP

| SN | Factor Affecting Adoption | Reachability Set | Antecedent Set | Intersection Set | Level |
|---|---|---|---|---|---|
| 1 | Government's Effort (**GE**) | 1,3,4,5 | 1 | 1 | |
| 2 | Financial Investment (**FI**) | 2,5 | 2 | 2 | |
| 3 | Empowered Customer (**EC**) | 3,4,5 | 1,3 | 3 | |
| 4 | Customer-Centric Care (**CC**) | 4,5 | 1,3,4 | 4 | |
| 5 | Data Revolution (**DR**) | 5 | 1,2,3,4,5 | 5 | Level-II |

## TABLE 13.10
## Iteration-3 for LP

| SN | Factor Affecting Adoption | Reachability Set | Antecedent Set | Intersection Set | Level |
|---|---|---|---|---|---|
| 1 | Government's Effort (**GE**) | 1,3,4 | 1 | 1 | |
| 2 | Financial Investment (**FI**) | 2 | 2 | 2 | Level-III |
| 3 | Empowered Customer (**EC**) | 3,4 | 1,3 | 3 | |
| 4 | Customer-Centric Care (**CC**) | 4 | 1,3,4 | 4 | Level-III |

## TABLE 13.11
## Iteration-4 for LP

| SN | Factor Affecting Adoption | Reachability Set | Antecedent Set | Intersection Set | Level |
|---|---|---|---|---|---|
| 1 | Government's Effort (**GE**) | 1,3 | 1 | 1 | |
| 3 | Empowered Customer (**EC**) | 3 | 1,3 | 3 | Level-IV |

Hence data revolution forms Level II. The level partitioning for the Iteration-2 is given in Table 13.9.

Since Factor-5 is used in Level-II, it was removed from all sets to find out the next level. Removal of Factor-5 resulted in the same reachability and intersection set for the Factor-2 and Factor-4. Hence, these factors were used in Level-III. The level partitioning for the Iteration-3 is given in Table 13.10.

Once the Level-III was decided, Factor-2 and Factor-4 were removed from all remaining sets. This resulted in the same reachability and intersection set for the Factor-3. Hence, the Level-IV contains Factor-3. The level partitioning for the Iteration-4 is given in Table 13.11.

Framework for Adoption of Healthcare 4.0

**TABLE 13.12**
**Iteration-5 for LP**

| SN | Factor Affecting Adoption | Reachability Set | Antecedent Set | Intersection Set | Level |
|---|---|---|---|---|---|
| 1 | Government's Effort (**GE**) | 1 | 1 | 1 | Level-V |

After removing Factor-3 from all sets, the only factor remaining was Factor-1, and it was included in the Level-V. Since none of the Factors remains, we stop the iteration. The level partitioning for the Iteration-5 is given in Table 13.12.

## 13.5 ISM MODEL FOR HEALTHCARE 4.0 ADOPTION

After completing the level partitioning step, the digraph was drawn. The relationship between various factors was established using the Initial Reachability Matrix (IRM). The model hierarchically organizes the factors affecting the adoption. The factors placed at a higher level help drive the factors at a lower level. The directional relationship decides the arrow direction between the factors. The model shows that the factors at Level-V, i.e. Government's Effort is the controlling factor that drives the factors at a lower level. The result of the model suggests that Government's Effort in building the policies and infrastructure is the most crucial factor out of the identified factors, and it drives the rest of the factors i.e. Empowered Customer, Data Revolution, Data Privacy, and Security. The Factor at the next level is the Empowered Customer, which drives factors such as Data Revolution, Data Privacy, and Security. The factors at Level –III are Financial Investment and Customer-Centric Care and these factors drive factors such as Data Revolution, Data Privacy, and Security. The factor Data Revolution is at Level-II and leads towards the establishment of another factor i.e. Data Privacy and Security, which is at Level-I. The ISM Model for the study is depicted with the help of the following diagram (Figure 13.3).

## 13.6 MICMAC ANALYSIS

The **MICMAC** Analysis is a graph plotted between the dependence power and the driving power of each factor. This analysis is shown in Figure 13.4. The graph illustrates the factors categorised in four cluster groups, namely Linkage, Independent, Autonomous, and Dependent. The factors in the Autonomous cluster have low driving and dependence power, and none of the factors falls entirely under this cluster. The factor Financial Investment falls at the boundary of Autonomous and Independent Clusters. The factors positioned under the Independent Cluster category have high driving power and low dependence power. The factors finding a place in this cluster are Government Effort, Empowered Customer, and Financial Investment. These factors are imperative as these factors drive the growth of other factors. The factors falling under the Linkage Cluster have high dependence and driving power and none of the factors falls under this cluster. The factor Customer-Centric Care has

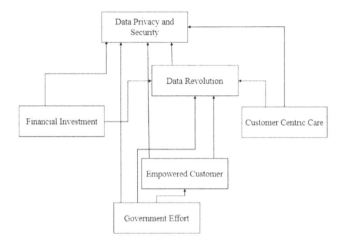

**FIGURE 13.3** ISM Model for the Study.

equal driving as well as dependence power. The last observed cluster is the Dependent Cluster, and the factors falling under this cluster are Data Revolution, and Data Privacy and Security. The factors falling under this cluster have high dependence and low driving power.

Figure 13.4 depicts MICMAC Analysis for factors responsible for adoption of Healthcare 4.0.

## 13.7 POLICY IMPLICATION

Healthcare in India faces the problem of inequity. Health equity is achieved when quality care is provided at an affordable cost. The adoption of technology can help in solving this perennial problem. The use of telemedicine has seen increased adoption after the outbreak of COVID-19, but the government needs to build infrastructure and guidelines for the continued use of telemedicine in healthcare. The adoption of Healthcare 4.0 requires financial investment.

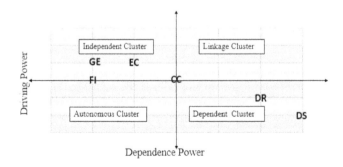

**FIGURE 13.4** Driving and Dependence Power Plot.

Considering the meagre expenditure of the government in this sector, it is hard to imagine that it will invest in the technological infrastructure requisite for Healthcare 4.0. Private players provide the majority of ambulatory care in India, and the government cannot achieve the set outcomes without involving private players in it. Alternative financing mechanisms like health insurance should be used for financing the investment required for the adoption of Healthcare 4.0. The government should build policies for data privacy and security to allay the apprehensions of the patient using Healthcare 4.0. Educating patients about the use of Healthcare 4.0 can further improve their adopting it.

## 13.8 CONCLUSION

The healthcare sector in developing countries like India needs a massive digitalization drive. The ubiquitous devices and data revolution have been strategic in driving the digitalization process, and this revolution is being termed as Healthcare 4.0 by healthcare experts and digital scientists. This study identifies six factors affecting the adoption of Healthcar4.0 in India. These factors are the Government's Effort, Empowered Customer, Customer-Centric Care, Financial Investment, Data Revolution, and Data Privacy and Security. Out of these factors, the Government's Effort in building policy and infrastructure for Healthcare 4.0 is most important. Factors such as Empowered Customer today is instrumental in making healthcare customer-centric, hence driving the adoption of technologies towards that goal. The study concludes that the adoption of Healthcare 4.0 is capital intensive, and financial Investment is required to drive Data Revolution. The study also observes that Data Privacy and Security is significant in the adoption of Healthcare 4.0.

## 13.9 LIMITATIONS AND FUTURE DIRECTION

The study identifies six factors affecting the adoption of Healthcare 4.0 in the country. The inclusion of extra keywords and databases can provide a more comprehensive list of the factors. The experts who participated in the study represent only two major cities. A future study can include experts from different parts of India to make the study more representative. A future study may be planned to explore further the factors identified in this study.

## ACKNOWLEDGEMENT

The authors are grateful for the support provided by FORE School of Management, New Delhi.

## REFERENCES

Abouelmehdi, K., Beni-Hessane, A., & Khaloufi, H. (2018). Big healthcare data: Preserving security and privacy. *Journal of Big Data*, 5(1), 1. 10.1186/s40537-017-0110-7

Aceto, G., Persico, V., & Pescapé, A. (2018). The role of Information and Communication Technologies in healthcare: Taxonomies, perspectives, and challenges. *Journal of Network and Computer Applications*, *107*, 125–154.

Aceto, G., Persico, V., & Pescapé, A. (2020). Industry 4.0 and health: Internet of things, big data, and cloud computing for healthcare 4.0. *Journal of Industrial Information Integration*, *18*, 100129. 10.1016/j.jii.2020.100129

Agarwal, R., Gao, G., DesRoches, C., & Jha, A. K. (2010). Research commentary—The digital transformation of healthcare: Current status and the road ahead. *Information Systems Research*, *21*(4), 796–809.

Anderson, C. L., & Agarwal, R. (2011). The digitization of healthcare: Boundary risks, emotion, and consumer willingness to disclose personal health information. *Information Systems Research*, *22*(3), 469–490.

Angst, C. M., & Agarwal, R. (2009). Adoption of electronic health records in the presence of privacy concerns: The elaboration likelihood model and individual persuasion. *MIS Quarterly*, *33*(2), 339–370.

Anshari, M. (2019). Redefining Electronic Health Records (EHR) and Electronic Medical Records (EMR) to promote patient empowerment. *International Journal on Informatics for Development*, *8*(1), 35–39.

Chanchaichujit, J., Tan, A., Meng, F., & Eaimkhong, S. (2019). An introduction to healthcare 4.0. In *Healthcare 4.0* (pp. 1–15). Palgrave Pivot.

Chen, C., Loh, E. W., Kuo, K. N., & Tam, K. W. (2020). The times they are a-Changin'– healthcare 4.0 is coming! *Journal of Medical Systems*, *44*(2), 1–4.

Ghassemi, M., Celi, L. A., & Stone, D. J. (2015). State of the art review: The data revolution in critical care. *Critical Care*, *19*(1), 1–9

Habran, E., Saulpic, O., & Zarlowski, P. (2018). Digitalization in healthcare: An analysis of projects proposed by practitioners. *British Journal of Healthcare Management*, *24*(3), 150–155.

Hathaliya, J. J., & Tanwar, S. (2020). An exhaustive survey on security and privacy issues in Healthcare 4.0. *Computer Communications*, *153*, 311–335.

Hathaliya, J. J., Tanwar, S., Tyagi, S., & Kumar, N. (2019). Securing electronics healthcare records in healthcare 4.0: a biometric-based approach. *Computers & Electrical Engineering*, *76*, 398–410.

Hillestad, R., Bigelow, J., Bower, A., Girosi, F., Meili, R., Scoville, R., & Taylor, R. (2005). Can electronic medical record systems transform health care? Potential health benefits, savings, and costs. *Health Affairs*, *24*(5), 1103–1117.

Jayaraman, P. P., Forkan, A. R. M., Morshed, A., Haghighi, P. D., & Kang, Y. B. (2020). Healthcare 4.0: A review of frontiers in digital health. *Wiley Interdisciplinary Reviews: Data Mining and Knowledge Discovery*, *10*(2), e1350.

Koster, J., Stewart, E., & Kolker, E. (2016). Health care transformation: A strategy rooted in data and analytics. *Academic Medicine*, *91*(2), 165–167.

Kumar, A., Krishnamurthi, R., Nayyar, A., Sharma, K., Grover, V., & Hossain, E. (2020). A novel smart healthcare design, simulation, and implementation using Healthcare 4.0 processes. *IEEE Access*, *6*(1), 118433–118471. DOI: 10.1109/ACCESS.2020.3004790

Kumari, A., Tanwar, S., Tyagi, S., & Kumar, N. (2018). Fog computing for Healthcare 4.0 environment: Opportunities and challenges. *Computers & Electrical Engineering*, *72*, 1–13.

Laurenza, E., Quintano, M., Schiavone, F., & Vrontis, D. (2018). The effect of digital technologies adoption in healthcare industry: A case-based analysis. *Business Process Management Journal*, *24*(5), 1124–1144.

Luthra, S., & Mangla, S. K. (2018). Evaluating challenges to Industry 4.0 initiatives for supply chain sustainability in emerging economies. *Process Safety and Environmental Protection*, *117*, 168–179.

Manogaran, G., Thota, C., Lopez, D., & Sundarasekar, R. (2017). Big data security intelligence for healthcare industry 4.0. In *Cybersecurity for Industry 4.0* (pp. 103–126). Springer.

Mettler, M. (2017). Focus on the end-user: The approach to consumer-centered healthcare. In *Health 4.0: How Virtualization and Big Data Are Revolutionizing Healthcare* (pp. 109–123). Springer.

Nokkala, T., & Dahlberg, T. (2018). Data Federation in the Era of Digital, Consumer-Centric Cares, and Empowered Citizens. *In International Conference on Well-Being in the Information Society* (pp. 134–147). Springer.

Ocloo, J., & Matthews, R. (2016). From tokenism to empowerment: Progressing patient and public involvement in healthcare improvement. *BMJ Quality & Safety, 25*(8), 626–632.

Papa, A., Mital, M., Pisano, P., & Del Giudice, M. (2020). E-health and wellbeing monitoring using smart healthcare devices: An empirical investigation. *Technological Forecasting and Social Change, 153*, 119226.

Piccarozzi, M., Aquilani, B., & Gatti, C. (2018). Industry 4.0 in management studies: A systematic literature review. *Sustainability, 10*(10), 3821.

Pousttchi, K., & Dehnert, M. (2018). Exploring the digitalization impact on consumer decision-making in retail banking. *Electronic Markets, 28*, 265–286.

Rantala, K., & Karjaluoto, H. (2016, October). *Value co-creation in health care: Insights into the transformation from value creation to value co-creation through digitization* [Conference session]. 20th International Academic Mindtrek Conference, ACM, pp. 34–41.

Tanwar, S., Parekh, K., & Evans, R. (2020). Blockchain-based electronic healthcare record system for healthcare 4.0 applications. *Journal of Information Security and Applications, 50*, 102407.

Thorun, C., Vetter, M., Reisch, L. A., & Zimmer, A. K. (2017). Indicators of consumer protection and empowerment in the digital world: Results and recommendations of a feasibility study. Institute of Consumer Policy, Available at: https://www.bmjv.de/G20/DE/ConsumerSummit/_documents/Downloads/Studie.pdf. (Assessed on January 12, 2019).

Thuemmler, C., & Bai, C. (Eds.). (2017). *Health 4.0: How virtualization and big data are revolutionizing healthcare.* Springer.

Tortorella, G. L., Fogliatto, F. S., Espôsto, K. F., Vergara, A. M. C., Vassolo, R., Mendoza, D. T., & Narayanamurthy, G. (2020). Effects of contingencies on healthcare 4.0 technologies adoption and barriers in emerging economies. *Technological Forecasting and Social Change, 156*, 120048. 10.1016/j.techfore.2020.120048

Tortorella, G. L., Fogliatto, F. S., Mac Cawley Vergara, A., Vassolo, R., & Sawhney, R. (2019). Healthcare 4.0: Trends, challenges and research directions. *Production Planning & Control, 31*(15), 1245–1260. 10.1080/09537287.2019.1702226

# Section C

*Impact of Disruptive Technologies in Society 5.0*

# 14 E-commerce Security for Preventing E-Transaction Frauds

*Reshu Agarwal, Manisha Pant, and Shylaja Vinaykumar Karatangi*

## CONTENTS

| | | |
|---|---|---|
| 14.1 | Introduction | 251 |
| 14.2 | Related Research | 252 |
| 14.3 | Proposed Approach | 254 |
| 14.4 | Features of E-Commerce Website | 254 |
| 14.5 | Security Considerations for E-Transactions | 255 |
| 14.6 | Types of Biometric Security | 256 |
| | 14.6.1 Physiological | 256 |
| | 14.6.2 Behavioral | 257 |
| 14.7 | Techniques Used in E-Commerce for Security | 257 |
| 14.8 | Data Analysis | 260 |
| 14.9 | Limitations and Challenges | 261 |
| 14.10 | Conclusion and Future Scope | 262 |
| References | | 262 |

## 14.1 INTRODUCTION

Electronic Commerce is goods and services exchanged for money on the Internet. Any seller or buyer of products through web is a participant of ecommerce. For example, versatile commerce, electronic subsidizes move, production network the executives, online exchange handling. Electronic Data Interchange (EDI) is a system for recording web-based business-to-business transactions like invoicing and ordering that makes paper-based recording outdated. There is a need to implement security measures to guard against fraud in online exchanges. With electronic commerce becoming more and more popular with customers, both B-to-C and B-to-B, and as people are moving more towards websites to purchase products given there are no "open" or "closed" signs dangling on websites, some extortion location models are proposed to forestall the misrepresentation of purchasers or vendors, and secure people and their trust to make the web a safer place to roam about. The current extortion location prototypes are utilized to check dealer respectability. Further, as many purchasers avoid using net banking, it becomes difficult to deal with ecommerce framework without its provisioning a wide range

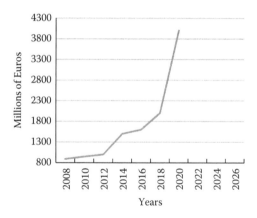

**FIGURE 14.1** E-Commerce Fraud Increases from 2008–2021.

of information. According to the Euromonitor International in the Europe, Middle East and Africa regions, as the number of fake transactions grows the amount of financial plans to secure against frauds grows too in euros from 2008 to 2021 as shown in Figure 14.1. To let a fraudster carry out a business transaction, it would suffice to tinker with the website's software underlying the business's operations but this could be successful in small network; yet, from the perspective of bigger network like a website having a worldwide reach, a fraudster will find it hard to carry out his nefarious activities without getting caught, since the fraud management process itself is divided and are seen as independent exercises. One, the significant verification issue in the traditional framework has a PIN and a secret password which can easily be fudged or tampered, so, to overcome this problem a proven model is suggested that can be useful for the readers. Internet business is big business since it has many advantages over brick and mortar stores though most of the latter have an online presence to catch on to the trend. It's easier to find products one needs on a website with quick and easy search functions to help us with our purchases. Self-ruling advisors, private endeavors, and gigantic associations have all benefitted by web based business seen on a scale that was impossible with conventional retail. In securing ecommerce against frauds, the fake avoidance prototype of ecommerce is utilized not just for securing the virtual stores and merchandise; but to forestall fraudsters and their use of fake credit cards and other activities. It not only has the option to discover fraud vendors functioning online, but fraud buyers too. It checks the physical address of the shippers that is known to the client, and helps to build client's trust and belief in online transactions by validating online stores and separating the wheat from the chaff.

## 14.2 RELATED RESEARCH

Several researchers have taken enormous strides in tackling e-commerce fraud security by utilizing Artificial Intelligence, deep learning, big data, and computational insight advancements. Scams during online payment has drawn a great deal attention and many researchers proposed proactive techniques for dealing with it (Aleskerov et al., 1997; Chiu & Tsai, 2004). Initially Ghosh and Reilly (1994) proposed a model for scam discovery by using neural network. Stolfo et al. (1997) have recommended a scam recognition framework utilizing various methods for detecting fake transactions. Caldeira et al. (2014) apply and assess computational insight strategies based on data mining and artificial intelligence to recognize scams in electronic exchanges, especially, in credit card activities. Guo and Li (2008) used a model to prove that neural network algorithms can be used for detecting scams on online exchanges. In this prototype, the sequence of tasks in master card exchange handling utilizing a confidence based neural network is proposed. A neural network is first prepared with artificial intelligence. In case, a prospective transaction using credit card isn't acknowledged by the prepared neural network prototype with adequately low confidence, it is viewed as fraud. Wang (2010) uses various mining algorithms such as decision tree, clustering, association rule mining for fraud detection. Mishra and Dash (2014) used data mining techniques such as decision tree and various complex algorithms to find out the pattern on complex and large data-sets. Based on this pattern, credit card fraud can be detected. Kataria and Nafis (2019) compare hidden markov model, deep learning, and neural network models used for detecting scam in online exchanges. Further, as online payment system through Alipay and WeChat Pay is becoming popular, fraud cases are proliferating. Wang et al. (2018) describe scam detection using Hidden Markov model where they utilized the k-means method to represent the exchange sum and recurrence order of a bank balance and was used to build and test the model. A HMM is at first trained with the typical conduct of a record. In the event that a processing credit card transaction isn't acknowledged by the prepared model with adequately high likelihood, it is viewed as fraudulent. Khan et al. (2012) used a clustering model to classify the legal and fraudulent transactions and also conducted experimental studies to show that their results are more valid. Sungkono and Sarno (2017) proposed a method for detecting fake recognition using Coupled Hidden Markov Model. The proposed strategy selects techniques using the exercises and structures a guide model of fraud detection utilizing probabilities of Coupled Hidden Markov Model. The test results show that the proposed technique gets a proper guide model of fraud detection. This chapter, in addition, exhibits an acquired model utilizing proposed strategy improving validity over an obtained model utilizing Map Miner Method. Further, false cases towards online insurance include different gatherings, for example, purchasers, merchants, and express organizations, and they could prompt heavy monetary losses (Abinayaa et al., 2020; Alsharif et al., 2018; Dalvi et al., 2016; Sailusha et al., 2020). Thus, it revealed the nexus of relations behind fraudsters and helped recognize fraudulent cases, Chen et al. (2019) built up a huge scope insurance fraud recognition framework. They examined various graphs to facilitate fraudster mining recognition. This model has detected 1000 fake claims and protected over tens of thousands of dollars daily. Michael et al. (2017) proposed a secured architecture for online banking. The model is characterized by features such as flexibility, reasonableness, dependability, availability, and processing time (Bay & Pazzani, 2001; Noto et al., 2012; Wei et al., 2013; Xu et al., 1999; Yamanishi et al., 2004). Saputra and Suharjito (2019) proposed a model for scam discovery in e-commerce using machine learning. This model applies a special technique to deal with class imbalances in the e-commerce exchanges dataset.

## 14.3 PROPOSED APPROACH

The proposed approach suggests a proven ecommerce model that can be more secure. Following steps will be followed by customers:

Step 1. At the very first step, a customer has to sign up to the website by providing their personal data like name, mobile number, email Id, and are required to set a password. Then, they should sign in to the website.
Step 2. After, the customers have to verify their mobile number through an OTP which was sent to them from the website.
Step 3. Then, the customer will choose some products to buy.
Step 4. Now, they can add the product to the cart.
Step 5. Then, they should complete their desired product's shipping address by entering their delivery address.
Step 6. After, some of the payment option will be shown to the customer, he/she can go for COD, payment by debit card/ credit card/ATM, netbanking, or though EMI. If they want to pay via card then they should add the card details. After that the order will be placed successfully.

## 14.4 FEATURES OF E-COMMERCE WEBSITE

The E-Commerce website should fulfill its customers' needs which is to buy a product they need in little time with the maximum satisfaction. The e-commerce website must have some features:

- Availability: The system should be available 24 ∗ 7.
- Flexibility: The system must be useable from everywhere; that is, for flexibility of the process the customer can access it anytime.
- Easy to use: Some easy methods should be there to help the user select his product, or the customer should adapt to the system easily.
- Trustable: The site should be trusted because if the customer is paying through e-payment, they are completely dependent on the site having to reveal confidential information that can be laundered and fudged, so, a basic requirement is the system should have security and made trusted.

Since Ecommerce is a rising business phenomenon, it is important to ensure its security by installing processes and checks in place to verify the installment procedure, the duration and timestamp of the hour of exchange, the purchaser, and vendor so that everything is authenticated and genuine.

- The purchaser must confirm his Aadhar Card, the universal identification of the card carrier verifying to his residency status as a permanent resident of India, which will anchor the installment procedure enough to confide in him as genuine. For forestalling scams, the client ought to be educated of the need to check if his/her record is "marked in" in another gadget. For "sign in" with another gadget, not the registered mobile, the client will get an OTP in his record, and he needs to punch it in the new gadget. At that point, no one but he can purchase things and pay through net banking.
- The authenticity of the purchaser as well as of the vendor can be checked through biometric security. Nowadays, there are several biometric processes that can be installed in the shops, Ecommerce websites, and on the vendors side, too, for validating the seller and vendors trustworthiness. Used during purchase, and before swiping the card after buying something online, a fingerprint scanning or eye recognition software could make identity thefts very difficult to manipulate and do, and should be implemented in the online shopping apps.

## 14.5 SECURITY CONSIDERATIONS FOR E-TRANSACTIONS

There are some vital safety considerations for e-transactions that internet business will do good to keep in mind while transacting on the Internet:

- The data for online exchanges should be confidential. Confidentiality means information pertaining to authentication of individual, assembly of people or an association, like individual's date of birth or first school's name, which are often used during two-step authentication process. If the client's name or registration information is purloined by a third-party or outside business contender, customer might face misfortunes. Cryptography assumes a vital importance here in keeping the information confidential. In this way, during data flows between interfaces, encryption is required to safeguard businesses.
- The data needed for online exchanges must be free of errors completely, or data integrity should be ensured and infallible. For data security to take effect, data integrity implies exact, precise, error-free data maintained throughout its lifecycle.
- Next challenge is to check the identity of members as real persons or not. Authentication guarantees that transactions taking place and money exchanging hands virtually are backed by real persons and real money so that exchanges and records are true. Likewise, for customers, it is imperative to know whether the websites are real.
- Non-repudiation is a very important concern and alludes to a situation where an individual can't deny his/her message, mark, reports, or any sort of agreements he entered while transacting over the internet where faceless transactions happen everyday on the basis of trust only. The exchange ought to set up the redundancy of explicit exercises like, affirming the buy orders and the method of registration.

## 14.6 TYPES OF BIOMETRIC SECURITY

### 14.6.1 Physiological

- Fingerprints: Fingerprint recognition is one of the traditional biometric methods that measures a finger's extraordinary ridges. Thus, basically it captures the print of a finger, then creates a unique digital biometric template by using some advanced algorithms, and lastly, the template is matched to existing fraudsters fingerprints to either verify or refute a match.
- Retina Scan: Retinal scan captures the eye vessels by using exceptional close-infrared cameras. Then the raw picture is pre-processed to improve the picture clarity, and last, processed as a biometric template to be used during log in by an individual so that only a genuine customer is let in the website.
- Facial Recognition: Face recognition alludes to the procedure by which a human or machine can distinguish between two people depending on their facial features. In reality, people can distinguish other individuals based on their countenances and we ace this ability at the age of just two months. As time progresses, the objective is to make machines do the tasks with a similar level of excellence. Since 2019, progress in this field has permitted organizations like Apple to change to facial recognition as an essential security strategy in their forthcoming iPhones (Bjelland et al., 2012). They utilized a variety of sensors to precisely delineate depths and isolate features of face in a computerized configuration. It is a technology for detecting a person from a video frame from a video source. It is the oldest way of biometric verification for, indeed, even newborn child will utilize facial recognition for the individuals nearest to them. Biometric facial acknowledgment programming works in the same way, but with progressively exact estimations. Actually, the software of facial recognition estimates the geometry of the face, including the gap between the two eyes, distance between the eyes, and the length from the jaw to the brows. After gathering the information, a sophisticated algorithm changes the data into an encrypted facial mark. Used in businesses such as payments methods, it makes payments easy for online shopping and contactless cards. As well as verifying a payment, face recognition can be integrated with devices and objects. In face recognition, no passwords are used.
- Iris Recognition: Iris is the shaded area of the eye that incorporates thick string type muscles that encourages the study to control the measure of light that enters the eye.
  By estimating the one of kind folds of these muscles, biometric validation devices can affirm eye features with exact accuracy. Liveness detection includes an extra layer of exactness and security.
- Hand Geometry: The hand geometry usually measures the hand like checking the length of the fingers, shape of the hand, and creates an encrypted data of that particular hand through some advanced algorithm. Nowadays, the fingerprint and facial recognition have taken the place of hand geometry as it's more relevant.

## 14.6.2 BEHAVIORAL

- Voice Recognition: This falls under both the physiological and behavioral biometric techniques. Behaviorally, the manner in which an individual says something is different from one person to another. Consolidating information from both physical and social biometrics makes an exact vocal mark. Sometimes because of sickness, mismatches can happen, though.
- Signature Recognition: It is a behavioral biometric that takes the measurement of how an individual puts a stroke on the paper, the pressure they put on that single pen, etc which creates a record of the signature's characteristics and which can be used during purchase transactions.
- Keystroke: Keystroke dynamics measures how an individual punches in the password using the keyboard, and records how much time that person takes to press each key, the timings between pressing each key, how many characters are typed per minute, and so on. By gathering all these information, it creates an impression of the keystroke which can be used to verify individual's genuineness during log in or checkout.

## 14.7 TECHNIQUES USED IN E-COMMERCE FOR SECURITY

As the authors have discussed in proposed approach, Aadhar card verification process should be used in which the purchaser must confirm his Aadhar Card, so that the user registration procedure will be confirmed on the buyer side helping to confide in him as authentic party. Secondly, the biometric processes can be implemented in E-Commerce websites for more security. Different techniques for ecommerce security are mentioned below:

- Fingerprints: The Fingerprint acknowledgment is one of the most established and efficient biometric strategy since fingerprints have for quite some time been perceived as a precise and essential ID verification technique, as the fingerprints of different people are never same. Given the various biometrics innovation on fingerprint recognition, the technique recognizes and confirms an individual's unique finger impression with fingerprint information gleaned in advance. It has always formed a part of criminological examination when fingerprint was discovered as having a unique mark among the people since the start of time and its verity as recognizable proof of person's identity. Unique finger impression acknowledgment has progressed significantly from that point onward as the procedure of unique mark as recognizable proof and coordinating its evidence progressed in criminological examination offices. These days, unique mark acknowledgment is broadly utilized from cell phones to entryway of bank vaults, and in any event, for high-security measures. ID and verification have been made conceivable by tiny yet productive unique mark sensors for cell phones. Ink and paper method is the oldest technique for fingerprints recognition but ink-less methods can be used to sense the ridges of a finger. Various devices that can be used for this purpose are Optical methods, CMOS capacitance, Thermal sensing, and

**FIGURE 14.2** Fingerprint.

Ultrasound sensing. The main benefit of this method is that it is cheap and easy to use. Moreover, finger print sensors are used everywhere. The main disadvantage of this method is that it is related with legal investigations and law procedure requirement. Moreover, sometimes when our hands are wet or made worse by skin disease, the sensor won't work. Figure 14.2 shows the finger print impression of a person's finger. The picture shows how a finger impression is capable of showing every ridge and whorl of a fingertip. Various methods like optical, capacitive and ultrasonic are utilized to gather fingertip design. The picture frame of a human's unique finger impression is improved to make it practical for use, and later, a biometric layout is created utilizing different refined calculations, that are constantly one of a kind for a person's finger. Thus, when an individual's information must be verified with this format, the biometric unique finger impression character can be leveraged for identity establishment. This layout is coordinated against the current sweeps, and the biometric framework establishes the genuineness of an individual, if a match is found; or rubbish, if there is no match. All the procedures portrayed above are at the client level, and the rest falling beneath unique mark acknowledgment framework.

- Facial recognition: It is the oldest way of biometric verification. Indeed, even newborn children utilize facial acknowledgment to recognize the individuals nearest to them. Biometric facial acknowledgment programming works in the same way, but with progressively exact estimations. Actually, the software of facial recognition estimates the geometry of the face, including the gap between two eyes, distance between the eyes, and the length from the jaw to the brows.

After gathering the information, a sophisticated algorithm changes the data into an encrypted facial mark. The main benefit of this technique is that it is easy to use and there is no need of any hardware. Moreover, it is easy to set up. The main disadvantage of this technique is that for identical twins, this system may fail; as well as skin disease will not let the system work. Facial acknowledgment innovation needs clear vision to set up. An advanced camera and facial acknowledgment programming are all you have to set up for this facial acknowledgment innovation. For slanted security applications, more equipment like the camera and infrared light producer, multi-camera arrangement, and so on are needed. Facial acknowledgment is one of the rapidly progressing and flexible biometric advances. The purpose for this development is the constant need of cell phones and individualized computing gadgets in everyday life. The modern day equipment convey two cameras, one at the front and the other at the back. Thus, it is simple to use face acknowledgment for client validation. These days gadget creators incorporate extra equipment like infrared producer for face acknowledgment and a camera to catch IR lit up 3D guide of facial structure. Apple was the principal organization to present this arrangement with iPhone X with Face ID moniker.

- Retina Scan: Retinal scan is a technique that captures the eye vessels by using unique close-infrared cameras. Then, the raw picture is pre-processed to improve the picture clarity; then, it is treated again as a biometric prototype to use during both enlistment and check. The main advantage of this technique is that everyone has a unique retinal structure and people can't copy it. The main disadvantage of this technique is that sometimes when there is some eye disease, the system won't work.
- Hand Geometry: The hand geometry basically measures the hand like the length of the fingers, or shape of the hand; then, it creates an encrypted data of that particular hand through some advanced algorithm. Nowadays, the fingerprint and facial recognition have taken the place of hand geometry as these are more relevant. The main benefit of this model is that everyone has a unique hand shape and people can't tamper with it. The main disadvantage of this method is that the hand sometimes is made worse by skin disease, and the system won't work. Figure 14.3 shows the hand geometry method.
- Signature Recognition: It is a behavioral biometric that takes the measurement of how an individual puts a stroke on the paper, the pressure they put on that single pen, and so on; then, the process creates a record of the signature. The main benefit of this technique is that it is simple, unique, and easy to prove. The main drawback is that anyone with wrong motives can copy it.
- Voice Recognition: It falls under both the physiological and behavioral biometric techniques. Physically, when a person says something the sound is produced by the shape of a person's vocal tract, the nose, mouth, and larynx. Behaviorally, the manner in which an individual says something is also different from person to person. Consolidating information from both physical and social biometrics makes an exact vocal mark. Sometimes because of ailments, mismatches can happen. The main advantage of this technique is that no one can copy it. The main disadvantage of this technique is that

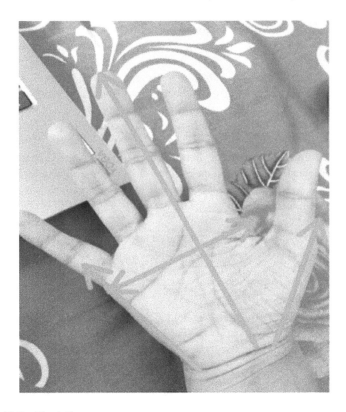

**FIGURE 14.3**  Hand Geometry.

nowadays there are many softwares that are able to make exact copies of voice. Moreover, if there is neck pain or cold, the sound quality may differ, so it will not work.

## 14.8 DATA ANALYSIS

"E-commerce security" can be assessed by:

- The use of electronic technology over the internet.
- Adding more components, fast and smooth transactions, and by marketplaces.
- Adding these security systems, the E-Transaction would be more secure.
- If all the transaction areas like, in app after paying the money, in shops after swiping the card can implement these security systems the E-Transaction would be fraud-less. Because if in every case people have to double check their fingerprints, face, retina, sound, shape of hand and the lot, or at least three of them, then the system can specify easily that any fraud is taking place or not.
- It is our obligation and duty as clients and web clients to protect our web-based business so that e-business can be more dependable in the coming future.
- To make sure about the online exchanges and be free from all sort of scams.

**TABLE 14.1 Comparison Between Current Methods and Proposed Method for Ecommerce Security**

|    | Basis              | Current Method | Proposed Method |
|----|--------------------|----------------|-----------------|
| 1. | Security           | Less secured   | More secured    |
| 2. | Biometric security | No             | Yes             |
| 3. | Trustable          | Less trustable | More trustable  |
| 4. | Reliability        | Less reliable  | More reliable   |
| 5. | Access             | Fast access    | Medium access   |

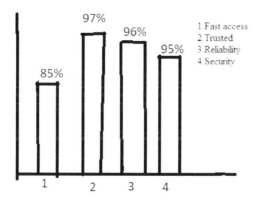

**FIGURE 14.4** Proposed Method on Various Parameters.

By utilizing this method, ecommerce becomes robust and efficient rather than a system riddled with defects. Table 14.1 shows a comparison between current methods and proposed work for ecommerce security. Based on Table 14.1, Figure 14.4 shows a bar graph representation of various parameters used on proposed method. Based on the analysis, it is found that proposed method is more secure, more trustable, more reliable, and easy to access. It turns out to be progressively dependable. Presently, individuals can trust ecommerce as it is ensured against frauds by biometric techniques implemented at various levels.

## 14.9 LIMITATIONS AND CHALLENGES

Some of the limitations and challenges are:

- Availability of data: Data should be collected from historic transaction database and with other related resources so that correct information can be collected for analysis.
- Skills shortage: Persons with various technical skills are needed for analyzing of data. Various data mining techniques can be used for analyzing of various patterns.

- Cost: Huge amount of resources is needed for implementation of special softwares and the person who will work on these softwares. This is the most important point to be taken care of.
- Time to implement the technology: Time is also an important factor that should be considered while planning for implementation of technologies for fraud detection.
- Customer Privacy: Customer data should not be exposed to anyone's view. It should be in safe hands.

## 14.10 CONCLUSION AND FUTURE SCOPE

In this chapter, the authors have discussed Ecommerce security in its various methods and how people should avoid frauds by using these methods. If a problem shows up, the customers might be disheartened about the reliability and trusted nature of E-Commerce. So from that point of view, if biometrics is used then transactions can be secured. Even if the card details of any customers are leaked then the Ecommerce practices shouldn't be abandoned and the CIA triad (Confidentiality, Integrity and Authenticity and Non-repudiation) should be followed. Besides, it is recognized that the payment methods security necessitates, and it additionally assists with, eliminating all the client concerns and its disadvantages. The effectiveness and safety can be given a greater scope for investigation in future. The exchange of cash and the conveyance of the item are the fundamental problems in each online transaction. Additionally, the cash transactions are completely reliant on the bank's management. The value-based data, common among client and between organizations, ought to be confidential. Utilization of new engineering has made the electronic exchange security dynamically productive. Future studies on this topic can be carried out by using various optimal computational techniques to control fraud in ecommerce. It is the need of the hour to build up a component with the mix of various methods and methodologies to alleviate the event of fake entries in the e-transaction by which the revenue loss can be controlled.

## REFERENCES

Abinayaa, S., Sangeetha, H., Karthikeyan, R. A., Saran, S. K., & Piyush, D. (2020). Credit card fraud detection and prevention using machine learning. *International Journal of Engineering and Advanced Technology (IJEAT)*, 9(4), April, 2020.

Aleskerov, E., Freisleben, B., & Rao, B. (1997, March). *Cardwatch: A neural network based database mining system for credit card fraud detection*. IEEE/IAFE 1997 computational intelligence for financial engineering (CIFEr), IEEE, pp. 220–226.

Alsharif, N., Dahal, K., Pervez, Z., & Sureephong, P. (2018, December). *Multi-dimensional e-commerce trust evaluation method* [Conference session]. Software, Knowledge, Information Management & Applications (SKIMA), 12th International Conference, IEEE, pp. 1–7.

Bay, S. D., & Pazzani, M. J. (2001). Detecting group differences: Mining contrast sets. *Data Mining and Knowledge Discovery*, 5(3), 213–246.

Bjelland, J., Canright, G., Engø-Monsen, K., Sundsøy, P. R., & Ling, R. S. (2012). *Social network study of the Apple vs. android smartphone battle* [Conference session]. Advances in Social Networks Analysis and Mining, IEEE/ACM International Conference, Istanbul, Turkey, pp. 983–987. 10.1109/ASONAM.2012.243.

Caldeira, E., Brandao, G., & Pereira, A. C. (2014, October). *Fraud analysis and prevention in e-commerce transactions* [Web Congress]. 9th Latin American Web Congress, IEEE, pp. 42–49.

Chen, C., Liang, C., Lin, J., Wang, L., Liu, Z., Yang, X., ... & Qi, Y. (2019, December). *InfDetect: A large scale graph-based fraud detection system for e-commerce insurance* [Conference session]. Big Data (Big Data), IEEE, pp. 1765–1773.

Chiu, C. C., & Tsai, C. Y. (2004, March). *A web services-based collaborative scheme for credit card fraud detection* [Conference session]. E-Technology, e-Commerce and e-Service, 2004. EEE'04. 2004, IEEE, pp. 177–181.

Dalvi, H. D., Joshi, A., & Shekokar, N. (2016, August). *Trustworthiness evaluation system in E-Commerce context* [Conference session]. Computing Communication Control and Automation (ICCUBEA), IEEE, pp. 1–6.

Ghosh, S., & Reilly, D. L. (1994, January). *Credit card fraud detection with a neural-network* [Conference session]. System Sciences, 1994, Twenty-Seventh Hawaii International Conference, IEEE, vol. 3, pp. 621–630.

Kataria, S., & Nafis, M. T. (2019, March). *Internet banking fraud detection using deep learning based on decision tree and multilayer perceptron* [Conference session]. Computing for Sustainable Global Development (INDIACom), 6th International Conference, IEEE, pp. 1298–1302.

Khan, A., Singh, T., & Sinhal, A. (2012, December). *Implement credit card fraudulent detection system using observation probabilistic in hidden markov model* [Conference session]. Engineering, 2012 Nirma University International Conference (NUiCONE), IEEE, pp. 1–6.

Michael, G., Arunachalam, A. R., & Srigowthem, S. (2017). Ecommerce transaction security challenges and prevention methods-new approach. *International Journal of Pure and Applied Mathematics, 116*, 285–289.

Mishra, M. K., & Dash, R. (2014, December). *A comparative study of Chebyshev functional link artificial neural network, multi-layer perceptron and decision tree for credit card fraud detection* [Conference session]. Information Technology, IEEE, pp. 228–233.

Noto, K., Brodley, C., & Slonim, D. (2012). FRaC: A feature-modeling approach for semi-supervised and unsupervised anomaly detection. *Data Mining and Knowledge Discovery, 25*(1), 109–133.

Sailusha, R., Gnaneswar, V., Ramesh, R., & Rao, G. R. (2020). *Credit card fraud detection using machine learning* [Conference session]. Intelligent Computing and Control Systems (ICICCS), 4th International Conference, Madurai, India, pp. 1264–1270. 10.1109/ICICCS48265.2020.9121114.

Saputra, A., & Suharjito (2019). Fraud detection using machine learning in e-commerce. *International Journal of Advanced Computer Science and Applications, 10*(9), 332–339.

Stolfo, S.J., Fan, D.W., Lee, W., & Prodromidis, A.L. (1997). *Credit card fraud detection using meta-learning: Issues and initial results* [Workshop]. AI Methods in Fraud and Risk Management, pp. 1–11.

Sungkono, K. R., & Sarno, R. (2017, October). *Patterns of fraud detection using coupled Hidden Markov Model* [Conference session]. Science in Information Technology (ICSITech), 3rd International Conference, IEEE, pp. 235–240.

Guo, T., & Li, G.-Y. (2008). *Neural data mining for credit card fraud detection* [Conference session]. Machine Learning and Cybernetics, Kunming, China, pp. 3630–3634. 10.1109/ICMLC.2008.4621035.

Wang, S. (2010, May). *A comprehensive survey of data mining-based accounting-fraud detection research* [Conference session]. Intelligent Computation Technology and Automation, IEEE, vol. 1, pp. 50–53.

Wang, X., Wu, H., & Yi, Z. (2018). *Research on bank anti-fraud model based on K-means and hidden Markov Model* [Conference session]. Image, Vision and Computing (ICIVC), 3rd International Conference, Chongqing, China, pp. 780–784. 10.1109/ICIVC.2018.8492795.

Wei, W., Li, J., Cao, L., Ou, Y., & Chen, J. (2013). Effective detection of sophisticated online banking fraud on extremely imbalanced data. *World Wide Web, 16*(4), 449–475.

Xu, X., Jäger, J., & Kriegel, H.P. (1999). A fast parallel clustering algorithm for large spatial databases. *Data Mining and Knowledge Discovery, 3*(3), 263–290.

Yamanishi, K., Takeuchi, J. I., Williams, G., & Milne, P. (2004). On-line unsupervised outlier detection using finite mixtures with discounting learning algorithms. *Data Mining and Knowledge Discovery, 8*(3), 275–300.

# 15 Botnet Forensic Analytics for Investigation of Disruptive Botherders

*Kapil Kumar, Shyla, and Vishal Bhatnagar*

## CONTENTS

| | | |
|---|---|---|
| 15.1 | Introduction | 265 |
| | 15.1.1 Problem Statement | 267 |
| | 15.1.2 Research Objective | 267 |
| 15.2 | Research Methodology | 267 |
| 15.3 | Related Work | 268 |
| | 15.3.1 Research Gap | 270 |
| 15.4 | Research Framework | 270 |
| 15.5 | Framework Deployment | 271 |
| | 15.5.1 Framework Phases | 271 |
| 15.6 | Data Description | 272 |
| 15.7 | Model Selection | 273 |
| 15.8 | Experiment and Result | 273 |
| | 15.8.1 Performance Analysis | 274 |
| 15.9 | Research Limitations | 275 |
| 15.10 | Conclusion and Future Scope | 275 |
| References | | 275 |

## 15.1 INTRODUCTION

Network Forensics is a portion of cyber forensics that makes investigative networking related digital evidence its focus of study. It includes noticing the soundtrack, scrutinizing it for anomalies, and construing complex congestion of information related to smart networks. Mechanisms of Complex Network Forensics include a packet of data captured, and its inspection, network device acquisition, and Incident response. The authors (Kaushik & Joshi, 2010) found that network forensics has become a vital constituent of the Information Technology business, given giant IT firms beleaguered by intruders all day are anxious to know the real-time information and status of devices. The occurrence of bouts has risen, which has become a problem not only for firms but also its clients.

The internet has grown exponentially with opportunities for expansion strewn about its way and the computer operator's life is going to be digital, from expenses to media, and from shopping to dining. The loads of information present on the network attract intruders stalking it to dent its armour. The network traffic is analyzed actively for the purpose of detection of malware and intrusions which involves monitoring of controlled packet flow in the network to troubleshoot the security constraints. In network forensics, the sleuths perform the monotonous job of finding these vulnerabilities in the network that could potentially attract muggers. One vulnerability is the information traveling in the form of data packets on the internet, and which are vulnerable to attack containing extremely important data as cradle, terminus, and stuffing.

Where an offensive attack takes place which signals the networks had missed taking note of, the sleuths need to examine these. The hints are named footprints. The investigation process, where the network mechanism is affected mostly, originates with noting traits, like checking the memory caches of the network after it fails for vital crash-related information. The caches need to be inspected by authorities and a timeline between stalking the network, attack, and crash takes shape.

The authors (Erbacher et al., 2006) describe the mining of data from caches of connected machines of a network as network data mining. It contains recognition, assortment, and scrutinizing the cached data. The examination of packets seized from a stream of traffic explicitly to regulate the traffic is caused by an authentic cradle or was arranged by the bots. Networking mechanisms contain critical information, which is valuable in an inquiry of cybercrime. Given every networking mechanism is robust and long-lasting, stages in its lifecycle are trailed to construct policies such as for firewalls, switches, and Intrusion Prevention Systems (IPS).

An approach for data mining needs to be created to lessen attacks and foresee attacks on IoT networks at a forensic level because IoT is catching up with both clients and intruders. With premonitions being sounded that the Internet needs to be watched on as not a harmless platform, but threatened by dangerous stalkers and thieves, loss in data from attacks having costs extending from simple troublesomeness to likely serious consequences are becoming more frequent. Further, Xu et al. (2018) gave a close up look of computer-generated exaction named botnets, estimated to be the single most damaging cause of network crashes.

The demarcation of a botnet as a set of internet-based usages, that include mainframe systems, waitrons, and IoT devices influenced, has been brought about through general style of bots. The current tools and techniques works on the set of protocols which includes procedures as Real-time monitoring, threat intelligence event correlation, Troubleshooting and In-depth analysis of possible threats. The researchers explore and critically examine the role of machine learning approaches to create a network forensic method based on the identification of ports which includes source addressing and destination addressing. The authors found that the network forensic is a segment of the computer forensics in the context of noting and checking link traffic aimed at outlining causes of defence plan abuses, and desecrations.

An approach made up of system comprehension, training, and test with schmooze data for sorting into sets of genuine and abnormal interpretations, and applied to network forensics for developing solutions is catching on. The foremost need of the study is the usage of the classification algorithm, and in the context of algorithm set up, to detect the origin of bots. The method is utilized to detect attack trajectories, whereas the derivation of bots associated with their stream is recognized as a forensic system.

### 15.1.1 Problem Statement

The Internet of Things-based (IoT) network of an entity has been improved with a few modules of counting to the extent of having enhanced its competency. However, with the flood of botnet attacks, the IoT is at a precarious situation given botnets are virulent with the capacity to destroy years of work. For instance, botnets are the reason for hefty safety risks and fiscal imbalances in companies across time.

The prevailing network forensic methods do not have the capability to classify and detect current slew of botnets with the mechanism refined to a sophistication that cannot be detected easily. The market's appliances largely depend on signature-based methods that are not skilled to detect a new pattern of a botnet. In the study, the authors proposed an approach based on the machine learning concept to evaluate the pattern of attacks.

### 15.1.2 Research Objective

The theme of the study is to enhance the network forensic analysis method for detecting bots while the investigation of IoT-related devices is an accessory to the study. The objectives include:

- Design a framework by using a support vector machine for including the cradle of botnets.
- Performance analysis of the designed framework for the IoT network by using python programming languages with a standard botnet dataset.
- The accuracy of classifying of attacks requires to be upgraded.

The remainder of the chapter is planned as follows: Segment 1 is Introduction, Segment 2 is about Research Methodology, Segment 3 explains the Related Work. Segment 4 is research framework, Segment 5 is about Framework Deployment, Segment 6 is Data Description. Segment 7 is Model Selection, Segment 8 is about the Experiment and Result, Segment 9 is Research Limitation and Segment 10 is Conclusion and Future Scope.

## 15.2 RESEARCH METHODOLOGY

Research methodology clearly defines the "how" in a research and it defines the systematic design of research to ensure the validity and reliability of research objectives.

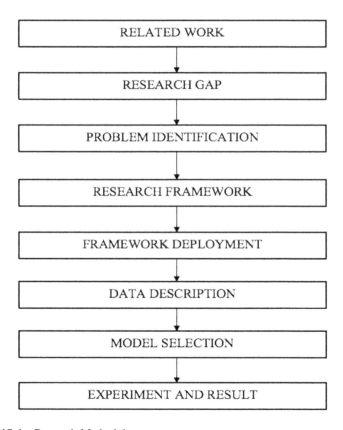

**FIGURE 15.1**  Research Methodology.

Figure 15.1 depicts the stages of the research which starts with identification of a research gap from surveying current literature. From the gap surfaces the problem and a framework to pattern the problem into research effort needed to solve it. Once deployed, the framework of research takes the direction of data description and modeling, algorithm selection, experimentation and results. Lastly, conclusions are made and future work projected for investigation.

## 15.3 RELATED WORK

The authors (Kaushik & Joshi, 2010) found that network classification in forensics is just all about seizing, soundtrack, and investigating network actions to discover the source of attacks during incident investigation. This paper faced difficulties in gathering, scrutinizing, and analyzing the process but not without finding ways out. The authors projected a framework by amassing network data, recognizing suspicious packets, observing protocol traits spoilt, and certifying the bots. The framework is put down with certain indications to defend bots on ICMP etiquette. The packets-seized folder is examined with the ICMP protocol traits in header to spot suspicious packets. The data contained inside the header is summarized in the packet detection folder ported to a databank. The guideline planned is for numerous ICMP bots queried in the databank to compute numerous statistical datasets. This data authenticates the occurrence of bots and will be highly rewarding during examination. The abridged packets contained data seize size simple to control as merely noticeable packets are measured. The etiquette traits typically operated via intruders are present in the database pattern for the following phase of examination and analysis. The authors (Erbacher et al., 2006), found that network forensics is the important next stage in the examination of network attacks, invasions, and abuses. It is the activity of forensics that will help in recognition of the attacks and process by which they occur. By explosion in amount and kinds of bots a novel method is built to help in the examination of supposed attacks. The latest extensive implementation of botnets is forensic methods that will let these botnets to be taken apart to evaluate their degree of harm, their competence, and their rheostat means. In this paper, the authors explain the visualization approaches developed around the analysis of network data and fitted to the scalability subjects key to such data. In combination with these approaches, the authors described fitting these methods into an analysis list, and their utilization for investigating the forensic process, and the process through which they will be used efficiently. The authors (Koroniotis et al., 2017) found a network of interrelated routine entity named as "things" that are expanded with a trifling degree of processing competencies. Recently, it was found that the IoT can be affected by numerous diverse botnet events. A system of interrelated ordinary entities called the internet of things (IoT) that have been enriched with a significant degree of figuring aptitudes. In not too distant past, the IoT was affected by diverse, dissimilar botnet events. In its role of adversary, botnets have been the root of significant safety risks and fiscal imbalance over the years. The standing Network Forensic Methods cannot classify and trail present knowledgeable approaches of botnets. This is why market equipment are mostly subjected to signature-based tactics that cannot notice new tendencies of the botnet. In related work, numerous reports have been in the forefront by using ML approaches to train and test the classifier. The study inspects function related to machine learning approach aimed at mounting a forensic assemblage based on network streaming data recognizer to deal with actions of botnets. The tentative fallouts using the botnet data with machine learning approaches with a streaming detector can detect the botnets effectively. The authors (Lubis & Siahaan, 2016) found that general cases that regularly appear on a cyber-network are a cause of vulnerability for computer networks. The network forensic is a method of examining movement, soundtrack, or to classify the network to discover digital signs from a cyber-delinquency. Internet behind network forensics and legal interference are significant advances for many organizations comprising small, medium commerce, investment, and economics industries. The archiving and restoration of internet data can be utilized for permissible signs in the context of disbalance in data maintenance. Administration and intelligence support use equipment to guard and preserve nationwide security. There are several ways to find a crime committed on a computer network. The use of several applications supported is to improve the success of network forensic processes in common cases. Geambasu et al. found the network intrusion detection systems (NIDS) utilize the ritual keys to record past network movements and sustain forensic examination by network managers. These keys are sophisticated, laborious, and inflexible. The author explores databank replenishment for co-operating network forensic investigation. The authors demonstrated that a different view about the (RDBMS) can upkeep sensible stream degree in a way that safeguards extreme probe recital. To allow provision for meaningfully advanced data rates, authors projected a method to propose on-demand opinion advent and indexing. The move towards the importance of objectives as frequency of growth pattern and bandwidth of network traffic is used to determine the intrusions attempt by tracking bottleneck and traffic spike.

### 15.3.1 Research Gap

- The primary problem related to network forensic methods are inability to detect and trace the existing approaches of botnets. Because of existing systems largely dependent on signature-based methods, the systems are not able to find novel form botnets.
- Controlling the colossal dimension of a complex network is a problem in the context of collecting, tracking, and examine data.
- Loss of information as collection of data does not complete the survey cycle.
- Tracing the routes of attacks is an issue, and there are high false alarm rates

## 15.4 RESEARCH FRAMEWORK

The model proposed by authors describe the methodological outlook of the path trailed by the SVM machine learning algorithm to develop an NFA. In the investigation, the data is assembled and refurbished into an appropriate shape by reforming the dataset into a standard pattern in the preprocessing stage to produce the training data.

Figure 15.2 shows the proposed network forensic analysis (NFA) framework with the inclusion of the machine learning concept to investigate the evidence present in IoT devices in a network that can be controlled by the attacker through

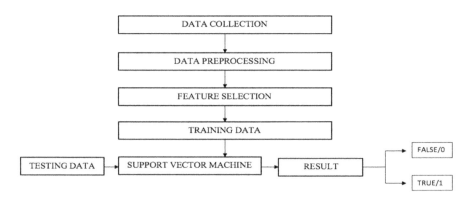

**FIGURE 15.2** Network Forensic Analysis.

botnet attacks. To find evidence for capturing the criminal, there is a need to find the origin of attacks and the creator of attacks. The authors in the paper used a concept of machine learning which includes support vector machines to investigate the IoT device. The authors used a botnet dataset which is concerned with IoT networks, downloaded it through the website of Kaggle. To train and test the classifier, there is a need to refine the dataset by preprocessing data by sensing the data for number of features, several classes, and categories of attacks in labels classes. There is a need to remove the inconsistencies in dataset and fill the null value with a constant value or by mean, median, or mode. There is a need to remove the noise that can create problems in obtaining the results by using wavelet transform and smoothing the data by using the binning methods. There is no need to convert the type of data into a standard type that suits the classifier by using the classifier named numeric. The next or last step of the data preprocessing is feature selection or select a subset of data by reducing dimensionality to make it representative of actual botnet data. This uses a suitable classifier which is known as an extra classifier to reduce the dimension of data which makes the classifier fast and reliable. The data must be split into a standard ratio of the train or test data which is defined as twenty: eighty ratio for test and train, respectively. The train data is utilized to train the classifier, and test data is used to refine the expectations. The prediction phase produced a condition matrix that divulges the variety of predictive results.

## 15.5 FRAMEWORK DEPLOYMENT

The projected model is instituted by using a suggested algorithm and assessing the enactment of classification. Researchers utilized the standard dataset for detection.

### 15.5.1 Framework Phases

The common stages of the process of a planned system can be split into three segments: pre-processing, training, and prediction.

Preprocessing
- **Loading the dataset**.
  Import<Dataset>
- **Transmuted the data in a suitable pattern to ease read.**
  csv->data frame.
- **Divide the dataset into independent and dependent features.**
  A<-data and B<-data.
- **If replicate features subsist then drip.**
- **Transform the entire data into binary.**
  Encoder->A[features].
- **Replace the null value with mean, median, or mode.**
  Null value<-simpleImputer (Null value<-np.nan, strategy->mode).
  Null value->Null value. Fit(A).
- **Divide data into train and test parts with a proportion of 0.25.**
  A_train, B_train, a_test, B_test<-split (A, B,0.20).

- **Do dimensional reduction to reduce the size of data.**
  Feature->Feature (No. of Element) and Feature. Fit(A_train).

Training
- **Consume the above-prepared data.**
- **Fix the feature, and target data set into the framework.**
  clf (A_train, B_train).
- **Perform.**
- **End.**

Prediction
- **Utilize the prepared test data.**
- **Envisage the consequence depends on fleeting the autonomous test dataset to the algorithm.**
  clf->Predict(A_test).

## 15.6 DATA DESCRIPTION

The dataset utilized in the trail is obtained by congregating the traits from the IoT network by uninterrupted scrutinizing of the packet by delivering through the network and by seizing the network traffic, which is extensively used for checking abnormality condition and detection. The dataset is settled on for use based on data seized from the IoT network that contains 42323 samples, with 22 attributes that are considered as true or false. The attacks are classified into one of the following categories:

- DDoS: The network or server attacks hinders system regularities with the objective of system blocking through anomalous traffic.
- Malware: In malware detection, attackers use a restrained executable code to run the network. The executable file amasses the ominous information and directs it to his chief to fix the cracks in the IoT network.
- Mutation: In mutation, close contact information is fetched from the network by becoming a pretender. The criminal can use imitate packets to acquire information from the network.
- Weakness Measures: The security manager machines do not exist in the devices that make the network susceptible to threats.

The authors analyzed the botnet that is known as a zombie that dislocates a target device. The bot is a plausible plug-in that is used to generate a network by reaching the state of the susceptibilities of the non-infected device. The bots focus on a fresh device for which specific techniques are needed to probe the device for getting undesirable entry to the network.

Non-P2P passage protocols such as HTTP, DNS, and SMTP with the loss of links is responsible for network outage; while in P2P traffic, the network disengages from the client-server network structure and the level of loss of links is enormous.

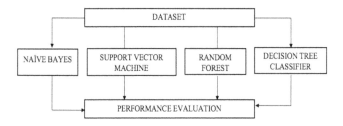

**FIGURE 15.3** Model Selection Framework.

The cause of elevated level of loss of links is because of the connectivity of a link with other comrade, and their re-linked to comrades that have been disengaged in P2P transmission.

## 15.7 MODEL SELECTION

The algorithm selection is the key part of creating the framework, It is required for accesing the targets. There are varied processes that are being utilized as Naive Bayes, Support Vector Machine, Random Forest, Decision Tree Classifier algorithm to inspect data. The investigators studied, the SVM, which is the best algorithm for detecting the android malware.

Figure 15.3 shows the model selection strategy where many algorithms present outcomes but SVM is revealed best of them.

## 15.8 EXPERIMENT AND RESULT

SVM is studied as a supervised machine learning algorithm that is utilized to scrutinize data for classification. The authors (Yang, 2019) found that SVM is a distinct classifier officially outlined by a detaching hyperplane. The algorithm yields the finest hyperplane which classifies given labeled training data. SVM are support vector machines where an algorithm is a description of data as points. The data in vector space is plotted into diverse data classes that are alienated by a clear distance. SVM can meritoriously accomplish linear and nonlinear classification by transforming inputs into dimensional trait plots. SVM learning mode is a type of supervised learning approach that brings in for classification. It is named as margin classifiers because SVM diminishes the experiential classification error and boosts the geometric mean instantaneously. The authors (Yan & Han, 2018) analyzed that in SVM two equivalent hyperplanes are structured to split the data entered for the creation of maximum separated hyperplanes to transform vectors into high dimensional spaces. The oversimplification miscalculation would vary on the minimal space between these hyperplanes.

Let us take into account the points $<(n1, m1), (n2, m2) \ldots (nn, mn)>$ and for two-dimensional vector w. $n + z = 0$ where w = direction vector and z is scalar constant and once the hyperplane is created then the hyperplane is used to make a prediction. The hypothesis function is defined as $P(n)$.

$$P(x) = \begin{cases} +1, & w.n + z \geq 0 \\ -1, & w.n + z \leq 0 \end{cases}$$

(Kim et al., 2003)

This shows the points above the plane for class +1 and below the plane for class −1. The goal of hyperplane is to separate the data accurately.

### 15.8.1 Performance Analysis

The expectation report proves the enactment of the model in trail operation and attained the result which divulges the accuracy, precision-score, and support.

Table 15.1 defines the accuracy, precision, recall, and f1-score for SVM with a macro average and weighted average.

Figure 15.4 shows the prediction result analysis using a graph that shows the accuracy, precision, and recall value for SVM.

**TABLE 15.1**
**SVM Performance**

|   | Precision | Recall | F1-Score | Support |
|---|---|---|---|---|
| 0 | 0.94 | 0.96 | 0.95 | 17613 |
| 1 | 0 | 0 | 0 | 8 |
| 2 | 0.97 | 0.94 | 0.96 | 10894 |
| 3 | 0.29 | 0.21 | 0.24 | 197 |
| 4 | 0.05 | 0.06 | 0.05 | 405 |
| Accuracy | N/A | N/A | 0.94 | 29117 |
| Macro Avg | 0.45 | 0.43 | 0.44 | 29117 |
| Weighted Avg | 0.94 | 0.94 | 0.94 | 29117 |

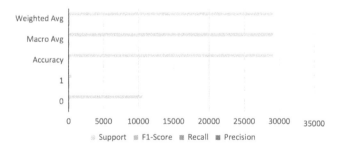

**FIGURE 15.4** Prediction Result Analysis.

## 15.9 RESEARCH LIMITATIONS

The research suffers certain drawbacks the authors faced in designing the NFA as follows:

- The researchers are powerless to use the sealed access tools and system that is a prerequisite to the trail to evaluate the result.
- The refusal of closed access journal deprived the authors of the learning to pinpoint papers of extremely valued journals.
- The research performed in a specific field by a repeated approach locked the doors of bestowing the results in an enhanced method.

## 15.10 CONCLUSION AND FUTURE SCOPE

A system of interrelated ordinary entities called the internet of things (IoT) that have been enriched with a significant degree of figuring aptitudes is undre consideration. Recently, the IoT has been affected by a diversity of dissimilar botnet events. In the role of adversary, botnets have been the root of serious safety risks and fiscal disbalance over the years. The standing Network Forensic methods cannot classify and trail present knowledgeable approaches of botnets. This is the cause of market equipment mostly subjected to signature-based tactics that cannot notice novel rituals of a botnet. In related works, numerous reports have shown use of machine learning approaches to train and test a classifier for such outbreaks, but still it generates high levels of incorrect classification for examining the routes of botnets. This study inspects the function of the machine learning approach for mounting a forensic contrivance over a network streaming data recognizer to detect events of botnets. The fallouts in loss of botnet databank with machine learning approaches with a streaming detector can detect the botnets effectively. The experiment result determines that the classifier gives optimal accuracy of 94% and shows the detection rate is 94%. Future research is suggested in the following:

- The future work includes the practice of artificial neural network (ANN) for investigating botnet data and the use of the optimal attributed to train the classifier.
- The research will be based on the blockchain supply method that contains an entire chain of information and updates the information in databases regularly as new malware occurs.

## REFERENCES

Erbacher, R. F., Christiansen, K., & Sundberg, A. (2006). *Visual network forensic techniques and processes* [Annual Symposium]. Information Assurance: Intrusion Detection and Prevention, pp. 59–72.

Kaushik, A. K., & Joshi, R. C. (2010). Network forensic system for ICMP attacks. *International Journal of Computer Applications, 2*(3), 14–21.

Kim, H. -C., Pang, S., Je, H. -M., Kim, D., & Bang, S. Y. (2003). Constructing support vector machine ensemble. *Pattern Recognition, 36*(12), 2757–2767.

Koroniotis, N., Moustafa, N., Sitnikova, E., & Slay, J. (2017). *Towards developing network forensic mechanism for botnet activities in the IoT based on machine learning techniques* [Conference session]. Mobile Networks and Management, pp. 30–44.

Lubis, A., & Siahaan, A. P. U. (2016). Network forensic application in general cases. *IOSR Journal of Computer Engineering, 18*(6), 41–44.

Xu, C., Shen, J., Du, X., & Zhang, F (2018). An intrusion detection system using a deep neural network with gated recurrent units. *IEEE Access, 6*, 697–707.

Yan, B., & Han, G. (2018). LA-GRU: Building combined intrusion detection model based on imbalanced learning and gated recurrent unit neural network. *Security and Communication Networks, 2018*, 1–13.

Yang, K. (2019). A novel research on real-time intrusion detection technology based on data mining. *Journal of Physics: Conference Series, 1345*(5), 016–052.

# 16 Design of a Toolbox to Give One Stop Solution for Multidimensional Data Analysis

*Dr. Prarthana A. Deshkar*

## CONTENTS

| | |
|---|---|
| 16.1 Introduction | 278 |
| 16.2 Review of the Existing System | 278 |
| 16.3 Design of the Toolbox | 280 |
|     16.3.1 Conceptual Design of a Toolbox | 280 |
|     16.3.2 Conceptual Design of a Query Generation Process | 281 |
|     16.3.3 Generation of Multidimensional Report and Visualization | 281 |
| 16.4 Statistical Analysis and Data Mining on Multidimensional Space | 284 |
|     16.4.1 Statistical Algorithms | 284 |
|         16.4.1.1 Single Series Algorithms | 284 |
|         16.4.1.2 Two Series Algorithms | 285 |
|         16.4.1.3 Multiple Series Algorithms | 286 |
|         16.4.1.4 Matrix Algorithms | 287 |
|     16.4.2 Advanced Techniques | 287 |
|         16.4.2.1 Regression Analysis | 287 |
|         16.4.2.2 XY-Graph | 288 |
|         16.4.2.3 Data Transformation | 288 |
|         16.4.2.4 K-Means | 288 |
| 16.5 Case study: Analyzing the Environment Using the Spatio-Temporal Data Generated by the Sensors | 288 |
|     16.5.1 Motivation | 288 |
|     16.5.2 Inputs and Workflow | 289 |
|     16.5.3 Data Modelling | 289 |
|     16.5.4 Multidimensional Analysis | 292 |
|     16.5.5 Outcome of Case Study | 296 |
| 16.6 Conclusion and Future Scope | 296 |
| References | 296 |

## 16.1 INTRODUCTION

Fast growth in technology also increases the variety of data sources. Multidimensional analysis of data is the basis for data analysis or decision making systems to get the more relevant and rational results. Traditional multidimensional data analysis systems are storing data in the multidimensional data cube. Currently good numbers of systems are available to perform the data analysis and helps in decision making. Most of them are to not design for multidimensional data analysis. Data analysis is an umbrella term having numerous statistical, data mining and machine learning techniques are under this concept. Available traditional systems do not provide complete data analysis solution under one roof. Multidimensional data model represents the data in terms of dimensions and measures also termed as fact values. Dimensions are the entities based on which information is analyzed. Dimension entity possess additional features, like sequential information, or the more complex relationship within each other, e.g. region dimension can have the hierarchy like country, state, region, city, area. Time dimension which is the essential aspect of any type of analysis can have both hierarchical and sequential features. Time can have year, quarter, month, week as hierarchical level structure. It also can possess the sequential property which represents the relationship like each month is associated with previous month and next month. Measures or fact values represent the actual transactional values which need to be analyzed.

As data analysis is a comprehensive term. It includes various statistical, data mining and machine learning algorithms to form a system which leads to the enhanced the decision making.

From the ancient time, statistical analysis plays a vital role in analyzing the data. Statistical analysis is used to collect and explore the data to identify the hidden patterns and trends from the source. The numerous techniques available in statistics helps research community, commercial researchers, government bodies to make more relevant and fact based decisions.

In this chapter the complete analysis toolbox which uses the on – the – fly query generation technique is presented. This chapter gives details of the numerous operations and customizations provided by the system to have efficient, accurate, and quick decision making. The chapter also focuses on the built in reporting and visualization techniques.

## 16.2 REVIEW OF THE EXISTING SYSTEM

As multidimensional analysis gains the importance in the building of decision making systems, leading brands in market are providing good range of such systems. Each system is offering some special characteristics to its user according to the analysis need. But most of the systems are working on the similar principle. One of the leading systems in the data analysis is Microstrategy which is a licensed product. A wide range of facilities are provided by the MicroStrategy which includes OLAP Services and Report Services which can be easily operated by using the unified Web interface. MicroStrategy stores data in a data warehouse. The architecture provides the separate data layer, presentation layer and the operational layer. Independent working of layers provides the hazel free internal working (Architecture for Enterprise Business Intelligence, an overview of the microstrategy platform architecture for big data, cloud bi, and mobile applications, n.d.).

Another popular licensed system is the IBM Cognos business intelligence solutions which facilitates the user with fast information usage for decision making. Cognos solutions can efficiently cater the requirements of all type of the users and provide the consistent data to work with in the collaboration. Also cognos provides the pool of products to serve the every functional level of organization. The Cognos system follows the web-based service-oriented-architecture. Issues in the complicated data modelling can be easily handled with the Cognos components like Framework Manager, Cube Designer, or Transformer. Its data modeler is responsible to populate the metadata for the multiple data sources in the form of package which can be used in the Cognos BI module. User of the Cognos can have maximum control on the usage of the memory because of the dynamic cubes technology. It also encourages the users to modify their deployments and supports the interactive analysis on the huge amount of data which may be in terabytes (IBM Cognos Business Intelligence V10.1 Handbook, 2010).

One more licensed product is BUSINESSOBJECTS (BO); it is a composite system which provides the tools for query processing, report generation and the data analysis mainly to the business professionals. It facilitates the users to directly work on the enterprise databases which are provided to them. Target users of the Business object are specially from the financial domain (Business Intelligence Platform Administrator Guide, 2016).

One more effective and popular option for data analysis is Tableau. It provides numerous techniques for the visually interactive reports mainly for the business analytics. This system provides limited functionality to the users in its free version and it is also having the more powerful and complete paid licensed version. With its large range of visualization facilities, Tableau facilitates its user with the better exploration of the temporal, spatial, topical, and network data. In the architecture of Tableau, the good weightage is given to handle the data by using the data warehouse, data marts, files and cubes (Rueter & Solutions, 2012 May).

This new era is experiencing the advancements in the technology and hence analytics needs to expand its techniques by adding univariate and bivariate methodology, time series statistical techniques and so on. Different machine learning algorithms, descriptive and predictive data mining algorithms are also the building blocks of the analytical solutions. As it is observed from the discussion in this section, maximum of the systems are providing all these mentioned techniques, but in maximum cases they are offering it as the separate components which need to be deployed separately on the users system.

Data mining techniques including descriptive and predictive data mining and various machine learning techniques are also pillars for complete data analysis solution. Most of the tools provide all of these techniques, but the user often needs to install separate components of the tools to get the required functionality. The non – technical users or researchers might not be comfortable with this approach.

Apart from these commercial players lot of researchers are also working on the architecture of the model and storage of the multidimensional data. It can be categorized in two aspects, as handling the cube efficiently and other different approaches to handle the multidimensional data.

One way to handle the data coming in the real time is dynamic cubing (Usman Ahmed, 2013). This approach uses the data partitions, and these partitions are having small cuboids as the materialized views generated at run time.

To have the reduced cube generation time, cubes can be generated in parallel. Multidimensional array can be used to create the cube (Jin & Tsuji, 2011).

Some research is also focusing the altogether different approach to perform the multidimensional data analysis. The system is considering Big Data as the source for the multidimensional data analysis, and the system stores data cube as the separate layer in the system (Sandro Fiore et al., 2013, 2014).

One more approach is the multidimensional data analysis system can be treated as software as service. In this approach predefined templates of data and the patterns to answer the user queries (Zorrilla & García-Saiz, 2012).

## 16.3 DESIGN OF THE TOOLBOX

### 16.3.1 CONCEPTUAL DESIGN OF A TOOLBOX

To perform the multidimensional data analysis, dimension structure is retrieved from the flat file of metadata. Figure 16.1 shows the conceptual design of the query generation process.

User events can be generated by selecting all the dimensions and measure entries. Dimension entities can be selected at any level of hierarchy. Higher level values will be calculated at the time of query execution and hence this architecture is following on – the – fly query generation process. According to design of the system, user events are generated with the help of various operators, which are used to make the analysis more customized and effective. Dimension filters are the operators which allow selecting the dimensional entities based on some predicate. Some standard and user defined functions can be applied on the measure value. Also any number of the standard aggregation function can be applied to measure value while generating the user event.

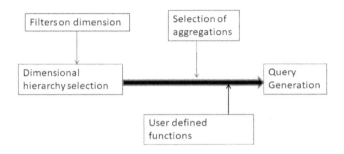

**FIGURE 16.1** Conceptual design of query generation process.

# Design of a Toolbox

## 16.3.2 Conceptual Design of a Query Generation Process

After generation of the user events, it is transferred to the query generation process to generate the multidimensional output. All the operators defined in the previous section are attached to the user events, which are further converted as the query to produce desired multidimensional matrix. Query generation process works with the two aspects, one is complete run time generation of the output, and other is generation of the output using hybrid cubes.

## 16.3.3 Generation of Multidimensional Report and Visualization

According to the design of the system, all the operators on dimension and measure are combined to form the input selection. This input selection is combined with some other output representation options to for the user event and this user event is then given to the system and system with the help of metadata generates the query.

The system does not support pre-aggregation; hence the system provides the various aggregation functions which can be calculated on multidimensional model at run time. System allows selecting more than one aggregation function at a time. The important benefit of this system is that, run time aggregation function can be decided, where as in the traditional multidimensional analysis system, the aggregations need to be decided at the time data absorption phase, and based on that cube is generated. The proposed system design allows row wise as well as column wise aggregation. Figure 16.2 shows the interface provided to have the aggregation and output representation design components.

Report generated shows the multidimensional analysis with the aggregations and the specific operators which are designed by the user. According to the selection of type of report, it is generated as web report or comma separated file or the pure graphical representation.

Output matrix and the graph is having the certain mapping in between. Y-axis is having the measure value. And the x-axis is plotted against each row from the output matrix. The columns from the output matrix are acting as legends to create overlapping plot. Figures 16.3 and 16.4 show the web report and the graphical representation respectively.

**FIGURE 16.2** System design components for aggregation.

58 results calculated in 1.094s

Date : 2018-02-27 13:58:44.994

| | | Canada | | | | | | | | | | | China | | | | | | | | | |
|---|---|---|---|---|---|---|---|---|---|---|---|---|---|---|---|---|---|---|---|---|---|---|
| | | 2002 | 2012 | 2003 | 2004 | 2005 | 2006 | 2007 | 2008 | 2009 | 2010 | 2011 | 2002 | 2012 | 2003 | 2004 | 2005 | 2006 | 2007 | 2008 | 2009 | 2… |
| Actual VALUE | Unemployment_total_(per_of_total_labor_force)_(modeled_ILO_estimate) | 7.70 | 7.30 | 7.60 | 7.20 | 6.80 | 6.30 | 6.00 | 6.10 | 8.30 | 8.10 | 7.50 | 4.41 | 4.47 | 4.30 | 4.30 | 4.14 | 4.00 | 3.76 | 4.36 | 4.29 | 4… |
| Actual VALUE | Unemployment_total_(per_of_total_labor_force)_(national_estimate) | 7.66 | 7.29 | 7.57 | 7.19 | 6.76 | 6.32 | 6.04 | 6.14 | 8.34 | 8.06 | 7.51 | 4.00 | 4.10 | 4.30 | 4.20 | 4.20 | 4.10 | 4.00 | 4.20 | 4.30 | 4… |

**FIGURE 16.3** Web report in the form of data matrix.

Design of a Toolbox 283

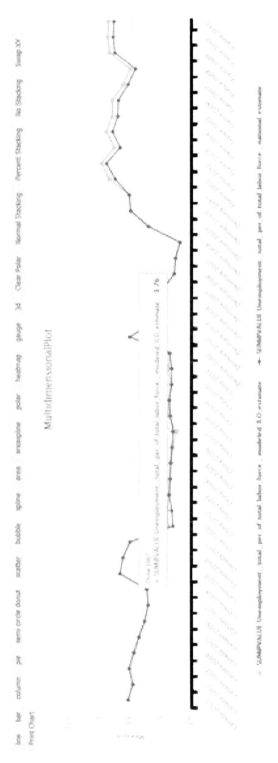

**FIGURE 16.4** Web report in the form of graphical representation.

Graphical representation of the report is designed to provide numerous facilities to make sophisticated visual representation of the results. Labels and legends are well defined. Almost all types of basic and advanced graphs are available for presentation. One more interesting and useful feature is added to remove some components from the heavily loaded graph so that user can comfortably analyse the results.

## 16.4 STATISTICAL ANALYSIS AND DATA MINING ON MULTIDIMENSIONAL SPACE

The system is designed to take the multidimensional output further to have the detailed analysis using numerous statistical techniques, and data mining algorithms. The matrix of values against the input selection is further given to the toolbox designed in system.

As data analysis is an umbrella term, various techniques are gathered together to analyse the data. Proposed system through this toolbox provides almost all type of functionality which fulfills the basic analysis need of any organization or research community. All the techniques and algorithms are designed to use the multi-dimensional output matrix. The algorithms can be applied on the matrix data with respect to row data or column data.

### 16.4.1 Statistical Algorithms

From ancient time statistical techniques are used to analyse the data. It helps to do the exploratory analysis of the data. It is used to discover the fundamental relationships between the data and helps to identify the trends in the data. Statistical analysis takes part in solving the problems from any domain. Business or commercial community can enhance the business by effectively planning marketing strategies, customer relationships, managing business needs, etc. Researchers will get benefited by using the predictions, finding out relationships between the data and using of hypothesis testing.

The toolbox, basically divides the statistical analysis space in four parts, single series algorithms, two series algorithm, multiple series algorithm, and matrix algorithm.

#### 16.4.1.1 Single Series Algorithms

*Test for Random Fluctuation*

Input: This test accepts single series time data as input. Consider 't' is the time series component and 'v' be the measure component, then input to this test will be, $t_1v_1, t_2v_2, t_3v_3, t_4v_4, \ldots t_nv_n$, where n is the count of components of time series data.

Functionality: This test predicts about the random element in the time series data.

Result: Hypothesis is considered in output, as seasonality is present in time series data or not. Test calculates the value for the test statistic and p-value as the regular test, the system from these values, output interprets whether evidence is there or not for the presence of seasonality in time series data. It also tells level of evidence whether small, moderate, and strong. [**Sample Output**: The evidence against the random element in fluctuations in the data is strong.]

## Design of a Toolbox

*Time Series Statistics*

Input: This test accepts single series time data as input. Consider 't' is the time series component and 'v' be the measure component, then input to this test will be, $t_1v_1, t_2v_2, t_3v_3, t_4v_4, \ldots t_nv_n$, where n is the count of components of time series data.

Functionality: It gives the basic statistics of the time series data.

Result: Result will display the values for basic statistics of the data like, mean, variance, 1st and 2nd partial autocorrelation, slop, and intercept.

*Anderson – Darling Test*

Input: This test accepts single series time data as input. Consider 't' is the time series component and 'v' be the measure component, then input to this test will be, $t_1v_1, t_2v_2, t_3v_3, t_4v_4, \ldots t_nv_n$, where n is the count of components of time series data.

Functionality: This test tells whether the given sample is drawn from the given said of distribution.

Result: Hypothesis is considered in output, as sample data follows normal distribution or not. Test calculates the value for the Anderson – Darling test as the regular output, the system interprets these values and give the descriptive output. From these values, output interprets whether the sample follows normal distribution or not. [**Sample Output**: At the level of Alpha of 0.05, the decision is to reject the null hypothesis that the Sample is normally distributed. Therefore the non-normality of the sample is significant.]

*Runs Test*

Input: This test accepts single series time data as input. Consider 't' is the time series component and 'v' be the measure component, then input to this test will be, $t_1v_1, t_2v_2, t_3v_3, t_4v_4, \ldots t_nv_n$, where n is the count of components of time series data. Along with the standard time series data, the test also asks for the function of central tendency.

Functionality: This test find out whether the sample is having random fluctuation or not. To find out the fluctuation, this test uses any of the central tendency function like mean, median or mode or any other customized function, according to the data set.

Result: The test calculates the value for selected central tendency function, based on the p-value with 0.05 significance it tell about whether the random fluctuation is present in the data or not. [**Sample Output:** Very strong evidence against randomness]

### 16.4.1.2 Two Series Algorithms

*Mann – Whitney U Test*

Input: Input to the algorithm is the multidimensional output. From that two series data is to be taken either row wise or column wise

Functionality: This test is used to find out whether the sample are equal or not. This is determined by testing whether the randomly selected number from the one sample is greater or smaller than the randomly selected from other sample.

Result: Hypothesis is considered for whether there is difference between the sample or not. It calculates the statistical parameter values like U-statistic, expected U-statistic, Z-value, etc. And with the help of these values, this test describes the chances of being the samples are equal or not. [Sample Output: At the level of Alpha of 0.05, the difference within the samples is significant. They have different means. Hence the decision is to reject the null hypothesis that the samples are not different.]

*Two Dependent Population Testing*

Input: Input to the algorithm is the multidimensional output. From that two series data is to be taken either row wise or column wise.

Functionality: This test is used to test the similarity between two populations, by testing mean and variance.

Result: Hypothesis is considered whether there is difference in mean or variance is similar to the claimed difference value. It is claimed that there is no difference in mean or variance. Mean and variance of sample 1 and sample 2, correlation coefficient, spearman rank correlation, all these values are calculated. As a result this test comments on the difference between means of two samples and variance of two samples. [**Sample Output:** (For mean) The evidence against the null hypothesis is small. (For variance) The evidence against the null hypothesis is strong.]

*Two Sample Comparison of Variance*

Input: Input to the algorithm is the multidimensional output. From that two series data is to be taken either row wise or column wise.

Functionality: This test is to find out the variance of samples is having large or small dissimilarity.

Result: Hypothesis is considered whether there is difference in variance is similar to the claimed difference value. It is claimed that there is no difference in variance. Mean and variance of sample 1 and sample 2, test statistic, F-statistics, p-value, all these values are calculated. As a result this test comments on the difference between variance of two samples. [**Sample Output:** There is some evidence against the null hypothesis.]

### 16.4.1.3 Multiple Series Algorithms

*One Factor ANOVA*

Input: Input to the algorithm is the multidimensional output. From which required data is to be taken either row wise or column wise.

Functionality: This test is used to find out whether there is any significant difference between means of more than two independent groups.

Result: Hypothesis is considered whether population means are same or at least one population mean differs significantly. Variance and other values are calculated, and with the help of these values system gives the descriptive result as whether the population mean is same or different. [**Sample Output:** The evidence supports null Hypothesis, i.e. population mean are almost same.]

# Design of a Toolbox

*Kruskal – Wallis Test*

Input: Input to the algorithm is the multidimensional output. From which required data is to be taken either row wise or column wise.

Functionality: It is rank based test. It is also used to find out whether there is difference between two or more independent groups.

Result: Hypothesis is considered as, whether there are differences among the samples or not. Test calculates the test statistic H value, P-value and with the help of these values, it talks about the differences among the samples. [**Sample Output:** At the level of Alpha of 0.05, there is no difference among the samples. They have similar means. The decision is to not reject the null hypothesis that there are no differences among the samples. The difference in the samples is not significant.]

### 16.4.1.4 Matrix Algorithms

*Rank Conversion*

Input: Based on user selection take the time series values from the row or column of the matrix. Consider column wise selection then values will be, $C_1R_1, C_1R_2, \ldots$, similarly for all the columns from matrix.

Functionality: Prepare the data based on the ranks.

Result: Displays the rank of each data point in the column or row.

*Transformations*

Input: Based on user selection take the time series values from the row or column of the matrix. Consider column wise selection then values will be, $C_1R_1, C_1R_2, \ldots$, similarly for all the columns from matrix. Along with data this test takes the various functions on the basis of which transformation will be performed. The functions are, LogN X, X * X, SQRT(X), and moving average.

Functionality: To transformed the data in specific format.

Result: Gives the transformed values for all the data values.

*Pearson Correlation Matrix*

Input: Based on user selection take the time series values from the row or column of the matrix. Consider column wise selection then values will be, $C_1R_1, C_1R_2, \ldots$, similarly for all the columns from matrix.

Functionality: It calculates the strength of the linear relationship between two variables.

Result: Gives the result matrix showing the strength of the relation

### 16.4.2 Advanced Techniques

#### 16.4.2.1 Regression Analysis

Input: First stage input taken by the system can be row wise or column wise from the multidimensional output. In the second stage input to this algorithm will be list of independent variables, and the dependent variable. Also the regression function needs to be provided.

Functionality: Regression analysis is done to find out the relation between the independent variables and dependent variable. Based on the relationship new value can be predicted.

Result: This algorithm gives various values as output. First the equation of regression is given, which tells about the relationship between the independent and dependent variable. System also gives the interpretation of mathematical regression equation, so that easily understood by user. It gives the error in the prediction, other parameter values like, p-value, t-ratio, and interpretation of all these parameters. Impact for all the independent variable is given separately. Multicollinearity analysis is also given. Result also talks about the heteroscedasticity analysis, whether it is present in the data or not. The output shows the descriptive interpretation of the result. [**Sample Output:** Impact Around Minimum (Actual SALE,MARATHAWADA,): 1% Change in (Actual SALE,VIDHARBHA,) = -0.5067976575183908% Change in (Actual SALE,MARATHAWADA,)]

#### 16.4.2.2 XY-Graph

Input: Variable which is to be plotted on X axis and variable to be plotted on Y axis need to be provided as input.

Functionality: To analyse the data by finding out relationship between the variables.

Result: As a result line graph is displayed between the variable on X axis and variables on Y axis.

#### 16.4.2.3 Data Transformation

Input: It requires the field to be transformed, and the expression to transform the selected field. System provides range of standard function along with the facility to provide customized transformation function.

Functionality: This technique is used to transform the data, to prepare it for analysis based on the analysis needs.

Result: It gives the transformed values for the selected data.

#### 16.4.2.4 K-Means

Input: K-means requires number of clusters as an input.

Functionality: This algorithm is used to group the based on similarity features present in the data.

Result: Gives the clusters along with its members. Also gives the graph for the clusters.

### 16.5 CASE STUDY: ANALYZING THE ENVIRONMENT USING THE SPATIO-TEMPORAL DATA GENERATED BY THE SENSORS

#### 16.5.1 MOTIVATION

Technological advancement is taking control of our daily life through various applications. In the current era systems are able to sense the surrounding and can act accordingly. Use of sensors in the systems is the means to achieve the awareness

# Design of a Toolbox

about the surrounding environment. Intelligence can be added in the devices using these sensors. Wide availability of internet infrastructure, motivate system developers to connect these intelligence devices through the internet can be accessed remotely. Hence Internet of Things (IoT) is the upcoming technology which helps in making the systems smart and intelligent. The concept of IoT is spreading throughout the available working domains from daily life, social life, security, commercial applications, research industry, etc. Such type of systems is generating the spatio-temporal data. The characteristic of this type of data is it is generated for particular location and having the temporal (time) factor. The data under consideration for this case study is generated by the sensors located at the various locations like dining room, outer area of the dining room, etc. and the data is generated for each second.

The given case study demonstrates how the system implemented using proposed architecture helps to absorb, model, and analyze the spatio – temporal data generated by sensors, which is generated continuously (stream data) and recorded after specific minutes.

### 16.5.2 Inputs and Workflow

The case study is taking the sensor data generated through a system which is mounted in the domotic house. It includes the daily monitoring data. This data set contains the data generated by 22 sensors. The sensors are mounted at various locations in the house. Following table shows the sensor names and the unit in which the data is generated by that sensor. Table 16.1

Along with all these sensors values, original data set contains the date and time attributes. The data set contains the sampled data that was originally collected for every minute and smoothed with 15-minute means. The data is collected from the UCI data repository (SML 2010).

### 16.5.3 Data Modelling

The first stage before the formal execution of the ETL process is to decide the dimension and measure values, and also decide the dimensional hierarchy from the data.

The dimensions and measure vale identified from the data set are:

- Dimension {Sensor} →{Sensor_Name},
- Dimension {Time} → {sdate},
- Measure {Value} → {Value}.

The multidimensional space with detailed dimensional hierarchy structure for the data under consideration is:

- Dimension {Sensor} → {All, SensorName}
- Dimension {Time} → {All, Year, Month, Day, Hour, Minute, Seconds}
Table 16.2

## TABLE 16.1
## Sensor information

| Sr. No. | Sensor name | Unit of data |
|---|---|---|
| 1 | Indoor temperature (dining-room) | Â°C. |
| 2 | Indoor temperature (room) | Â°C |
| 3 | Weather forecast temperature | Â°C. |
| 4 | Carbon dioxide (dining room). | ppm |
| 5 | Carbon dioxide (room). | ppm |
| 6 | Relative humidity (dining room), | %. |
| 7 | Relative humidity (room), | % |
| 8 | Lighting (dining room). | Lux |
| 9 | Lighting (room). | Lux |
| 10 | Rain, the proportion of the last 15 minutes where rain was detected | |
| 11 | Sun dusk. | |
| 12 | Wind, in | m/s |
| 13 | Sun light in west façade | Lux |
| 14 | Sun light in east façade | Lux |
| 15 | Sun light in south façade | Lux |
| 16 | Sun irradiance\ | W/m2 |
| 17 | Enthalpic motor 1 | 0 or 1 (on-off). |
| 18 | Enthalpic motor 2 | 0 or 1 (on-off). |
| 19 | Enthalpic motor turbo | 0 or 1 (on-off). |
| 20 | Outdoor temperature | Â°C. |
| 21 | Outdoor relative humidity | % |
| 22 | Day of the week | 1 = Monday, 7 = Sunday |

## TABLE 16.2
## Data points in the data set

| Attribute | Number of data points |
|---|---|
| Sensor_Name | 22 |
| Time | 1359 |
| Value | 60412 |

The data is collected at the second level but for analysis requirement, it may be possible to have aggregation at the level of day or month level, hence time hierarchy is created from seconds to year level.

# Design of a Toolbox

| | | Temperature_Corriedor_Sensor | Temperature_Habitacion_Sensor | Temperature_Exterior_Sensor |
|---|---|---|---|---|
| Actual VALUE | 18-Apr-12 0:15 | 20.16 | 19.76 | 16.04 |
| Actual VALUE | 18-Apr-12 0:30 | 20.07 | 19.66 | 16.02 |
| Actual VALUE | 18-Apr-12 0:45 | 19.98 | 19.58 | 15.94 |
| Actual VALUE | 18-Apr-12 1:0 | 19.89 | 19.50 | 15.86 |
| Actual VALUE | 18-Apr-12 1:15 | 19.81 | 19.43 | 15.68 |
| Actual VALUE | 18-Apr-12 1:30 | 19.72 | 19.37 | 15.56 |
| Actual VALUE | 18-Apr-12 1:45 | 19.64 | 19.27 | 15.51 |
| Actual VALUE | 18-Apr-12 2:0 | 19.56 | 19.20 | 15.52 |
| Actual VALUE | 18-Apr-12 2:15 | 19.47 | 19.11 | 15.71 |
| Actual VALUE | 18-Apr-12 2:30 | 19.39 | 19.04 | 15.95 |
| Actual VALUE | 18-Apr-12 2:45 | 19.33 | 18.98 | 16.12 |
| Actual VALUE | 18-Apr-12 3:0 | 19.30 | 18.94 | 16.31 |
| MIN | 18-Apr-12 0 | 80.41 | 78.80 | 64.07 |
| | | 19.30 | 18.94 | 15.51 |
| MAX | | 80.41 | 78.80 | 64.07 |

**FIGURE 16.5** Report for maximum and minimum temperature parameters at various locations till 3 pm.

### 16.5.4 MULTIDIMENSIONAL ANALYSIS

The data set contains the temperature sensors which detect the temperature from dining room, other room, and the exterior environment. If the user wants to analyse the values of all these temperature parameters, then the system generates the report in few seconds, though the data is at the second's level as shown in Figure 16.5.

Sensors are detecting the temperature throughout the day, the analysis system can talk about the random fluctuation present in the collected data or not. The statistical algorithm 'Test for Random Fluctuation', talks about the possibility of presence of randomness in the data set. According to this test there is strong evidence against random element in fluctuation as depicted in Figure 16.6.

System provides one more test to check whether there is randomness in data set or not. This allows us to compare the results. The non – parametric 'runs' test is used to test the presence of random fluctuation in the data set as shown in Figure 16.7.

The non – parametric runs test also states that there is strong evidence against the randomness.

The data set contains the temperature parameter which is forecasted. If user wants to check whether forecasted and actual temperature is similar or different. Series of non – parametric tests are available to test the samples. Using this it is easy to predict whether the actual and forecasted temperature is similar or not as shown in Figure 16.8.

The test indicates that the samples are significantly different. Hence it can be concluded that the forecasted temperature and the actual temperature recorded is significantly different.

If user wants to detect the reason of increase in the humidity in the environment, there might be any factor which is affecting the humidity. Here we have tested the emission of $CO_2$ in the dining room, lightning experienced, temperature, and the humidity at the same location. Regression can be used to find out the relationship between these variables. Figure 16.9 shows the report of regression analysis.

The report generated by the system tells that increase in emission of $CO_2$ increases humidity. Increase in temperature will decrease the humidity, whereas increase in lightening increases the humidity in the atmosphere.

Output-

| Headers | Test Statistic | P-Value | Conclusion |
|---|---|---|---|
| Temperature_comedor_sensor | 64.6863 | 0.0003 | The evidence against the random element in fluctuations in the data is strong |
| Temperature_habitacion_sensor | 54.4821 | 0.0003 | The evidence against the random element in fluctuations in the data is strong |
| Temperature_exterior_sensor | 4.6907 | 0.0007 | The evidence against the random element in fluctuations in the data is strong |

**FIGURE 16.6** Report indicating the presence of randomness in the values for temperature parameters.

**Runs Test**

Ho: Samples is having random Fluctuation

Ha: Samples is not having random Fluctuation

Output:

| Headers | Test Value (Mode) | Cases < Test Value | Cases >= Test Value | Total Cases | Number of Runs | Z Value | P Value | Alpha | Conclusion | Note : Multiple modes have been observed, therefore, highest value within the series will be considered. |
|---|---|---|---|---|---|---|---|---|---|---|
| Temperature_comedor_sensor | 20.1600 | 11.0000 | 1.0000 | 12.0000 | 1.0000 | -2.6833 | 0.0073 | 0.0500 | Very strong evidence against randomness | Multiple modes have been observed, therefore, highest value within the series will be considered. |
| Temperature_habitacion_sensor | 19.7600 | 11.0000 | 1.0000 | 12.0000 | 1.0000 | -2.6833 | 0.0073 | 0.0500 | Very strong evidence against randomness | Multiple modes have been observed, therefore, highest value within the series will be considered. |
| Temperature_exterior_sensor | 16.3100 | 11.0000 | 1.0000 | 12.0000 | 1.0000 | -2.6833 | 0.0073 | 0.0500 | Very strong evidence against randomness | Multiple modes have been observed, therefore, highest value within the series will be considered. |

**FIGURE 16.7** Report to display result of runs test.

**Mann-Whitney U Test**

Applied on Variables: Temperature_Comedor_Sensor and Temperature_Habitacion_Sensor

Ho: There are no differences among the samples
Ha: There are differences among the samples

Output:

| Headers | U Statistic | Alternate Formula for U Statistic | Expected U Statistic | Z (Observed Value) | Z (Two-tailed Critical Value) | Probability Value of Z (Two Tailed) | Alpha | Standard Error of U Statistic | Conclusion |
|---|---|---|---|---|---|---|---|---|---|
| Temperature_comedor_sensor | 72.0000 | 72.0000 | 72.0000 | 0.0000 | -100.0000 | 0.3989 | 0.0500 | 17.3205 | At the level of Alpha of 0.05, the difference within the samples is significant. They have different means. Hence the decision is to reject the null hypothesis that the samples are not different. |

**FIGURE 16.8** Report of Mann-Whitney test showing samples are equal or not.

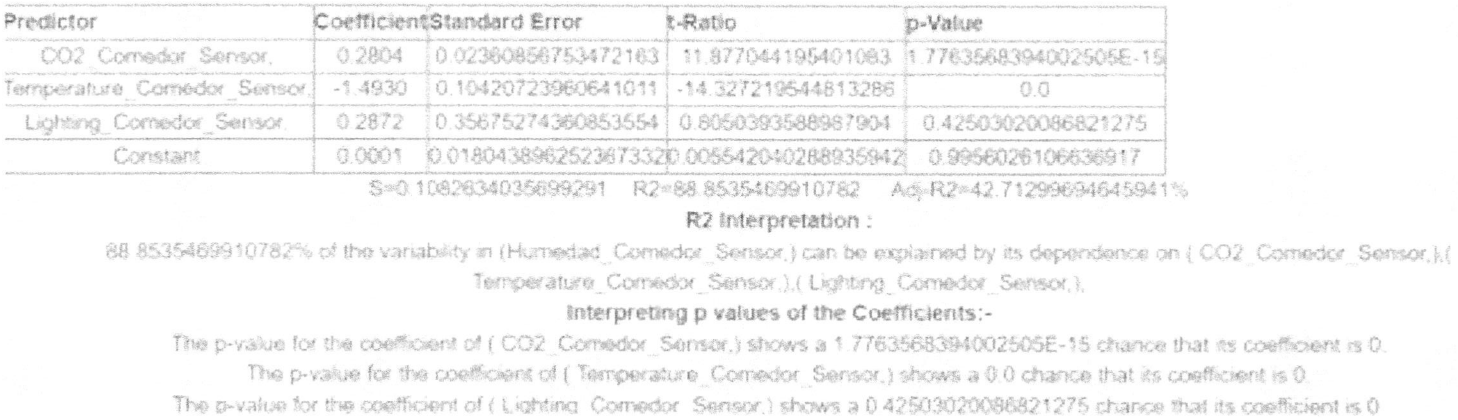

# Design of a Toolbox 295

| Predictor | Coefficient | Standard Error | t-Ratio | p-Value |
|---|---|---|---|---|
| CO2_Corredor_Sensor | 0.2804 | 0.023608575347283 | 11.877044194401083 | 1.7763568394002505E-15 |
| Temperature_Corredor_Sensor | -1.4930 | 0.104207239606410 | -14.327210544813286 | 0.0 |
| Lighting_Corredor_Sensor | 0.2872 | 0.356752743608355 | 0.805039358986790 | 0.42503020086821275 |
| Constant | 0.0001 | 0.0018043896252367332 | 0.0055442002889359 | 0.99560261063691 |

S=0.10628340356990291   R2=88.85354699910782   Adj.R2=42.71294634645941%

R2 Interpretation :

88.85354699910782% of the variability in (Humedad_Corredor_Sensor.) can be explained by its dependence on ( CO2_Corredor_Sensor.),( Temperature_Corredor_Sensor.),( Lighting_Corredor_Sensor.).

Interpreting p values of the Coefficients :-

The p-value for the coefficient of ( CO2_Corredor_Sensor.) shows a 1.7763568394002505E-15 chance that its coefficient is 0.
The p-value for the coefficient of ( Temperature_Corredor_Sensor.) shows a 0.0 chance that its coefficient is 0.
The p-value for the coefficient of ( Lighting_Corredor_Sensor.) shows a 0.42503020086821275 chance that its coefficient is 0.

**FIGURE 16.9** Regression analysis between emission of $CO_2$ and humidity.

## 16.5.5 Outcome of Case Study

The motive behind discussing this case study is to demonstrate how the system is helping to analyse the data which is coming with the level of highest granularity (at the second level). With the help of automated and customized ETL process the data set gets easily absorbed in the system, and the metadata is generated. User can analyse the data with respect to the different sensors and time.

The analysis techniques provided by the system helps to extract the hidden information and trends in the dataset. System reports talk about the similarity or dissimilarity in the data sets, relationship in terms of dependency between the attributes, etc. Hence within few seconds system helps to understand and analyse the data.

## 16.6 CONCLUSION AND FUTURE SCOPE

This chapter describes the functionality of a toolbox which provides modelling and analyzing data. To perform the multidimensional analysis, user needs to generate the events. These events are fully loaded with the customized functionalities attached in terms of filters and functions to the dimensional hierarchy and measure values. Query builder component generates the query from these events and fired on the data to generate the desired multidimensional analysis report, with grid and graph layout. Once event is generated for specific report, it can be stored, so that same report can be generated at any time or can be modified. Events generated are demanding the aggregated dimensional information, whereas no such information is separately stored. Aggregations are generated dynamically at the time of execution of query. The output of multidimensional analysis is given to the analysis system. It includes the good range of statistical algorithms for single series to multiple series and matrix series algorithms. Some advanced algorithms are also provided for analysis. The specialty of this system is all the algorithms taking multidimensional grid as the input; it gives unique feature of advanced analysis.

## REFERENCES

Architecture for Enterprise Business Intelligence – Microstrategy http://www2.microstrategy.com/download/files/whitepapers/open/Architecture-for-Enterprise-BI.pdf

Business Intelligence Platform Administrator Guide. (2016). SAP BusinessObjects Business Intelligence Platform Document Version: 4.2 SP6 – 2018-07-18, available at: https://help.sap.com/doc/ec7df5236fdb101497906a7cb0e91070/4.2.6/en-US/sbo42sp6_bip_admin_en.pdf

Cognos, B. I. (2010). IBM Cognos Business Intelligence V10. 1 Handbook. [Online], [Retrieved December 10, 2012], http://www.ibm.com/redbooks.

Fiore, S., D'Anca, A., Palazzo, C., Foster, I., Williams, D. N., & Aloisio, G. (2013). Ophidia: Toward big data analytics for escience. *Procedia Computer Science, 18*, 2376–2385.

Fiore, S., D'Anca, A., Elia, D., Palazzo, C., Williams, D., Foster, I., & Aloisio, G. (2014, July). Ophidia: A full software stack for scientific data analytics. In 2014 International Conference on High Performance Computing & Simulation (HPCS) (pp. 343–350). IEEE.

Jin, D., & Tsuji, T. (2011, November). Parallel data cube construction based on an extendible multidimensional array. In 2011 IEEE 10th International Conference on Trust, Security and Privacy in Computing and Communications (pp. 1145–1139). IEEE.

Rueter, A. M., & Solutions, S. (2012). Tableau for the enterprise: An overview for IT. *Tableau Software*.

SML. (2010). https://archive.ics.uci.edu/ml/datasets/SML2010

Usman Ahmed. (2013, February). Dynamic cubing for hierarchical multidimensional data space. PhD thesis.

Zorrilla, M., & García-Saiz, D. (2013). A service oriented architecture to provide data mining services for non-expert data miners. *Decision Support Systems*, *55*(1), 399–411.

# 17 IoT Based Intelligent System for Home Automation

*Navjot Kaur Sekhon, Surinder Kaur, and Hatesh Shyan*

## CONTENTS

17.1 Introduction .................................................................................................. 299
    17.1.1 History ............................................................................................. 302
    17.1.2 Working of IoT ................................................................................ 302
    17.1.3 Importance of IoT ............................................................................ 303
    17.1.4 IoT Benefits to Organizations .......................................................... 303
    17.1.5 Advantages and Disadvantages of IoT ............................................ 304
    17.1.6 Privacy and Security Issues ............................................................. 304
    17.1.7 Major Components of IoT ............................................................... 305
    17.1.8 Applications of IoT ......................................................................... 306
17.2 Architecture ................................................................................................... 307
17.3 Literature Survey ........................................................................................... 308
17.4 Experimental Setup ....................................................................................... 310
17.5 Conclusion ..................................................................................................... 312
17.6 Future Scope .................................................................................................. 312
References ............................................................................................................... 313

## 17.1 INTRODUCTION

The day-by-day changing trend of the technology leads to the emergence of some new ideas from the existing ones, and the same can be said about the trendy field of IoT (Said et al., 2017). The crucial idea behind the IoT network is ruling out human intervention in its everyday functioning because connected physical objects of everyday use, known as things, with embedded sensors, micro-controllers, and protocol stacks can communicate with each other and with remote computers to pass on specific information to and receive instructions from so as to facilitate actions on the objects, all happening with Internet connection. Due to increasing demand, day by day, the main advantage of using an IoT application is from enhanced protocols, drivers, and middleware services all developed with efficient coding skills (Pandiyan et al., 2020). Given the field of network of networks is the backbone of IoT, the connected things may be placed in a remote location, and if there is need to access the object without stepping into the location physically, to

**FIGURE 17.1** Physical Objects in IoT Concept.

control the thing, then wireless remote sensors are the only option (Sathyadevan et al., 2019; Singh & Sachan, 2015a, 2015b, n.d., 2015c, 2019). This implies that wireless sensors or a sensor network performs a vital role in activating the idea behind IoT. Figure 17.1 represents some of the physical objects that can directly be considered as things in IoT. The following section of the study gets data from a remote object with the help of Internet. The data related to IoT are generated in huge quantities as any data, however tiny, be it minute to minute change of state in an object or hourly change of state, has to be recorded (Reddy & Krishnamohan, 2018). To deal with this kind of application requirements, huge storage is needed.

It's necessary to introduce the role of cloud computing for the implementation of an IoT network. Cloud storage with its virtually limitless capacity and robust availability has to be used to meet the needs of storing and analyzing of bulk information. To capture, store, analyze and interpret IoT data from millions of connected objects in the context of day-to-day existence of human beings is a challenge because it needs infallible infrastructure and cutting-edge technology to carry on the activities of IoT, mainly sensing, measuring, interpreting, connecting, analyzing and learning, acting and optimizing, and sensing again all in minute-level gaps, and to send it across the network for the system to work. Figure 17.2 represents the position of physical objects in the coming years or in near future, we can say.

Figure 17.3 shows the data from year 2010 to 2020. During 2010, the number of physical devices recorded as things in IoT were 6.8 billion (Al-Mohair et al., 2015; De et al., 2015; Jain & Sahu, 2016; Jin et al., 2015; Silva & Miyasato, 1997; Wu et al., 2001) and within the time span of five year these devices were recorded with an increase rate of 0.4 billion i.e. total recorded devices in 2015 were 7.2 billion. But the right side of the graph shows the challenge, which is the expectation of an approximate 50 billion devices as things in IoT network. To deal with these kinds of

**FIGURE 17.2** State of Physical Devices Over Time.

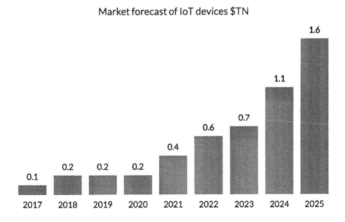

**FIGURE 17.3** Data Generation by IoT Devices.

expectations, the IPv6 addressing scheme is capable but one cannot imagine the scenario over the next five years or so. Figure 17.3 represents the amount of data produced by IoT devices over the years. Given everyday the devices can generate over 20 quintillion data, the amount of storage needed by an IoT network can be easily predicted. This strengthens the concept of cloud computing as a backbone of IoT field. The data introduced by these devices can be considered in the form of normal textual information or in the form of some multimedia data such as video, speech signals etc. Internet of things (IoT) is a set of objects/devices, which are assigned unique IP Addresses, and are able to transmit data over internet/network without the need of human interference. A thing in IoT may be a person with Heart Monitor Implant or a vehicle having sensors to alert the rider when the tires are low on pressure, OR any other object which may be assigned an IP address and can pass on the information over Internet.

### 17.1.1 History

The Internet of Things was first included in a presentation by Kevin Ashton in 1999 made to Procter & Gamble at the Massachusetts Institute of Technology (MIT), though the idea of connected devices has been in use since 1970. Kevin Ashton wanted to draw P&G's management's attention to RFID while calling the presentation "Internet of Things".

A coke-vending machine in early 1980s was the first Internet appliance, which updated the status of coke temperature in stock and also the information about the last refills made. It has been evolving over the years since then and, now, we have billions of devices connected to the Internet. At a very early stage of its development, Internet was used to assign the connection between people and things. In 2008, the number of connections between devices by means of the Internet was more than the population of world; hence, it led to the invention of IoT. To put it simply, IoT deals with the real-world objects or physical objects that are used with various sensing and technological devices to get sense data from the environment and to share information by means of the Internet. The main physical object of IoT is RFID-based, which collects information in microchips and makes that data transferable to the user via wireless communication channel. The main objective of IoT is to optimize the state of affairs within an organized whole by connecting various physical objects, and to perform various functions like analysis of data, sensing incoming information from other objects, applying machine learning algorithms to track sensor location, as well as identifying, monitoring, managing, and deciding, in collaboration with an autonomous distributed system of sensors, information on other devices.

### 17.1.2 Working of IoT

The IoT system consists of intelligent objects having embedded processors, sensors, and connection components to detect a change of state or measure a quantity from the object, analyze, and pass on the information they collect from their environment as an input signal for a computer via an IoT gateway. Also, such objects can transfer data to other devices and process the information they share between themselves. These devices are capable of handling their change of state without human intervention, though users can interact with these devices, maybe to set it up or give instructions or get data. The protocols used with these devices mainly depend on the IoT applications that are deployed. Artificial Intelligence and Machine learning can further aid to make the data collection process easier and dynamic.

Intelligent IoT System

### 17.1.3 IMPORTANCE OF IoT

IoT networks help people in making their lives smart and to get better control over their lives with the help of smart devices/homes. IoT has become essential to businesses as well. IoT helps businesses in getting a real-time picture of how their setup will actually work, give insights into their performance, and connect with the supply chains and logistics operations, all in one, connected, Internet-enabled business setup. IoT helps the companies in not only automating the processes and reducing the labor costs but having enabled connectivity to a level where siloed datasets and processes are replaced with superior connected machines and systems, all connected to a main system to monitor and take action when need. Moreover, it helps in reducing the waste and in making service delivery efficient, for example, by reducing the inventory time gap between inventory request and replenishment so that service delivery costs fall, which also brings transparency in customer transaction handling. IoT has thus assumed significance as a technology of everyday use and we will see it gaining more limelight once more businesses start using it to be more competitive.

### 17.1.4 IoT BENEFITS TO ORGANIZATIONS

IoT is useful in numerous ways to organizations, both industry-specific and from multiple industries. Few of the usual benefits of IoT to organizations include: monitoring business processes and implementing improvements, where need; improving customer experience by streamlining the service delivery process; providing time-and-cost effective solution to organizations and their customers by cutting down on wastage, shrinkage, and time taken to complete service delivery; enhancing employee productiveness by implementing machines for repetitive work and leaving essential workflows for human intervention; integrating new models in business that infuse new ideas in outdated ways of working; making timely decisions to coordinate with workflow change, and thus, help in generating new income. IoT helps businesses to revisit the possible ways to carry on their work by offering insights into siloed processes, and helps improvise business approach and strategies with the help of various tools. It's frequently used in manufacturing and logistic organizations, which make use of various IoT sensors and devices, leave alone being used in agriculture, infrastructure, and home automation. For example, in agriculture, IoT helps farmers by collecting data on weather (rain, temperature, humidity etc.) and also on soil content, especially soil containing water, nutrients, plant and animal waste, and other related features, which helps with automating the farming methods and connecting it to a central server monitored by humans reducing labor and time taken to respond to critical events. Similarly, the infrastructure industry uses IoT to keep an eye on operations from production to warehousing not only saving time and cost but also keeps the workflow paperless, improving quality of workflow, while sensors monitor the changes in building structures, bridges etc and report anomalies. The latest use of IoT is in home automation industry, which in monitoring and manipulating the mechanical and electrical systems in the buildings, helps keep homes safe and liveable. Such smart homes/towns can contribute in reducing waste and power consumption. So, several modern businesses/industries of most sectors (health, finance, manufacturing or retail) are trying to reap the benefits of IoT.

### 17.1.5 Advantages and Disadvantages of IoT

- IoT helps to access information anywhere, on any device, and in real-time.
- Improves communication among various inter-connected objects.
- Saves time and money by transferring the data over Internet in real-time.
- Automates jobs; thus, improving the quality of service with least human errors from doing away with human intervention.
- Security concerns arise as the devices increase and a lot of information is shared over the network, thus increasing the risk of hacking/information stealing.
- As the systems get larger – with maybe millions of devices involved – managing the data from such large number of devices may become a challenge in itself.
- A bug in the system can corrupt every device that is connected to the IoT network.
- Due to lack of any international standard for IoT devices, it may become tough for devices to communicate with each other, when acquired from different vendors.

### 17.1.6 Privacy and Security Issues

The major concern faced by IoT is in the field of security and privacy, as IoT may involve millions of devices connected to the Internet, which further involves billions of data points open to inappropriate use by sinister forces, if the data falls in the hands of data thiefs and hackers.

As IoT has a network of closely connected devices. So, a hacker can preempt a vulnerable point in the network to get access to all of the network's data and tamper its data integrity to make it unusable. The manufacturers and agriculturists, or other stakeholder in the industry, need to update their devices with the latest upgrades in the software incorporating the latest virus detection procedures, and frequently, so that they are less vulnerable to hackers. As in many cases, the connected devices may transmit personal information of the end users, which raises a serious privacy concern for the IoT users, given the valid concern of customers that the companies could use those devices to sell user's personal information. Below are the few privacy and security concerns in wider IoT usage:

    i. **Authentication and Authorization in IoT**: Authentication is the process of verifying the user's access to the resource in question that checks whether he is allowed access or not. According to the research, we are going to identify how to authenticate the object using authentication methods such as ID's, Passwords or other types of credentials. As objects belong to different class, and location, therefore based on their complexity, traditional methods are not used for authentication and authorization of objects.

ii. **Object Identification and Location in IoT**: Object Identification is the process of uniquely identifying the object using some object identification method. Whatever method is used to identify the object, it not only identifies the object but also identifies its properties. However, as the object is associated with its location; so, location identification should also be considered. The recently used location method is Internet protocol based i.e. IPv4/IPv6.
iii. **Cryptosystem and Security Protocols**: Cryptosystem is essentially a method, way, or a technique to implement cryptographic algorithm to make information, data, and communication channels secure. Whenever we talk about secure channel, we need to consider the protocol suite, which consists of a series of protocols to perform secure transmission among various nodes. As cryptosystem and protocols are considered stable and rigorous, therefore, these are not used for resource-constrained devices.
iv. **Vulnerability and Analysis in IoT**: Together with authentication and authorization, or privacy and security, the use of vulnerability and analysis is equally important in IoT security. When any software is in the development phase, a software developer has to check its vulnerabilities to make it fullproof against failures. Before releasing any software project to the market, it undergoes product analysis, which identifies the software's vulnerabilities.
v. **Attacks/Malwares/Viruses in IoT**: Malware is a single name to club together malicious software variants present in the Internet, like spywares, ransomware, or phishing attacks. Succinctly, malwares infect the security system of the IoT defense chain taking control of anything, from the objects' IP to the browser, to gain the rights to access the network's resources and data. In 2013, Symantec discovers the first malware in IoT.

### 17.1.7 Major Components of IoT

Major components of IoT have the SMART factor, that is SMART is synonymous with the things considered under IoT concept. Every physical object linked with IoT must be smart enough to share information, to generate events, and then, perform necessary actions against these events (Bharath & Madhvanath, 2009; Graves et al., 2009; Holzinger et al., 2010; Pratikakis et al., 2012; Konno & Hongo, 1993; Kumar et al., 2011; Lee & Kim, 1994; Lehal, 2009; Lehal et al., 2001; Lehal & Singh, 2000; Razzak et al., 2010; Sharma et al., 2008; Sharma et al., 2009; Seiler et al., 1996). Following are some categories specifying the major components of IoT in Figure 17.4:

- Sensors or Actuators.
- Platforms for Servers.
- Servers or Middleware platforms.
- Engines for Data analysis.
- Applications or Aaps.

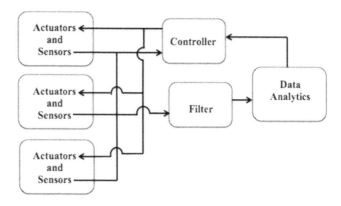

**FIGURE 17.4** Components of IoT.

Sensors or actuators sense information like a change of state requiring action in the device leading to triggering of events, that is, the data about certain states to generate actions. The data sensed must be sent via the network to be recorded on some server or middleware as a part of cloud computing storage. Given the needs for a suitable server platform to enable storage, therefore Internet connection with the cloud server must be available. The enormous quantities of information stored after being sensed by various sensors must be analyzed with various data analytics tools and the results offering recommendations about the event to happen and the actions needed from the applications at user end are updated by the user to create an efficient IoT network.

### 17.1.8 Applications of IoT

Following are some of the applications of IoT which, directly or indirectly, have an impact on human lives by providing a level of comfort:

- **Logistics:** RFID technology is used to track the movement of goods from one location to another.
- **Energy Savings:** Smart buildings, smart grids, smart meters etc. are the examples of the application directly impacting the cause of energy saving.
- **Security and safety:** Smart building, smart homes, Smart cities are the examples where devices are used for the purpose of security and surveillance.
- **Industrial Applications:** Automatic monitoring of industrial equipment to reduce the maintenance cost, automatic fault diagnosis, etc.
- **People tracking and inventory control:** Objects, persons, vehicles can be tracked by using RFID tags, among other ways.
- **Smart Surroundings:** Existence of common sensors like temperature, humidity, smart soil monitoring etc. constitute Smart Surroundings.
- **Smart Home:** This deals with the concept of automated home environment, home appliances, etc.

- **Traffic Monitoring:** traffic information updates of different locations, smart parking sensors, and information on available parking areas, etc.

## 17.2 ARCHITECTURE

The cloud computing model has a three-layered architecture, with the three layers represented as follows as Figure 17.5 depicts:

i. **Application layer:** This layer is in charge of handling all the applications throughout the processing of objects' information in the middle layer.
ii. **Network Layer:** This layer is responsible for transferring information securely collected from various sensing devices on the network layer to the information processing system. Also called "Transmission Layer", the network layer's transmission media can be connection-oriented or connectionless, depending on the devices at the layer.
iii. **Perception Layer:** Sensor layer is used to gather the information from sensors, which includes RFID's, scanning codes, like bar codes based on the unique identification number of objects (Said et al., 2017). The collected data is then sent to the network layer to be securely transmitted to the Application Layer above for further processing. Sensor layer is also known as perception layer or device layer as it is made up of sensors on devices perceiving information from the environment and collecting data from the devices. Sensor layer will have sensors based on the application it will service. Given sensors are used to collect information required to analyze data of interest, and with the Network layer forming a kind of bridge between sensors planted on devices and the cloud storage used to store the bulk of information, it's the application layer that will plough back the results of analysis as recommendations for improving the sensors in the network for a more efficient network. At the level of application layer, the type of processing as per the demands of application under consideration has been performed and the author of the above said literature was also considering to work on the specific layer of the model. Figure 17.5 shows the kind of work performed by each layer in IoT model along with the applicable technology on each layer of the model.

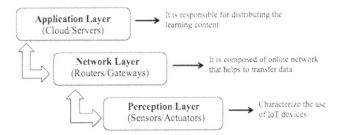

**FIGURE 17.5** Three Layer Architecture of IoT.

## 17.3 LITERATURE SURVEY

Nguyen et al. (2016), in their study, have developed the framework for FRASAD which was used to enhance the MDA method and also upgrades the utilization, sustainability, and adaptability of sensor networks. The proposed architecture was a rule-based model and a Domain Specific Language (DSL) is adopted to delineate the applications. The MDA method is implemented to layout the presented architecture. In this architecture, each and every layer can interact with its upper and lower layers using interfaces (Nguyen et al., 2016).

Mehta et al. (2017) recommendations were implemented in 2017 on Calvin: A Frame for IoT applications which had distributed resources connected in a heterogeneous manner. This research mainly focuses on the thin resources that have very short available memory and limited speed for processing data. The Calvin provides the concept of server-less architecture through which we were able to see function as service or actor as service (Mehta et al., 2017).

Saleem et al. (2017), in their work, have presented a framework for Social IoT requirements and services between the varieties of IoT applications, and also prescribe an IoT network for an example that highlights the relationship among various devices and between people and various things. The researcher proposed the new concept in which objects are able to communicate via people-to-people and people-to-things interfaces, where this concept is called context awareness and capable of creating its own networking infrastructure. This research shows that a person was able to share his information using IoT services with his fellows or things (Saleem et al., 2017).

Laubhan et al. (2016), in this research paper, mainly focused on four-layered architectures for IoT application i.e. node layer, sensor layer, hub layer, and cluster layer. In this architecture, the complete process is expatiated on as to how to collect data from the patients and its environmental factors using the low power-consuming sensor networks. Then the manner in which that data was sent to the central station using the highly upgraded automation systems to make its access easy is presented. The users can check the data using the REST web services of a web (Laubhan et al., 2016).

Mehta et al. (2018), in their survey paper, described the various key challenges faced by IoT, its vision, applications, and various aspects to make IoT a reality for the modern era. As per this research, the author also describes the five-layered IoT architecture which includes layers like Business Layer, Application Layer, Middleware Layer, Network Layer, and Device Layer (Mehta et al., 2018).

Zhang et al. (2014), in their work, describe information security for IoT, the various challenges the IoT has to face (like framework, cryptography, viruses, malwares) to come into existence, and the privacy and security as playing a major role in IoT. So, it defines the very first aspect of security that is to identify the object and its location before considering any other aspect. Whenever we were able to identify the object using proper object identification method, then it not only identified objects but also described the properties of that object (Zhang et al., 2014).

Han et al. (2012), in their survey paper, briefly describe the significance of IoT with technologies and methods in SSME field for asymmetric information collection. This research shows the SSME Theory based IoT application framework has multiple layers: the lowest layer is entity layer, which consists of objects serviced and service provider; application interface layer to communicate between the layers, and which is also called transition layer; and, the topmost layer is called core layer or SSME layer (Han et al., 2012).

Pan et al. (2018), in their research work based on edge chain, described the edge chain as a collection of cloud platforms, that carry out resource management which defines how much resources a device can use from the edge servers. The main focus of this study was on how to manage and control the provision of resources so that when it's time to allocate resources and to decide which device to allocate to, a decision is made backed up by the decision of where to allocate when the resources are free for the use of other devices (Pan et al., 2018).

Sun and Memon (2016), in their research work, explained how to overcome the challenges faced by IoT framework, and also proposed the system layout and related services for maintaining, and scaling up or down, the communication from service layer to physical layer in a way that handles their core challenges in the most expeditious way. By providing such services, our system design was capable of supporting interoperability and handle the heterogeneous objects efficiently. This layout supports many applications like home automation, artificial intelligence, remote sensing, data analytics, geo-sensing services, traffic control systems, smart cities, and remote health monitoring, to name a few (Sun & Memon, 2016).

Datta et al. (2016), in their research work, explained how IoT helps to communicate with peers using sensors, actuators, and RFID tags. This IoT world makes system communication (like home monitoring, remote monitoring, traffic control systems, and automation systems) feasible using Internet. To achieve the smooth functioning of the network, this research proposes a machine-to-machine based MSC model in smart cities for IoT framework which mainly considers three aspects: i) Data Conversion and communication, ii) Data Filteration for abstraction, and iii) Data Distribution to the interested consumers (Datta et al., 2016).

Sri Harsha et al. (2017), in their study, proposed the enhanced automation system with 100% efficiency, low power consumption, and enhanced reliability. The major components that were used to design this system are portable microcontroller, raspberry pi, and Arduino; plus various android-based applications for testing the system were used. The raspberry pi acts as main component or heart of the system that was used to handle the requests/response sent by the android applications, and also worked as the server to store the data collected by the sensing devices like RFID's (Sri Harsha et al., 2017).

Dey et al. (2016), in their research, proposed a home automation system, which has three layers: sensor layer, network layer, and the application layer. The Application layer being the top layer provides communication between applications and system. The Network layer being the middle layer, and also called the device layer, provided data transmission facilities between application and sensor layer. The last and the bottom layer was Sensor layer which operated the sensing devices like RFID's bar coder reader devices used to collect the information (Dey et al., 2016).

Wu et al. (2020), in their research article, recognized the human activity based on IoT to validate the efficiency of the customized learning for effective IoT application (Wu et al., 2020). The authors define the various challenges an IoT network has to face in the fast-paced environment of technology advancement, and also describe the features or factors that can impact future smart applications. In their research work, the authors describe how blockchain has become essential or closely associated with IoT. The author also mentioned how blockchain can overcome the challenges faced by IoT (Pavithran et al., 2020).

Chattopadhyay et al. (2020) implements the SDN with the help of aloe. Aloe framework was designed to reduce the flow setup delay and to handle the fault tolerance of the microcontroller.

Pandiyan et al. (2020), in their work, presented the dynamic cloud framework for IoT application to handle the storage challenges; and the data handling requirements in IoT by using cloud techniques and cloud algorithms.

## 17.4 EXPERIMENTAL SETUP

For the purpose of implementation, a programmable hardware Arduino is used. The Arduino is basically a microcontroller board, works as an open source electronics platform, is easy to use as hardware as well as software, can also be used as sensing device to read inputs, lights on sensors, or fingers on buttons, to generate our own data set. The basic knowledge of C++ or Java programming is sufficient to operate this hardware device. For experimental setup, the Arduino board is required along with Ethernet shield. This Ethernet shield is required to communicate with Internet or other connected networks using Ethernet library while SD library is used to read and write content to SD card. The other alternative for experimenting with IoT is Arduino Yun pack. Two main functions or Arduino environments have been used as setup () and loop(). The function setup () worked as an event generator and Loop () function has been used for representing different cycles. Apart from these, the main function number of other library functions is also available with Arduino environment.

1. **Arduino board (microcontroller):** Arduino is basically a microcontroller board, works as an open source electronics platform, is easy to use as hardware as well as software, and can also be used as sensing device to read inputs, lights on sensors, or fingers on buttons to generate our own data set.
2. **Ethernet shield board:** For the experimental setup, the Arduino board is required along with Ethernet shield. This Ethernet shied is required to communicate with Internet or other connected networks using Ethernet library; and the SD library is used to read and write content to SD card.
3. **Support of Internet:** For providing communication between Ethernet shield and Arduino IDE, an Internet connection of very good capacity bandwidth and of bandwidth quality must be available. Without Internet support, our system will be unable to share the data amassed to generate some sort of output.
4. **A breadboard with some jumper wires:** Breadboard is basically an electronic chip with tiny holes to connect various electronics components to

design, generate, or temporarily prototype an electronic circuit. Jumper wires are simple wires that have connectors at both the ends and are used to provide connection to the various electronic components to complete circuits.
5. **Thermistor Circuit 10 K Ohms:** The thermistor circuit is basically a temperature-sensing circuit that uses 10 K Ohms of resistance to record the room temperature for the given experimental setup. This type of thermistor circuit is used to form a voltage divider, which must produce output voltage.
6. **10 K Ohms Resistor:** Resistors have 4-color band color-coding concept through which we are able to calculate the resistor tolerance.
7. **Door movement monitoring circuit (Ultrasonic distance sensor):** The ultrasonic sensors are used to sense the door movements and to keep track of the change in distance of the object.
8. **Arduino IDE:** Arduino is basically a microcontroller board, works as an open source electronics platform, is easy to use as hardware as well as software, and which can also be used as sensing device to read inputs, lights on sensors, or fingers on buttons to generate our own data set (Figure 17.6).

A 4-pin ultrasonic sensor is used for temperature data generation. The Thermistor Circuit reads the data by using the function analogRead(A0). A0 represents the analog pin 0 of the circuit. A Thermistor resistor will give voltage output which passes through it based on its resistance value. So what we get in "temperature" variable will be just some voltage references. With the help of certain library

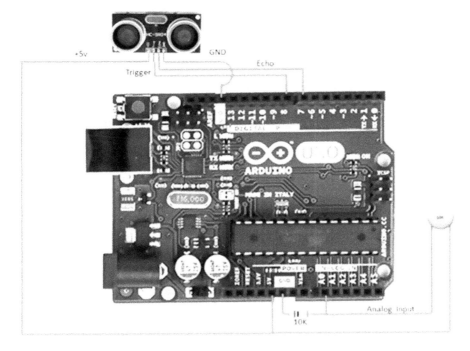

**FIGURE 17.6** Arduino Hardware Kit.

functions, Arduino has the ability to connect to the Ethernet client, and the connection will raise an HTTP request to connect to the specific service. The model used for the purpose of communication is Client-Server model. Various services are requested by the clients after the identification of a suitable server. Arduino IDE also consists of a text editor for writing a program code, a message area, a toolbar with buttons for implementing the common functions, and a text console to check output of an executed code. The main part of this experimental setup is to record room temperature or the object movement activities. Then this information is passed on with the help of Internet to the network. The following equipment's are required according to the experimental setup.

## 17.5 CONCLUSION

This paper elaborates the concept of IoT. It covers different aspects of the research by showing the use of Wireless Sensor Networks (WSN) and cloud computing as backbone to IoT networks. Also the major part of the study has to do with the data analytics being carried out on information gleaned from the IoT. To understand the enormous quantities of data produced by the sensors, it must be converted into useful information. The three-layered architecture of IoT network has also been discussed with the use of applications of IoT in different fields having ensuing benefits to the society. This research study mainly focuses on home automation system using wireless sensor network on the cloud. In the study, we define the procedure of working and managing various types of electrical and electronic appliances using remotely controlled systems. This manner of working and managing appliances is called automation system, which serves to complete our daily routines, nowadays. The proposed system is user-friendly, is based on IoT platform, are connected to wireless sensor networks, and needs cloud computing. The outcomes-based system is very economical, 100% efficient, and consumes little power. To make the system workable throughout, its database is stored on the cloud, where all the files or records are updated and synchronized on a daily basis, and are available to be accessed by authorized users only. By using such systems, we tend to enhance living standards, provide secure and safe environment to the users, make handling devices very fast, and do away with worries of handling and coping with any types of mishap.

## 17.6 FUTURE SCOPE

An extension to this work can be in implementing speech-to-text recognition module for users with visual difficulties to interact with such systems. For implementation, it needs the installation of inexpensive cameras for facial recognition to enhance the user security and user privacy. This work can also be implemented for different platforms i.e. iOS and Android. To integrate more IoT features like tracking the user sleep patterns, health conditions, or enhancement in model lifestyle of the users, the capabilities of artificial intelligence and machine learning algorithms can be implemented.

# REFERENCES

Al-Mohair, H. K., Mohamad Saleh, J. M., & Suandi, S. A. (2015). Hybrid human skin detection using neural network and K-means clustering technique. *Applied Soft Computing*, *33*, 337–347. 10.1016/j.asoc.2015.04.046.

Bharath, A., & Madhvanath, S. (2009). *Online Handwriting Recognition for Indic Scripts*. 10.1007/978-1-84800-330-9.

Chattopadhyay, S., Chatterjee, S., Nandi, S., & Chakraborty, S. (2020). Aloe: Fault-tolerant network management and orchestration framework for IoT applications. *IEEE Transactions on Network and Service Management*, *17*(4), 2396–2409. 10.1109/TNSM.2020.3008426.

Datta, S. K., Da Costa, R. P. F., Bonnet, C., & Harri, J. (2016). *OneM2M architecture based IoT framework for mobile crowd sensing in smart cities* [Conference session]. EUCNC 2016 – European Conference on Networks and Communications, ii, pp. 168–173. 10.1109/EuCNC.2016.7561026.

De, A., Saha, A., & Pal, M. C. (2015). A human facial expression recognition model based on eigen face approach. *Procedia Computer Science*, *45*(C), 282–289. 10.1016/j.procs.2015.03.142.

Dey, S., Roy, A., & Das, S. (2016). *Home automation using Internet of Thing* [Conference session]. 2016 IEEE 7th Annual Ubiquitous Computing, Electronics and Mobile Communication Conference, UEMCON 2016, pp. 1–6. 10.1109/UEMCON.2016.7777826.

Graves, A., Liwicki, M., Fernández, S., Bertolami, R., Bunke, H, & Schmidhuber, J. (2009). A novel connectionist system for unconstrained handwriting recognition. *IEEE Transactions on Pattern Analysis and Machine Intelligence*, *31*(5), 855–868. 10.1109/TPAMI.2008.137.

Han, K., Liu, S., Zhang, D., & Han, Y. (2012). Initially researches for the development of SSME under the background of IoT. *Physics Procedia*, *24*, 1507–1513. 10.1016/j.phpro.2012.02.223.

Holzinger, A., Schlogl, M., Peischl, B., & Debevc, M. (2010). *Preferences of handwriting recognition on mobile information systems in medicine: Improving handwriting algorithm on the basis of real-life usability research* [Conference session]. E-Business (ICE-B), pp. 1–8. 10.1007/978-3-642-25206-8_6.

Jain, C., & Sahu, R. (2016). Study and Comparison of Face Recognition based on PCA, *3*(2), 37–41.

Jin, J., Xu, B., Liu, X., Wang, Y, Cao, L., Han, L., Zhou, B., & Li, M. (2015). A face detection and location method based on Feature Binding. *Signal Processing: Image Communication*, *36*, 179–189. 10.1016/j.image.2015.06.010.

Konno, A., & Hongo, Y. (1993, October). Postprocessing algorithm based on the probabilistic and semantic method for Japanese OCR. In *Proceedings of 2nd international conference on document analysis and recognition (ICDAR'93)* (pp. 646–649). IEEE.

Kumar, M., Jindal, M. K., & Sharma, R. K. (2011). *k-nearest neighbor based offline handwritten Gurmukhi character recognition* [Conference session]. Image Information Processing (ICIIP). 10.1109/ICIIP.2011.6108863.

Laubhan, K., Talaat, K., Riehl, S., Morelli, T., Abdelgawad, A., & Yelamarthi, K. (2016). A four-layer wireless sensor network framework for IoT applications. *Midwest Symposium on Circuits and Systems*, *0*(October), 16–19. 10.1109/MWSCAS.2016.7870142.

Lee, S. W., & Kim, E. S. (1994). Efficient postprocessing algorithms for error correction in handwritten Hangul address and human name recognition. *Pattern Recognition*, *27*(12), 1631–1640. 10.1016/0031-3203(94)90082-5.

Lehal, G. (2009). A survey of the state of the art in Punjabi language processing. *Language in India*, *9*(10), p11–p11.

Lehal, G. S., Singh, C., & Lehal, R. (2001). *A shape based post processor for Gurmukhi OCR* [Conference session]. Document Analysis and Recognition, ICDAR, 2001-January, pp. 1105–1109. 10.1109/ICDAR.2001.953957.

Lehal, G.S., & Singh, C. (2000). *A Gurmukhi script recognition system* [Conference session]. Pattern Recognition, International Conference, vol. 2, pp. 557–560. 10.1109/ICPR.2000.906135.

Mehta, A., Baddour, R., Svensson, F., Gustafsson, H., & Elmroth, E. (2017). *Calvin constrained – A framework for IoT applications in heterogeneous environments* [Conference session]. Distributed Computing Systems, pp. 1063–1073. 10.1109/ICDCS.2017.181.

Mehta, R., Sahni, J., & Khanna, K. (2018). Internet of Things: Vision, applications and challenges. *Procedia Computer Science*, *132*, 1263–1269. 10.1016/j.procs.2018.05.042

Nguyen, X. T., Tran, T. T., Baraki, H. T., & Geihs, K. (2016). *Frasad* [World forum]. Internet of Things, WF-IoT 2015, pp. 387–392.

Pan, J., Wang, J., Hester, A., Alqerm, I., Liu, Y., & Zhao, Y. (2018). EdgeChain: An edge-IoT framework and prototype based on blockchain and smart contracts. *ArXiv*, *6*(3), 4719–4732.

Pandiyan, S., Lawrence, T. S., Sathiyamoorthi, V., Ramasamy, M., Xia, Q., & Guo, Y. (2020). A performance-aware dynamic scheduling algorithm for cloud-based IoT applications. *Computer Communications*, *160*(April), 512–520. 10.1016/j.comcom.2020.06.016.

Pavithran, D., Shaalan, K., Al-Karaki, J. N., & Gawanmeh, A. (2020). Towards building a blockchain framework for IoT. *Cluster Computing*, *23*(3), 2089–2103. 10.1007/s10586-020-03059-5.

Razzak, M. I., Anwar, F., Husain, S. A., Belaid, A., & Sher, M. (2010). HMM and fuzzy logic: A hybrid approach for online Urdu script-based languages' character recognition. *Knowledge-Based Systems*, *23*(8), 914–923. 10.1016/j.knosys.2010.06.007.

Reddy, M., & Krishnamohan, R. (2018). Applications of IoT: A study. Special issue published in *International Journal of Trend in Research & Development (IJTRD)*, pp. 86–87. 10.13140/RG.2.2.27960.60169.

Pratikakis, I., Gatos, B., & Ntirogiannis, K.. (2012, September). ICFHR 2012 competition on handwritten document image binarization (H-DIBCO 2012). In *2012 international conference on frontiers in handwriting recognition* (pp. 817–822). IEEE.

Said, O., Albagory, Y., Nofal, M., & Al Raddady, F. (2017). IoT-RTP and IoT-RTCP: Adaptive protocols for multimedia transmission over Internet of Things environments. *IEEE Access*, *5*, 16757–16773.

Saleem, Y., Crespi, N., Rehmani, M. H., Copeland, R., Hussein, D., & Bertin, E. (2017). *Exploitation of social IoT for recommendation services* [World forum]. Internet of Things, WF-IoT 2016, 3rd World Forum, pp. 359–364. 10.1109/WF-IoT.2016.7845500.

Sathyadevan, S., Achuthan, K., & Doss, R. (2019). Protean authentication scheme – A time-bound dynamic KeyGen authentication technique for IoT edge nodes in outdoor deployments. *IEEE Access*, *7*, 92419–92435. 10.1109/ACCESS.2019.2927818.

Seiler, R., Schenkel, M., & Eggimannn, F. (1996). *Off-line cursive handwriting recognition compared with on-line recognition* [Conference session]. Pattern Recognition, vol. 4, pp. 505–509. 10.1109/ICPR.1996.547616.

Sharma, A., Kumar, R., & Sharma, R. K. (2008). *Online handwritten Gurmukhi character recognition using elastic matching* [Congress session]. Image and Signal Processing, CISP 2008, 1st International Congress, vol. 2. 10.1109/CISP.2008.297.

Sharma, Anuj, Kumar, R., & Sharma, R. K. (2009). *Rearrangement of recognized strokes in online handwritten Gurmukhi words recognition* [Conference session]. Document Analysis and Recognition, ICDAR, pp. 1241–1245. 10.1109/ICDAR.2009.36.
Silva, L. C. D. E., & Miyasato, I. T. (1997). *Facial emotion recognition using multi-modal information* [Conference session]. Information, Communications and Signal Processing ICICS, September, pp. 9–12. 10.1109/SMC.2015.387.
Singh, G., & Sachan, M. (2015a). A framework of online handwritten Gurmukhi script recognition. *International Journal of Computer Science and Technology (IJCST), 6*(3), 52–56.
Singh, G., & Sachan, M. (2015b). *Multi-layer perceptron (MLP) neural network technique for offline handwritten Gurmukhi character recognition* [Conference session]. Computational Intelligence and Computing Research, IEEE ICCIC 2014. 10.1109/ICCIC.2014.7238334.
Singh, G., & Sachan, M. K. (2015c). *Data capturing process for online Gurmukhi script recognition system* [Conference session]. Computational Intelligence and Computing Research (ICCIC), pp. 518–521. 10.1109/ICCIC.2015.7435778.
Singh, G., & Sachan, M. K. (2019). A bilingual (Gurmukhi-Roman) online handwriting identification and recognition system. *International Journal of Recent Technology and Engineering 2277-3878, 8*(1), 2936–2952.
Singh, G., & Sachan, M. K. (n.d.). *An Unconstrained and Effective Approach of Script Identification for Online Bilingual Handwritten Text*. 10.1007/s40009-020-00889-0.
Sri Harsha, S. L. S., Chakrapani Reddy, S., & Prince Mary, S. (2017). *Enhanced home automation system using Internet of Things* [Conference session]. IoT in Social, Mobile, Analytics and Cloud, I-SMAC 2017, pp. 89–93. 10.1109/I-SMAC.2017.8058302.
Sun, L., & Memon, R. A. (2016). *Sun2017*, pp. 154–162.
Wu, J., & Zhang, X. (2001). *A PCA classifier and its application in vehicle detection* [Conference session]. Neural Networks, IJCNN'01. International Joint Conference (Cat. No.01CH37222), vol. 1, pp. 600–604. 10.1109/IJCNN.2001.939090.
Wu, Q., He, K., & Chen, X. (2020). Personalized federated learning for intelligent IoT applications: A cloud-edge based framework. *IEEE Open Journal of the Computer Society, 1*, 35–44.
Zhang, Z. K., Cho, M. C. Y., Wang, C. W., Hsu, C. W., Chen, C. K., & Shieh, S. (2014). *IoT security: Ongoing challenges and research opportunities* [Conference session]. Service Oriented Computing and Applications, SOCA 2014, IEEE 7th International Conference, pp. 230–234. 10.1109/SOCA.2014.58.

# 18 Digital Learning Acceptance during COVID-19: A Sustainable Development Perspective

*Praveen Srivastava, Shelly Srivastava, S.L. Gupta, and Niraj Mishra*

## CONTENTS

18.1 Introduction ..................................................................................................... 317
18.2 Problem Statement ........................................................................................... 318
18.3 Literature Review ............................................................................................. 319
18.4 Research Methodology .................................................................................... 320
18.5 Findings ........................................................................................................... 320
18.6 Discussion ........................................................................................................ 322
18.7 Managerial Implication ................................................................................... 323
18.8 Limitation ........................................................................................................ 323
18.9 Future Research ............................................................................................... 324
References ................................................................................................................ 324

## 18.1 INTRODUCTION

E-learning, as the term implies, is the use of the computer in order to provide schooling or online programs. Since the emergence of e-learning, the term has been sublet as a generic term for any online course (Hemming, 2008). E-learning has paved way for the evolving concept in education, i.e. hybrid learning. It can be understood as the deliberate combination of conventional and digital training in order to provide educational resources that optimize the advantages of each medium and, therefore, to more efficiently promote student learning (Ayaia, 2009). Today, online education or e-learning is a strong and innovative medium to expand conventional forms of learning and develop capability in the field of education (Kemp & Jones, 2007; Mikhaylov & Fierro, 2015). With the increasing popularity of online education, higher education organizations, globally, have made the transition from traditional frameworks to modern Learning Management Systems (LMS) (Diep et al., 2017).

Online education methods implemented by several universities, globally, allowed students to participate in their independent study without concern for their job obligations, location and distance from educational site, and insufficient prelearning interactions which could be hurdles, if present, leading to students being refused full-time conventional learning (Khatib, 2013). Hence, it gives opportunity to the student community to avail the benefit of learning without physically attending the classes. However, to avail the full benefit of online learning, several dimensions of effective learning need attention from its practitioners, specially in the developing countries. These countries face the issue of infrastructural advancement, pedagogic concerns, as well as the need to connect the utility of technology to student life thereby improving the learning experience (O'neill et al., 2004). Given the complexity of online learning, study into this emerging area of teaching will take a number of forms and answer a number of concerns (Ahern & Repman, 1994).

In the future, this technical direction will theoretically improve the learning method, though, it will not substitute the lecturer or mentor. When scholars teach or students learn, the ramifications of ed-tech are extensive. Universities are being more dependent on tailoring their courses to student's expectations from learning courses in an attempt to accommodate the varied educational needs of students, which would eventually enable courses evaluated as good or bad based on comparisons between courses to change to match the student's needs and not the educational institution's needs (O'neill et al., 2004).

The COVID-19 pandemic forced all the educational institutions to close their gates to the students. A lock-down was imposed in many countries and, hence, physical classes were not possible. Digital mode of learning was the only option left to students with physical classes not possible during the pandemic because hybrid learning as practised before the pandemic, where students had the option to physically interact with their mentor, was ruled out. This led to global increase in the use of e-learning platforms. Though the acceptance of any technology differs from one individual to another, due to scarcity of time, every student was forced to adopt it within the same time frame.

## 18.2 PROBLEM STATEMENT

Though e-learning is not a new phenomenon for students, its adoption among student community needed more time and should have been a gradual process rather than a sudden one. Since the undergraduate students were forced to adapt to it as soon as pandemic hit the globe, the acceptance of digital learning among students remained a neglected area. The institutions were not concerned since there was no other way to continue the learning-and-teaching process and still adapt to the pandemic's conditions and, hence, they had to continue with online learning. This implies that the knowledge was to be disseminated digitally irrespective of the preparedness of the students to imbibe the learning that the teachers disseminated through the online mode. However, whether online learning was accepted has not been much explored in research previously. Hence, it becomes imperative to investigate the acceptance of digital learning among students by delving into the preparedness of students and teachers towards e-learning. With this backdrop, the present study is an attempt to explore the acceptance of digital learning among undergraduate students during pandemic.

## 18.3 LITERATURE REVIEW

Online readiness in the learning community has been assessed by numerous previous studies which make an effort to explore how much students are ready to accept e-learning. (Hung, 2016; Smith et al., 2003). However, there are scarce studies which focus on acceptance of technology enabled e-learning after it has been used by the student. There is a plethora of research focusing on technology acceptance, and various theories have assessed the technology acceptance with a pioneering study being the Technology Acceptance Model (TAM) proposed by Fred D Davis (Davis, 1989). This model hypothesized that as people are confronted with a new technology, a variety of considerations will affect not only their choice but also when and how they will use it. Perceived usefulness and Perceived ease of use were identified as two important constructs influencing the technology acceptance. External influences such as social influence or other societal viewpoints also affect the way the individual thinks about the technology in question. A model was later proposed by Venkatesh (Venkatesh et al., 2003), highlighting the Unified Theory of Acceptance and Use of Technology (UTAUT). The UTAUT attempts to clarify the intentions surrounding use of an information system and the resulting user actions. According to this model, there are four main constructs of technology acceptance namely performance expectancy, effort expectancy, social influence, and facilitating condition. Several previous studies have successfully applied this model in various technology acceptance situations. The same model can also be used to identify the construct of technology enabled e-learning acceptance.

In recent years, a number of research efforts has been conducted on the role of online learning in higher education (Al-Rahmi et al., 2018; Hollis & Was, 2016; Tess, 2013). A study investigated student views of online instruction in higher education courses. The study concluded that effective communication is one of the most important elements of a successful online course (Young, 2006). In another study, it was found that effective e-learning is influenced by course content which is appropriately designed, has depth of interaction between the instructor and learners, takes into account instructors preparedness level and support, is conducive to development of online learning community, and spurs on rapid advancement of technology (Sun & Chen, 2016)

A study done on higher education sector in Korea found that learning satisfaction was related to learning achievement. The study has further found that certain psychological variables such as "achievement goal orientation, self-regulation, test-anxiety, self-efficacy for designing, and managing online learning environments" are crucial for success of online learning (Im & Kang, 2019).

## 18.4 RESEARCH METHODOLOGY

Data for present study was obtained from undergraduate students Pan-India. An electronic questionnaire was used to collect the data. The questionnaire was framed by modifying the item stated in the UTAUT model and making it e-learning specific. A total of 196 data was received. However, after scrutiny, 160 data was found to be usable. Smart PLS was used to analyse the obtained data. A modified scale to measure the four constructs was used. Measurement model metrics were verified and, after establishing reliability and validity of the construct, the structural model was analysed using Smart PLS.

## 18.5 FINDINGS

The measurement model gave the result for reliability and validity. The proposed model and the value obtained after running PLS Algorithm is given in Figure 18.1.

The model suggests that the outer loading value of all the constructs is above the minimum threshold of 0.6 which explains that there is no problem in the items selected for construct. Further, the $R^2$ value obtained is 0.714, which implies that all the construct explains 71.4% of the behavioural intention toward e-learning acceptance.

Further, Table 18.1 displays the results of calculations for reliability and validity of the data.

Table 18.1 suggests that all the values of Cronbach's alpha and Composite Reliability are above the prescribed minimum value of 0.7. Hence, the reliability

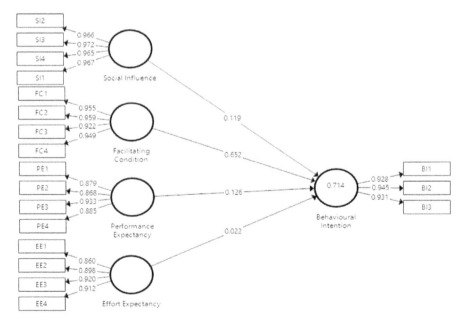

**FIGURE 18.1** Research model.

## TABLE 18.1
### Reliability and Validity of Data

| Column1 | Cronbach's Alpha | rho_A | Composite Reliability | Average Variance Extracted (AVE) |
| --- | --- | --- | --- | --- |
| Behavioral Intention | 0.928 | 0.93 | 0.954 | 0.874 |
| Effort Expectancy | 0.92 | 0.925 | 0.943 | 0.806 |
| Facilitating Condition | 0.961 | 0.963 | 0.972 | 0.895 |
| Performance Expectancy | 0.914 | 0.921 | 0.939 | 0.795 |
| Social Influence | 0.977 | 0.979 | 0.983 | 0.936 |

is established. The value of Average Variance Extracted (AVE) is above 0.5, this indicates that the items explain more than 50% of the construct, which is above the prescribed minimum for convergent validity (Henseler et al., 2009). Hence convergent validity is established. In order to establish discriminant validity, the (Fornell & Larcker, 1981) criteria was applied. The result is depicted in Table 18.2 which suggests that all the diagonal values are greater than corresponding off-diagonal values in the table. Hence, the discriminant validity is established.

After validating the reliability and validity, the structural model was analysed by bootstrapping method. The result of the bootstrapping is shown in Table 18.3.

The result indicates that the relationship of the three constructs i.e. Performance Expectancy, Social Influence, and Facilitating Condition with Behavioral Intention is significant ($t > 1.96$, $p < 0.05$). Hence, these variables have positive and significant impact on the digital learning acceptance. However, the relationship between effort expectancy and behavioral intention is not found to

## TABLE 18.2
### Discriminant Validity

| | Behavioural Intention | Effort Expectancy | Facilitating Condition | Performance Expectancy | Social Influence |
| --- | --- | --- | --- | --- | --- |
| Behavioral Intention | 0.935 | | | | |
| Effort Expectancy | 0.64 | 0.898 | | | |
| Facilitating Condition | 0.833 | 0.703 | 0.946 | | |
| Performance Expectancy | 0.61 | 0.715 | 0.626 | 0.892 | |
| Social Influence | 0.669 | 0.582 | 0.726 | 0.501 | 0.967 |

## TABLE 18.3
### Bootstrapping Result

| | Original Sample (O) | Sample Mean (M) | Standard Deviation (STDEV) | T Statistics (\|O/STDEV\|) | P Values |
| --- | --- | --- | --- | --- | --- |
| Effort Expectancy -> Behavioral Intention | 0.022 | 0.018 | 0.083 | 0.27 | 0.787 |
| Facilitating Condition -> Behavioral Intention | 0.652 | 0.651 | 0.081 | 8.034 | 0 |
| Performance Expectancy -> Behavioral Intention | 0.126 | 0.128 | 0.052 | 2.439 | 0.015 |
| Social Influence -> Behavioral Intention | 0.119 | 0.124 | 0.06 | 2 | 0.046 |

be significant ($t < 1.96$, $p > 0.5$), hence, it can be concluded, given effort expectancy does not impact the behavioral intention toward acceptance of digital learning during COVID-19, thus it implies this relationship does not influence the e-learning acceptance.

## 18.6 DISCUSSION

From the findings, we can conclude that various factors influence the acceptance of digital learning among undergraduate students during COVID-19. It is understood that during the pandemic, e-learning was the only option and students were forced to accept it, but the actual acceptance of digital learning differs from the picture where they were forced to accept it. Though several students have found e-learning to be more convenient as they can continue learning being at the comfort of their home, still, students may not be able to concentrate on the lectures as they are in their home environment and that there are disturbances to learning cannot be denied. Additionally, the findings suggest that *Facilitating Conditions* significantly impact the acceptance of e-learning. Hence, factors like poor internet connectivity, lack of proper machines (computer, laptop, mobile etc.), or poor audio-video reception will lead to instances of poor acceptance among students. This factor has been given maximum weightage as its strength of relationship with intention behaviour is highest. Hence, it can be concluded that the learning infrastructure from the student or mentor end has a huge impact on instances of technological acceptance. In a developing country, where internet has not reached everyone's home and connectivity issue is a frequent problem, and where personal computers, laptops, or smart phones are not cheaply available, the acceptance of digital learning will be an issue. Next, the *Social Influence* i.e. peer pressure, which is social pressure created by friends, relatives, teachers, or batchmates to accept e-learning, plays a significant role in accepting the e-learning as the prevalent mode of study. Students will make an effort to understand the technology and adapt it, if the social group which holds it in importance wants them to do so. However, if the social group is of the opinion that e-learning is of no use and does not create the pressure on students to accept it, students may not make an effort to accept digital learning. Additionally, results suggest that *Performance Expectancy* plays a significant role in e-learning acceptance. This implies the degree to which students believe using the digital-learning will help them. If a student is of the opinion that e-learning will improve their skill and will encompass additional knowledge, they will show an inclination to accept it. However, if they believe otherwise, then they may not accept this form of education.

Lastly, the finding suggests that *Effort expectancy* does not impact the acceptance of e-learning.

This indicates that the amount of effort required to accept e-learning is not of importance for its acceptance. Students are not afraid of making an effort to accept e-learning, provided they believe that it will improve their performance, that the facility provided is good, or that there is social pressure among them to accept the same. This finding is not in agreement to the findings in several previous studies of technology and e-learning acceptance (Abushanab & Pearson, 2007; Tarhini et al., 2016) and, hence, this suggests that the amount of effort required for acceptance of e-learning is not important during the COVID-19 pandemic.

## 18.7 MANAGERIAL IMPLICATION

The finding is of use to the educational institution as there are many takeaways for them, which is listed below.

- An effort is required to provide good internet connectivity to the teachers so that there is no technical issue from their end. Institutions should work to find a way for the same.
- Teachers should make an effort to make e-learning more informative. The latest pedagogy like Flip classroom, or providing out-of-class activity etc. should be tried to ensure that students understand the importance of e-learning in improving their skills.
- Teachers and management should interact with students online before starting the session. This interaction will be of help to ensure that students understand the e-learning process and the importance of it. Also, it may act as a social influence for students to accept e-learning.

## 18.8 LIMITATION

Data is self-reported by the students and, hence, the chance of personal bias cannot be denied. Also, the response may have been influenced by the peer group or a last experience with digital learning.

## 18.9 FUTURE RESEARCH

Present study will encourage the researchers to investigate the results by analysing the moderating role of several variables on the identified relationships. Moderators like gender, or experience with digital learning etc. may provide an interesting result. Additionally, the continuation of digital learning after the pandemic provides a relevant research area. Institutions are planning to open their doors again to the students. Hence, to investigate students' intention to continue with this mode of learning, once the compulsion is no longer there, may reveal some surprise output. This may provide us with the preferred mode of learning after lockdown i.e. Digital, Hybrid or Offline. Hence, these directions in future studies will be of great help to researchers and institutions alike.

## REFERENCES

Abushanab, E., & Pearson, J. M. (2007). Internet banking in Jordan: The unified theory of acceptance and use of technology (UTAUT) perspective. *Journal of Systems and Information Technology*, *9*(1), 78–97. 10.1108/13287260710817700.

Ahern, T. C., & Repman, J. (1994). The effects of technology on online education. *Journal of Research on Computing in Education*, *26*(4), 537–546. 10.1080/08886504.1994.10782109.

Al-Rahmi, W. M., Alias, N., Othman, M. S., Marin, V. I., & Tur, G. (2018). A model of factors affecting learning performance through the use of social media in Malaysian higher education. *Computers and Education*, *121*, 59–72. 10.1016/j.compedu.2018.02.010.

Ayaia, S. J. (2009). Blended learning as a new approach to social work education. *Journal of Social Work Education*, *45*(2), 277–288.

Davis, F. D. (1989). Perceived usefulness, perceived ease of use, and user acceptance of information technology. *MIS Quarterly: Management Information Systems*, *13*(3), 319–339. 10.2307/249008.

Diep, A. N., Zhu, C., Struyven, K., & Blieck, Y. (2017). Who or what contributes to student satisfaction in different blended learning modalities? *British Journal of Educational Technology*, *48*(2), 473–489. 10.1111/bjet.12431.

Fornell, C., & Larcker, D. F. (1981). Evaluating structural equation models with unobservable variables and measurement error. *Journal of Marketing Research*, *18*(1), 39–50. 10.2307/3151312.

Hemming, H. (2008). E-Learning, in a world with too much information. *Legal Information Management*, *8*(1), 43–46. 10.1017/s1472669608000008x.

Henseler, J., Ringle, C. M., & Sinkovics, R. R. (2009). The use of partial least squares path modeling in international marketing. *Advances in International Marketing*, *20*, 277–319. 10.1108/S1474-7979(2009)0000020014.

Hollis, R. B., & Was, C. A. (2016). Mind wandering, control failures, and social media distractions in online learning. *Learning and Instruction*, *42*, 104–112. 10.1016/j.learninstruc.2016.01.007.

Hung, M. L. (2016). Teacher readiness for online learning: Scale development and teacher perceptions. *Computers and Education*, *94*, 120–133. 10.1016/j.compedu.2015.11.012.

Im, T., & Kang, M. (2019). Structural relationships of factors which impact on learner achievement in online learning environment. *International Review of Research in Open and Distance Learning*, *20*(1), 112–124. 10.19173/irrodl.v20i1.4012.

Kemp, B., & Jones, C. (2007). Academic use of digital resources: Disciplinary differences and the issue of progression revisited. *Educational Technology & Society, 10*(1), 52–60. http://www.ifets.info/index.php?http://www.ifets.info/issues.php?id=34.

Khatib, N. M. (2013). Students attitudes towards the web based instruction. *Gifted and Talented International, 28*(1–2), 263–267. 10.1080/15332276.2013.11678421.

Mikhaylov, N. S., & Fierro, I. (2015). Social capital and global mindset. *Journal of International Education in Business, 8*(1), 59–75. 10.1108/jieb-09-2014-0018.

O'Neill, K., Singh, G., & O'Donoghue, J. (2004). Implementing eLearning programmes for higher education: A review of the literature. *Journal of Information Technology Education: Research, 3*, 313–323. 10.28945/304.

Smith, P. J., Murphy, K. L., & Mahoney, S. E. (2003). Towards identifying factors underlying readiness for online learning: An exploratory study. *Distance Education, 24*(1), 57–67. 10.1080/01587910303043.

Sun, A., & Chen, X. (2016). Online education and its effective practice: A research review. *Journal of Information Technology Education: Research, 15*, 157–190. http://www.informingscience.org/Publications/3502.

Tarhini, A., Teo, T., & Tarhini, T. (2016). A cross-cultural validity of the E-learning Acceptance Measure (ElAM) in Lebanon and England: A confirmatory factor analysis. *Education and Information Technologies, 21*(5), 1269–1282. 10.1007/s10639-015-9381-9.

Tess, P. A. (2013). The role of social media in higher education classes (real and virtual)-A literature review. *Computers in Human Behavior, 29*(5), A60–A68. 10.1016/j.chb.2012.12.032.

Venkatesh, V., Morris, M. G., Davis, G. B., & Davis, F. D. (2003). User acceptance of information technology: Toward a unified view. *MIS Quarterly: Management Information Systems, 27*(3), 425–478. 10.2307/30036540.

Young, S. (2006). Student views of effective online teaching in higher education. *American Journal of Distance Education, 20*(2), 65–77. 10.1207/s15389286ajde2002_2.

# 19 A Framework for Real-Time Accident Prevention using Deep Learning

*Pashmeen Kaur, Dr. K.C. Tripathi, and Prof. (Dr.) M.L. Sharma*

## CONTENTS

19.1 Introduction ........................................................................................................327
19.2 Drowsiness Detection ........................................................................................328
    19.2.1 Ways to Detect Drowsiness ...................................................................328
    19.2.2 Related Work ..........................................................................................329
19.3 Proposed Model ..................................................................................................330
    19.3.1 System Requirement ..............................................................................330
    19.3.2 Proposed System Algorithm ..................................................................331
    19.3.3 Pre-Processing .......................................................................................332
    19.3.4 Classification and Feature Extraction using CNN ................................332
19.4 Results .................................................................................................................334
19.5 Conclusion and Future Scope .............................................................................335
References ....................................................................................................................336

## 19.1 INTRODUCTION

One of the basic needs of a Human Being is considered to be the need for sleep. When a human body doesn't get enough sleep, it responds inefficiently to brain stimulus requiring activity. Lack of sleep reduces both the reflex action speed and wakefulness in a person. It also leads to loss of concentration to finish an activity he has started and low alertness to surroundings. Different activities based on care also reduces the person's ability to perform, especially in the case when a person is driving a vehicle.

    Nowadays, many people use automobiles to commute to their destinations comfortably within time constraints, and to keep to their living standards. This lifestyle behavior has led to the rise in the volume of traffic in urban areas as well as on connecting highways. With rising traffic, the chances of road accidents, resulting in deaths of and physical injuries in people, has increased multifold to four times equaling to over a lakh people affected every year. A report for the last decade shows the average number of deaths because of road accidents in India to be 136118 per year. A major count of this was accounted to be from the age group of 18–35 in 2016, which grossed 60% of the total deaths.

The data released by the Indian Road Transport in 2017, in the presence of highways minister Nitin Gadkari, showed that around 147913 people were killed in road accidents. Uttar Pradesh, with the largest road accident fatalities of 20124 accidents, was followed by Tamil Nadu with 16157 people. About 40% of accidents were caused by people dozing off at the wheel while driving, and shows research done by the Central Road Research Institute (CRRI) who based their analysis on accidents taking place on the 300-km Agra-Lucknow Expressway. The occurrence of these accidents was found to be mostly between 2 am and 5 am, the unusual time when people are expected to be asleep rather than driving on roads.

It is found in studies that the major reason for these accidents, and accompanying death in many cases, is more likely to be from deprivation of sleep and drowsy driving than from texting while driving, drunk driving, and negligent driving combined. Thus, to reduce driver falling asleep at wheels and, therefore, to reduce the number of accidents taking people's lives, a sleep detection system is needed to alert the driver with the help of an alarm.

Here, we are going to discuss one of the techniques introduced to beat this challenge, which makes use of Deep Learning and Computer Vision. The model proposed uses Convolutional Neural Networks to extract the desired features indicating sleep that meets our purpose, giving us good accuracy of alerting drivers when they are feeling drowsy.

## 19.2 DROWSINESS DETECTION

As discussed, the requirement for a regulatory system has been rising to preempt the instances of deaths from occurring, many because of their lack of care on the road while driving. The way to conquer this challenge is to detect drowsiness as it infiltrates the driver's wakefulness and to put him/her on alert with a proctored alarm system installed in the car for the same. Hence, the primary step is to detect the drowsiness in a person accurately, so let's consider the ways of doing so below.

### 19.2.1 Ways to Detect Drowsiness

Many different factors can be observed to diagnose drowsiness in a person, such as the state of one's eyes, one's stance, lane detection (whether a person is in a lane or not), frequent yawning while driving, and so on.

- As fullproof indication of vigilance for monitoring the driver's operations, Soares et al. (2020) found his physiological characteristics, and vehicle responses, supported by sensing his actions to be nearly accurate.
- The measuring of physiological signals such as heart rate, blinking of an eye, and brain waves; or, that of physical changes such as state of eyes or position of the head is monitored (Mardi et al., 2011; Noori & Mikaeili, 2016).
- For the real-world driving condition, usage of video cameras for detection of

physical changes is preferred over usage of sensing electrodes, which is not considered realistic as it needs special attachments of some equipment to the driver's body, which could be distracting and can be felt sometimes unnecessary too.
- In this work, too, we are using physiological signals or behavioral measures as it is a more effective and realistic way to assess a person's state of drowsiness/fatigue. This detection of signals can be done by analyzing the motion of the eyes, mouth (yawn or not yawn), or the position of the face.

Here, in our model, we detect the state of eyes of a person as they tend to be good indicator of fatigue or sleepy state of a person, and as soon as a microsleep nap is detected, the sleep indicator points to sleepiness as having occurred in the driver.

In computer vision, the face of a person is detected by making use of face-detection technique. After detecting a face of a person, the region of interest, i.e. our eyes, are detected by processing the image collected, and feeding it into a Haar cascading algorithm enabling us to distinctly identify the feature of interest. In the third step, we pass this image of the eyes into our classifier made to detect its state. The classifier is made of a convolutional neural network as it's best for processing signal detection over an image. The image processed through it, is then classified as closed or open to classify whether a person is sleeping or is awake. This label is returned to us, for which we keep its count and track the number of times a person's eyes are found closed, and if the score exceeds the threshold score, then the person is awakened with an alarm.

Figure 19.1 given below is the flow diagram of the Driver Drowsiness/Consciousness Checker System.

### 19.2.2 Related Work

To manage the drowsiness of driver based on its motion, eyes' state, path detection, etc., many approaches have been taken up to solve its repurcussions. Friedrichs and

**FIGURE 19.1** Flowchart of driver drowsiness checker system.

Yang (2010) proposed that drowsiness can be detected under experimental conditions or simulators, in which the experiment is capable of measuring the driver's eyes, as measures of prediction of fatigue are monitored on the basis of eye-tracking simulators. If the blink detection tracking drowsiness works properly, then the result when assessed statistically on real road drivers for 90 hours of dataset can be seen with utmost accuracy. Flores et al. (2011) also described that to cut back the number of such loss of life, to compute the index for drowsiness, to analyze both eyes and face of the driver, and to process the visual information, the need for a sophisticated driver assistance system was formed into a module using AI algorithms (Ravi et al., 2020). A testing drive was also conducted wherein images of real drivers in the vehicle at night were captured to double check the algorithm proposed. For monitoring the bus drivers, development of a vision-based fatigue system was proposed by Mandal et al. (2016) which detected the head and shoulder movements, using SVM and HOG respectively. The model used OpenCV face detector to detect one's face, and a new method was used to estimate eye openness after recognizing the shape of an eye using Spectral Regression Embedding and OpenCV eye detector. Zhang et al. (2012) introduced a robust eye-tracking and image processing system to deal with issues like illumination and changing stances of the driver. An independent index was extracted and the decrease in co-relations was made combined with a stepwise method (Fisher's linear discriminated functions). These included factors such as opening and closing velocity of eyes, minimum duration of eyelid closure, blink frequency etc. As a result these experiments gave 86% accuracy.

Another eye-tracking based system was proposed by Said et al. (2018) which ran on driver's drowsiness and alerted the driver by ringing an alarm. Viola-Jones model was used bearing 82% and 72.8% of accuracy in indoor and outdoor environments, respectively. For continuous monitoring of alertness of a vehicle driver, Mbouna et al. (2013) also proposed visual analysis working on driver's eye state and head-nodding angles. The scheme proposed made the use of different visual features such as pupil activity (PA), head pose (HP), and eye index (EI) to assess the inattentiveness of the driver (Isola et al., 2017). A sequence of the video sections was categorized into alert and no-alerting driving states by a support vector machine (SVM). The experiment resulted in high accuracy even for the people of varied gender and ethnic backgrounds, that too with low false alarms frequency.

## 19.3 PROPOSED MODEL

The most feasible and efficient method observed was to measure physical changes of the driver i.e. by making use of an eye-tracking based system. Here, we are using deep learning and computer vision to extract the desired features of a person, i.e. eyes, our region of interest, requiring the least software requirements and feasible to experiment upon. So, the system requirements and the proposed algorithm for the system are defined below.

### 19.3.1 SYSTEM REQUIREMENT

For the proposed project, we need the following Software Requirements in a System:

# Real-Time Accident Prevention Framework

1. Windows 10
2. Webcam
3. Python 3.6.12
4. Tensorflow
5. Keras
6. OpenCV
7. Anaconda Navigator
8. Jupyter Notebook

### 19.3.2 Proposed System Algorithm

i. Recognize the face of a person by making use of OpenCV, a python library.
ii. After identifying the face, use Haar Cascades to detect the person's eyes as it is our Region of Interest.
iii. Pass the extracted eyes into the CNN classifier determining the state of eyes as open or closed.
iv. CNN classifier uses layers to detect the deep eyes features and will pass them further into a fully connected neural network layer to determine the eye state.
v. The last layer, softmax layer, outputs the image as a closed or open eye.
vi. If the output labeled image is found as "closed", system will alert the user by playing an alarm.

The flowchart of the proposed system algorithm of Driver Drowsiness/Consciousness Checker is shown in Figure 19.2, which depicts the different phases in the System: extracting the image, then, detecting ROI, next, passing the image into the classifier, and lastly, predicting the state of eyes.

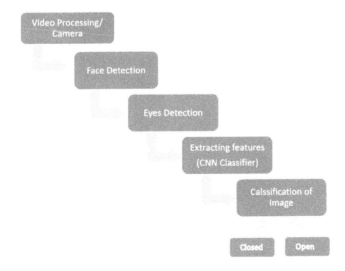

**FIGURE 19.2** Proposed system algorithm.

### 19.3.3 Pre-Processing

The real-time video of the person is captured while driving, but our requirement to detect drowsiness in the person is fulfilled by detecting the person's eyes. To do so, firstly, by the use of Haar cascade, a famously known framework used for object detection, we detect the driver's face. With the help of this, we detect the eyes of a person, left eye and right eye and, then, pass them into further steps for processing. This Haar Cascade/Viola-Jones object/face detection algorithm was developed in 2001 (Huynh et al., 2016; Kim et al., 2017) which accurately detects objects in the images and human faces.

To make system fast with improved detection accuracy, Haar-like features, Integral images, and the AdaBoost Algorithm-like algorithms are combined in the Viola-Jones object detection framework. In this work too, this algorithm with Haar cascade classifier, used with OpenCV, is implemented to detect the driver's face and eyes. Figure 19.3 shows the images of the identified right eye and left eye, respectively, when closed; and Figure 19.4 shows the images of the identified right and left eye, respectively, when open, extracted from the face detected from the user's real-time video.

### 19.3.4 Classification and Feature Extraction using CNN

When there is a need of extracting the interesting parts of the image from a large pool of pixels of the image, reducing its dimensionality, it is called feature extraction which is commonly used in face and object detection. It builds the features

  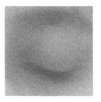

*Right eye*     *Left Eye*

**FIGURE 19.3**  Closed eyes.

*Right Eye*     *Left Eye*

**FIGURE 19.4**  Open eyes.

of image by starting from a certain set of pre-initialized data, and by subsequent learning and generalization occurring at the backend, to lead to the fulfillment of desired requirement (Jensen, 2008; Rahman et al., 2015).

In the work presented here, Convolutional Neural Network is used to abstract the features from a person's eye image, which is further used to classify the state of eyes as closed or open.

**Convolutional Neural Network:** To perform image recognition or classifications, Convolutional Neural Network (CNNs or ConvNets) is the most famously used neural network used for varied experimental purposes. It is widely used in object detection, face recognition, image processing etc. For example, for image processing, it takes the image as an input, processes it, and then classifies it under defined categories as required.

Computers see images as an array of pixels based on its resolution. It recognizes it as H × W × D (height × width × dimension), for example a 5 × 5 × 3, a 3D array is recognized as RGB image as 3 refers to RGB configuration, similarly, 6 × 6 × 1 array of matrix would be identified as a grayscale image.

In Convolutional Neural Networks, an input image passes through many different layers performing different functionalities like, first, it goes through convolutional layers (with filters), then to pooling layer, next, fully connected layers, and finally, through softmax layer to predict the class of image depending upon its probability.

Here, in the method described below, we are making use of three convolutional layers and, then, two fully connected dense layers are used.

The images of eyes with the size 24 × 24, are passed as input into the first convolutional layer. In this first layer, we have 32 filters each of size 3 × 3, transformed by ReLU function, and has a Max pooling over 1 × 1 cell to be added, requiring 896 parameters. The output of the first convolution layer is then passed as an input into the second convolution layer. In the second convolutional layer, the input is convolved again with 32 filters each of size 3 × 3, requiring 9284 parameters. When the image is convolved after passing the two layers, it is then transformed non-linearly by ReLU function and its MaxPooling is performed over 1 × 1 cell with a stride. The output of the second convolution layer is further fed into the third convolution layer. Here, in this layer, the input is convolved with 64 filters with size 3 × 3, each. After this convolution, non-linear transformation ReLU and MaxPooling over 1 × 1 cell with stride is applied. This layer required 18496 parameters, marking the last convolutional layer.

After processing through convolutional layers, its output is passed further into fully connected two layers, which are followed by a dropout of 25% and 50%, respectively. Finally, a binary output is desired; so, softmax function is applied as a last step to predict the eye state.

Thus, our classification model produces the state of eyes as output giving us the accuracy of 97% on running over 436 validation test samples. The model was trained over 1016 training samples. Thus, it gives us a pretty good accuracy of 97% on our sample images.

## 19.4 RESULTS

In the proposed model, the classifier used to classify the state of eyes was trained over 1016 training samples and validated on 436 samples containing images of the closed and open eye. Among these, closed and open eyes are uniformly distributed to give a fair prediction.

The classifier, first, did some pre-processing of data with data augmentation etc. and, then, fitted over these 1016 training samples. The testing dataset predicted output with an accuracy of 97%. The graphs in Figure 19.5 and Figure 19.6 depict the accuracy and loss metrics of the training set and validation set varying with the no. of epochs our model executed, respectively. Classification Report and Confusion Matrix of y_test and predictions are shown in Figure 19.7.

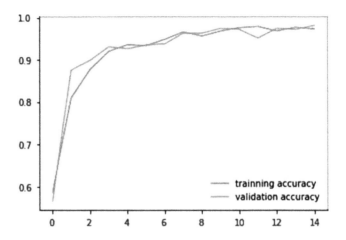

**FIGURE 19.5** The training accuracy and validation accuracy over no. of epochs.

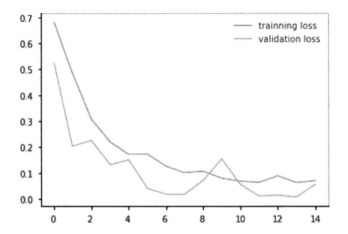

**FIGURE 19.6** The training loss and validation loss over no. of epochs.

```
Classification Report :
              precision    recall  f1-score   support

           0       0.99      0.96      0.98       228
           1       0.96      0.99      0.97       208

    accuracy                           0.97       436
   macro avg       0.97      0.98      0.97       436
weighted avg       0.98      0.97      0.97       436


Confusion Matrix :

[[220   8]
 [  3 205]]
```

**FIGURE 19.7** Classification report and confusion matrix of y_test and predictions.

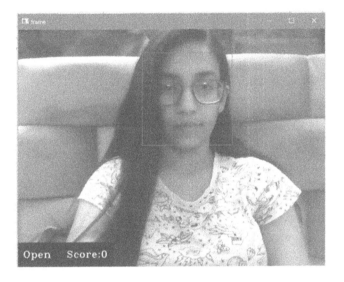

**FIGURE 19.8** Response for open eyes.

Figures 19.8 and 19.9 depict the frames of our model in working state where it detects the state of one's eyes as open and closed, respectively, with the score calculated for each state.

## 19.5 CONCLUSION AND FUTURE SCOPE

Nowadays, more and more professions require long-term concentration, and drivers must keep an eye on the road so that they can react immediately to uncertain situations like accidents etc.

This system will help us to improve road safety and prove to be a proactive measure to prevent accidents to some extent. Alerting the driver about his health condition, while he is driving, and making him attentive, while driving, does improve his consciousness and decreases his accident-prone driving drowsiness rate.

**FIGURE 19.9**  Response for Closed Eyes

In this proposed system, the method used for drowsiness detection is based on the detection of the eye state of the driver. Here, the state of the eye can be drowsy or not-drowsy depending on whether the eyes of the driver are closed or open. This system registered the accuracy of detection to be 97%. In future, for improving this system, we can include the inspection of more features like head positioning, steering movement, etc. to detect the same and timely alert the concerned before there is any mishap.

## REFERENCES

Flores, M. J., Armingol, J. M., & de la Escalera, A. (2011). Driver drowsiness detection system under infrared illumination for an intelligent vehicle. *IET Intelligent Transport Systems*, 5(4), 241–251.

Friedrichs, F., & Yang, B. (2010, June). *Camera-based drowsiness reference for driver state classification under real driving conditions* [Symposium]. 2010 IEEE Intelligent Vehicles Symposium, IEEE, pp. 101–106.

Huynh, X. P., Park, S. M., & Kim, Y. G. (2016, November). *Detection of driver drowsiness using 3D deep neural network and semi-supervised gradient boosting machine* [Conference session]. Computer Vision, Asian Conference, Springer, Cham, pp. 134–145.

Isola, P., Zhu, J. Y., Zhou, T., & Efros, A. A. (2017). *Image-to-image translation with conditional adversarial networks* [Conference session]. Computer Vision and Pattern Recognition, IEEE Conference, pp. 1125–1134.

Jensen, O. H. (2008). *Implementing the Viola-Jones face detection algorithm* [Master's thesis], Technical University of Denmark, DTU, DK-2800 Kgs, Lyngby, Denmark.

Kim, K. W., Hong, H. G., Nam, G. P., & Park, K. R. (2017). A study of deep CNN-based classification of open and closed eyes using a visible light camera sensor. *Sensors*, 17(7), 1534.

Mandal, B., Li, L., Wang, G. S., & Lin, J. (2016). Towards detection of bus driver fatigue based on robust visual analysis of eye state. *IEEE Transactions on Intelligent Transportation Systems, 18*(3), 545–557.

Mardi, Z., Ashtiani, S. N. M., & Mikaili, M. (2011). EEG-based drowsiness detection for safe driving using chaotic features and statistical tests. *Journal of Medical Signals and Sensors, 1*(2), 130.

Mbouna, R. O., Kong, S. G., & Chun, M. G. (2013). Visual analysis of eye state and head pose for driver alertness monitoring. *IEEE Transactions on Intelligent Transportation Systems, 14*(3), 1462–1469.

Noori, S. M. R., & Mikaeili, M. (2016). Driving drowsiness detection using fusion of electroencephalography, electrooculography, and driving quality signals. *Journal of Medical Signals and Sensors, 6*(1), 39.

Rahman, A., Sirshar, M., & Khan, A. (2015, December). *Real time drowsiness detection using eye blink monitoring* [Conference session]. *2015 National software engineering conference (NSEC)*, IEEE, pp. 1–7.

Ravi, A., Phanigna, T. R., Lenina, Y., Ramcharan, P., & Teja, P. S. (2020, July). Real time driver fatigue detection and smart rescue system. Electronics and Sustainable Communication Systems (ICESC), International Conference, IEEE, pp. 434–439.

Said, S., AlKork, S., Beyrouthy, T., Hassan, M., Abdellatif, O., & Abdraboo, M. F. (2018). Real time eye tracking and detection-a driving assistance system. *Advances in Science, Technology and Engineering Systems Journal, 3*(6), 446–454.

Soares, S., Monteiro, T., Lobo, A., Couto, A., Cunha, L., & Ferreira, S. (2020). Analyzing driver drowsiness: from causes to effects. *Sustainability, 12*(5), 1971.

Zhang, W., Cheng, B., & Lin, Y. (2012). Driver drowsiness recognition based on computer vision technology. *Tsinghua Science and Technology, 17*(3), 354–362.

# 20 Multi-Modality Medical Image Fusion Using SWT & Speckle Noise Reduction with Bidirectional Exact Pattern Matching Algorithm

*Kapil Joshi, Minakshi Memoria, Laxman Singh, Parag Verma, and Archana Barthwal*

## CONTENTS

| | | |
|---|---|---|
| 20.1 | Introduction | 340 |
| 20.2 | Medical Modalities | 343 |
| | 20.2.1 X-Ray | 343 |
| | 20.2.2 Ultrasound | 343 |
| | 20.2.3 Computed Tomography (CT) | 344 |
| | 20.2.4 Magnetic Resonance Imaging (MRI) | 344 |
| | 20.2.5 Positron Emission Tomography (PET) | 345 |
| | 20.2.6 Speckle Noise Based Model | 345 |
| |     20.2.6.1 Types of Speckle Noise Filter | 346 |
| 20.3 | Strategies of Image Fusion | 347 |
| | 20.3.1 Spatial Fusion Domain | 347 |
| |     20.3.1.1 Average Methodology | 348 |
| |     20.3.1.2 Principal Component Analysis (PCA) | 348 |
| |     20.3.1.3 Intensity Hue Saturation (IHS) | 348 |
| |     20.3.1.4 High Pass Filter (HPF) | 348 |
| |     20.3.1.5 Brovey Transforms (BT) | 349 |
| | 20.3.2 Transform Fusion of Domain | 349 |
| |     20.3.2.1 Discrete Wavelet Transform (DWT) | 349 |
| |     20.3.2.2 Methodology of Pyramids | 349 |
| |     20.3.2.3 Stationary Wavelet Transform (SWT) | 349 |

DOI: 10.1201/9781003154686-20

| | | |
|---|---|---|
| 20.4 | Proposed Work | 351 |
| 20.5 | Result and Discussion | 351 |
| | 20.5.1 Average Pixel Intensity (API) | 352 |
| | 20.5.2 Standard Deviation (Sd) | 352 |
| | 20.5.3 Coefficient of Correlation (Cc) | 353 |
| | 20.5.4 Average Gradient (Agr) | 356 |
| | 20.5.5 Entropy (En) | 356 |
| 20.6 | Conclusion | 357 |
| References | | 357 |

## 20.1 INTRODUCTION

The term fusion normally implies the state of being combined into one pattern. Image fusion is a method for generating combos of images of a minimum of two given pictures (Sahu & Parsai, 2012). The aim of blending footage is to incorporate and compliment the information multiple footage in an extremely secluded image. Image fusion has had a serious impact at intervals in the sector of Medical Sciences. In current days, medical modality fusion has become an active area of research (Ramandeep, 2014). The foremost thought methodology of blending medical footage is to fuse multiple pictures such as CT, X-ray, MRI, Ultra-sound, PET, etc. in extremely solitary images (Mandhare et al., 2013). Medical specialty uses different types of fusion techniques with healthful footage and each image provides meaningful information (Shalima & Virk, 2015). As an example, Magnetic Resonance Images provide information on soft-tissues, although CT pattern footage offers information on dense-tissues. As a result, individual footage obtained from different medical imaging methods may not offer enough clinical conditions for conclusions to be drawn. Therefore, it's become necessary to consolidate the pictures provided by different modalities into completely different combos generating a single image (Sahu & Parsai, 2012). The generated single image will meet the suitable clinical desires from medical point of view. The fusion of healthcare footage will enhance the common features more profoundly (Balachander & Dhanasekaran, 2016) and will make an image more readable (Deshmukh & Malviya, 2015).

Figure 20.1 illustrates the general process of image fusion methodology.

Image de-noising is the subset of image processing (Malviya & Bhirud, 2009) and it is the biggest problem in this field. An image is a collection of pixels or a function denoted as f(x, y) where x is the number of pixels in a row and y is the number of pixels in a column. Before discussing speckle noise, firstly, we must discuss about the noise. It is random variations (Mishra & Bhatnagar, 2014) of color and brightness or increasing and decreasing the value of pixels. Noise can be added to an image due to digital camera problem (sensor problem), climate change, adding of unwanted signals. In an image, noise can be added externally also. Noise is divided into Salt and Pepper Noise, Gaussian Noise, Poisson Noise, and Speckle Noise. In this work, we explain speckle noise. Speckle noise mostly appears in medical images (Ultrasound images), SAR (Synthetic Aperture Radar) and Active Radar Images. 2-D images are used in satellite imaging. SAR image is a kind of

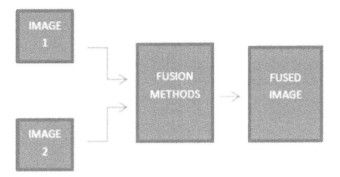

**FIGURE 20.1** General Process of Image-Fusion.

radar used for creating 2-D and 3 D images of landscape and object. Its advantage is that it works in all-weather conditions (Yang et al., 2016).

SAR Image gets affected by multiplicative speckle noise. Bayesian Based Algorithm can be used for de-noising the image, while preserving the data (details) (Qu et al., 2001). Removal of Noise (Wang & Ma, 2008) is considered critical for medical images to extract the important medical information or yield an accurate result. Speckle noisy images (James & Dasarathy, 2014) includes MRI (Medical Resonance Image), CT (Computed Tomography) and X-ray. The performance of any algorithm for these multi-resolution images can be determined in term of PSNR, MSE, and accuracy. The most important thing behind applying any algorithm is to recover actual image information from hidden data (Yang et al., 2008). In case of discrete wavelet transform algorithm (DWT), if an input image contains speckle noise, then, DWT (Jaffery et al., 2017) is applied to produce a de-noised image. In this algorithm, a noisy image is decomposed into sub-parts by applying DWT filter (Bhatnagar et al., 2013). After removal of noise, an image can be re-stored and its performance can be analyzed. For de-noising an image, wavelet filtering technique (Singh & Jaffery, 2018; Zhou et al., 2020) can be used with soft and hard thresholding. DTCT may also be used for de-noising the image. It is an improvement over CWT (Complex Wavelet Transform). CWT is used to filter the signal into real and imaginary parts, while wiener filtering is a non-linear filter that de-noises the image in spectral domain (Zaheeruddin et al., 2012). DTCWT for DT-CWT and DT-WT with wiener filter tend to increase the performance of an algorithm as compared to SWT and DWT for low noisy images. However, for high noisy images SWT, DTCWT, and DWT do not yield good result as compared to DT-CWT with wiener filter.

Ultrasound images are corrupted by Speckle noise, which creates a problem in diagnosing a disease; and thereby, creates a problem in providing a proper health treatment to the patient. In Ultrasound images (US), reduction in resolution or quality creates a noise; and thereby, the noise causes blurring of an image (Constantinos et al., 2001). Speckle reduction has two approaches, first is computing approach and second is post processing approach. In computing approach, we acquire the data and convert it into an appropriate form. However, many different filtering techniques can be used for post processing (Calhoun & Adali, 2008). Numerous speckle noise reduction algorithms are available in literature for de-noising the medical images (Aguilar & Garrett, 2001). Speckle noise occurs due to monochromatic radiations. In Adaptive Weighted Median Filter (AWMF), which add (combine) edge preserve details of the Weighted Median, only AWMF filter is more suitable than other non-linear space varying filters as it is able to preserve the minute details of images (Kirankumar & Shenbaga Devi, 2007).

SRAD derivation for speckle images corrupted with additive noise using anisotropic diffusion with Lee and Kahn Filters form final output image.

Satellite based hyper spectral imaging in Nov 2000 successfully launched, which is a kind of image with spectral bands (Anbarjafari et al., 2015). Fuzzy logic can also be employed for reduction of speckle type noise. It is a soft computing technique that adds an effectiveness, robustness, and tractability on currently de-noising methods. However, it is mostly used in Ultrasound and SAR Image (Liu et al., 2006). In illumination model, there is the difficult task of removing multiplicative noise by using homographic framework which gets converted into additive form (Achim et al., 2003). A variation model is considered the fastest algorithm for image restoration, and for de-noising the speckle noise in image, thereby preserving the finer details of an image. Using this algorithm, PSNR, and, SSIM value can be improved, which in turn tend to improve the image quality of the restored image. The output results demonstrate that the proposed algorithm offers reasonable speed and less computation time (Loupas et al., 1989). Adaptive filter achieved better results owing to usage of low pass smoothing filters that help in preserving sharpness and content (Yu & Acton, 2002). A new Filtering Technique (Katiyar & Santhi, 2017) for de-noising speckle noise from medical pictures supports adaptive and allotropic diffusion filter. The downside of pattern matching is often divided into real pattern matching and approximate pattern matching. Details of Real Pattern match algorithm is described in Intrusion Detection Method (Huang et al., 2019). Text Editor, plagiarism, and many more areas find use of pattern match (PM) algorithm. Approximate pattern matching is outlined in Bioinformatics (Gloor et al., 1992). Video Retrieval uses the Pattern window in text string with edit distance. Conjointly (Charras & Lecroq, 2004) divided the pattern matching in such a way to assist the framework in the form of single PM or multi PM. Machine learning algorithms offer an ability to search out without being explicitly programmed, and the pattern recognition and automation theory in AI (SaiKrishna et al., 2012) have been developed by using this concept. Such algorithms overcome the strictly static programming by making knowledge-driven choices (Knuth et al., 1977). These construct a model from sample inputs, and are capable to learn and predict from the given data (Boyer & Moore, 1977). Many activities like email filtering, identification of network (Sunday, 1990), intruders' detection, and optical character recognition are based on machine learning algorithms. Machine learning and data mining algorithms learn and establish baseline behavioral profiles for varied entities to detect and realize the anomalies in any network or image (Horspool, 1980).

An ADPS is claimed to discover from aptitude E according to any class of movement T and execution live P if its presentation at T errands, as surveyed by P, increments with skill E. Tom Mitchell cited" (Raita, 1992). This definition of machine learning is followed by Alan Turing's paper "Computing Machinery and Intelligence", where he queries "Can a machine do what I'm willing to do". Machine learning can perform the various types of tasks such as:

**Regulated learning:** The pc is given by an "instructor" with model data sources and their ideal yields and, furthermore, the goal is to sort out an overall principle that guides contributions to yields.

**Solo learning:** No unit of the zone names given to the preparation recipe follows up on its own to search for structure in its information. Unattended learning is targeted in itself (disclosure of mystery information designs) or an approach to achieve a related objective.

Semi-managed learning is considered among directed and unattended learning (Blum & Langley, 1997). Any place the educator offers a connected fragmented training signal: an instructing set with none or a few of the objective yields. Transduction (Cover & Hart, 1967) might be a unique instance for such type of hypothesis.

**Reinforcement learning:** A malicious programme communicates with a complex environment through which an explicit purpose would be accomplished (such as driving vehicle or taking part in a game against associate opponent). Feedback is basically provided to the programmer in terms of incentives and penalties, when it navigates the area of disadvantage.

The other classes of machine learning problems can be defined as:

(a) Learning from the new knowledge from its own inductive bias. (b) Learning in Development. (c) For AI learning, elaborate. (d) Socio-technical learning and so on. Machine Learning sets its goal from achieving AI to confronting resolvable issues through sensible approach. It moved from symbolic family approaches from AI to custom-made methods and models from statistics and applied mathematics. Machine learning and data processing work with similar techniques, but machine learning focuses on prediction, knowledge-supported properties, and data processing, and also on the disclosure of unknown knowledge properties.

## 20.2 MEDICAL MODALITIES

### 20.2.1 X-Ray

The X-Ray is electromagnetic influx of a related degree that recognizes the body's inward constructions and takes pictures of those designs on a photographic material or a fluorescent screen. Such pictures are called representative radiographs or symptomatic X-beams. Chest visualization is explained through Figure 20.2.

### 20.2.2 Ultrasound

A therapeutic imaging watch that utilizes high redundancy sound waves to catch living pictures from inside the anatomy might be an ultrasound. Figure 20.3 shows the details of new born baby.

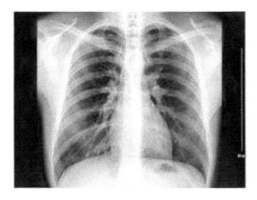

**FIGURE 20.2** Chest Picture of Human Body.

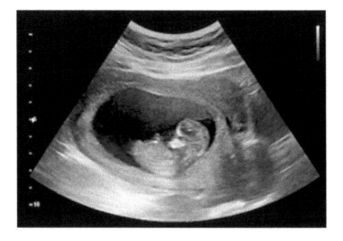

**FIGURE 20.3** Ultrasonic New Born Baby Imaging.

### 20.2.3 COMPUTED TOMOGRAPHY (CT)

It is a connected insightful degree check of clinical pictures used to take expounds pictures of inside organs, bones, sensitive tissues, and nerves (Calhoun & Adali, 2008). The best strategy for routinely secluding tumors is CT imaging. Figure 20.4 presents the modality of CT scan image of brain.

### 20.2.4 MAGNETIC RESONANCE IMAGING (MRI)

MRI radiography can be a clinical imaging approach that commonly shapes photos of the body's life frameworks and physiological cycles in any wellbeing and unwellness. The handling of pictures by imaging and CT could yield a particular demonstrative data, so it is essential to join the various pictures to initiate exact and unmistakable indicative data (Kirankumar & Shenbaga Devi, 2007).

Figure 20.5 presents the modalities of MRI of brain.

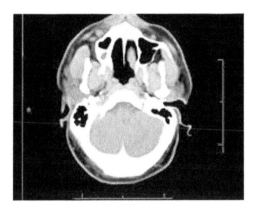

**FIGURE 20.4** Brain CT Modality.

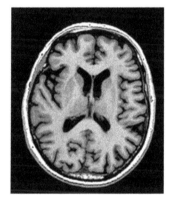

**FIGURE 20.5** Brain MRI Modality.

### 20.2.5 POSITRON EMISSION TOMOGRAPHY (PET)

It is a clinical imaging assessment (Kirankumar & Shenbaga Devi, 2007) technique that bolsters the working of anatomical tissues and organs. To show certain exercises, a PET output utilizes a hot medication (tracer). All in all, this output will take a gander at sicknesses prior to turning up on other clinical imaging contemplates. A scanned image of brain using PET modality is shown in Figure 20.6.

### 20.2.6 SPECKLE NOISE BASED MODEL

Images from active radar, synthetic aperture radar, medical ultrasound, and optical coherence tomography are all examples of this imaging modality. Speckle is a granular interference and degrades the image quality drastically. Speckle Noise and the distribution noise can be expressed as:

$$g(n, m) = f(n. m) \times u(n, m) + \xi(n, m)$$

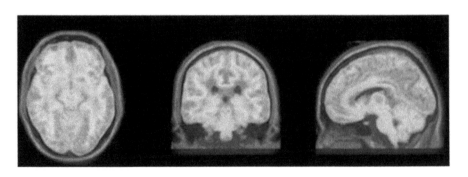

**FIGURE 20.6** PET Modality Brain Scanned Image.

Where $g(n, m)$ represents the observed image, $u(n, m)$ represents the multiplicative component, and $\xi(n, m)$ represents the additive component of the speckle noise.

#### 20.2.6.1 Types of Speckle Noise Filter
Speckle is a granular interference that occurs in active radar, synthetic aperture radar (SAR), medical ultrasound, and optical coherence tomography images and it degrades their effectiveness. There are two types of filters.

*20.2.6.1.1 Wiener Filter*
The Wiener filter (Zaheeruddin et al., 2012) is a linear time-invariant filter used in signal processing to filter an observed noisy process and estimate a desired or target random process, assuming known stationary signal and noise spectra and additive noise. The Wiener Filter was invented by Wiener in 1940. The noise in an image is typically minimized using this filter.

It is used in spatial domain and linear filter. Due to linear motion within the signal processed region, it is treated as an effective technique to minimize the blurring effect in medical images as well as other normal images.

The Wiener Filter is known to have a non-linear spectral domain that might be clubbed with different techniques such as DT-CWT and DT-CWT DTCWT.

The Wiener filtering presents results in terms of mean square error. In other words, it minimizes the total mean square error during the inverse filtering and noise smoothing operation. Wiener filtering is a linear estimate of the original image. The strategy is built on a stochastic basis. Fourier domain of Wiener filter can be expressed as follows:

$$W(f_1, f_2) = \frac{H \times (f_1, f_2) S_{xx}(f_1, f_2)}{|H(f_1, f_2)|^2 S_{xx}(f_1, f_2) + S_{\eta\eta}(f_1, f_2)}$$

Where $S_{xx}(f_1, f_2) + S_{\eta\eta}(f_1, f_2)$ power spectra of the original image. The additive is respectively noise. $H \times (f_1, f_2)$ is the blurring filter. It is easy to see that there are two different parts of the Wiener filter, an inverse filtering component and a noise smoothing portion. It not only performs de-convolution by inverse filtering (high-pass filtering), but also reduces noise with a compression process (low-pass filtering).

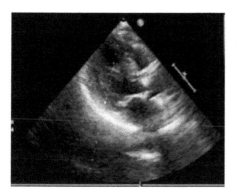 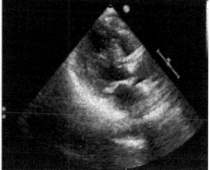

**FIGURE 20.7** Source Image of New Born Baby with Speckle Noise.

*20.2.6.1.2 Median Filter*

Non-linear optical filtering is a technique for removing noise from an image or signal, and it is treated as the median filter. This type of noise reduction is common in pre-processing to enhance the outcomes of subsequent processing, when need.

The median filter is usually used, much like the mean filter, to decrease noise in a picture. It often does a better job, however, than the average filter of retaining useful information in the picture (Singh & Jaffery, 2017).

The median is calculated with, first, arranging all of the pixel values in the surrounding neighborhood via numerical order, and afterwards supplementing the middle pixel value only for considering the pixel.

The reduction of multiplicative and additive noise is achieved by:

$$G(m_1) = f(m, n) * u(m, n) + Ig(m, n) - Ig(m, n) - (m, n)(m, n)$$
$$= f(m, * u(m, n)(2)$$

Figure 20.7 illustrates the source image of a new-born baby with speckle noise image. Figure 20.7 shows the gamma distribution using the following equation:

$$f(g) = \frac{g^{\alpha-1}}{(\alpha - 1)!a^\alpha} e^{\frac{-g}{a}}$$

Figure 20.8 presents the source image of Lena corrupted with speckle noise.

## 20.3 STRATEGIES OF IMAGE FUSION

### 20.3.1 SPATIAL FUSION DOMAIN

The term spatial refers to the term when any picture is represented in spatial domain. Within the image, the spatial domain based method manages the pixels. In order to get ideal outcomes, constituent estimates of unit area must be regulated.

**FIGURE 20.8** Source Image of Lena with Speckle Noise.

### 20.3.1.1 Average Methodology
This is considered as one of the powerful techniques for image fusion, in which the weighted conventional pixels of unit area of an input image are directly calculated and replicated on the merged output image.

### 20.3.1.2 Principal Component Analysis (PCA)
PCA can be a vector house modification that turns more relevant factors into several unrelated factors. PCA employs an orthogonal transformation technique to transform a set of observations of possibly correlated variables into a set of values of linearly uncorrelated variables known as Principal components (PC). PCs are used for image pattern and image level characterization. The PCA theory is to minimize the quantity of variable information while retaining actual information in term of principal components. The first PC will have the highest variance and followed by each succeeding components' variance falling, provided that they are orthogonal to each other (Singh & Jaffery, 2018).

### 20.3.1.3 Intensity Hue Saturation (IHS)
Intensity-Hue-Saturation is a well-known shading hybrid technique. The RGB based image is transformed into power segments using this approach (i.e. Intensity (I), Hue (H), and Saturation (S)) thereby subjecting the intensity to a panchromatic representation of the upper spatial goals. At that point, a reverse transformation is applied to get RGB back as an output.

### 20.3.1.4 High Pass Filter (HPF)
The high pass frequency component basically represents the edges, whereas the low pass frequency component indicates the smooth regions. In the field of Image Processing, Ideal High Pass Filter (IHPF) is used to yield a sharpened image in the frequency domain. Image Sharpening refers to the technique that enhances the fine details and highlights the edges in a digital image. It has the ability to eliminate low-frequency components from the digital image and preserves high-frequency components. This ideal high-pass filter performs the reverse operation of the ideal low-pass filter.

### 20.3.1.5 Brovey Transforms (BT)

The BT was proposed and developed by American scientist, named Brovey. This method is also known as the color normalization transform as it involves a red-green-blue (RGB) color transform method. The Brovey transformation method avoids the disadvantages of the multiplicative method and is considered to be a very simple method for combining data from different sensors. It performs the combination of arithmetic operations and, thereby, normalizes the spectral bands their multiplication with the panchromatic images. It helps in retention of the corresponding spectral feature of each pixel, and converts all the luminance information into a panchromatic image possessing high resolution.

### 20.3.2 TRANSFORM FUSION OF DOMAIN

Image Fusion method can be broadly categorized into two methods: spatial domain fusion method and Transform domain fusion method. In the spatial domain method, we directly deal with the pixels of the input image. In this method, the pixel values are altered to achieve the desired result. In the transform domain methods, the Fourier transform of the image is computed first. Thereafter, all the Fusion operations are computed on the Fourier transformed image followed by application of the Inverse Fourier transform to get the resultant image.

#### 20.3.2.1 Discrete Wavelet Transform (DWT)

Discrete wavelet transform (DWT) is known as a wavelet transform in which the wavelets are discretely sampled. As with other wavelet transforms, a key advantage of DWT Fourier transforms is that it provides temporal resolution. It means DWT captures both frequency as well as location information (location in time) (Jaffery et al., 2017).

#### 20.3.2.2 Methodology of Pyramids

The picture pyramid will be addressed as an adjustment of a copy of a periodic or high band pass of one picture with as little as could really be expected. The most critical task of mixing images is to create a pyramid illustration for the primary images to use the rules of role choice, by betting on the strategy of the pyramid. The picture can be obtained by rotating the pyramid backwards.

#### 20.3.2.3 Stationary Wavelet Transform (SWT)

SWT is a wavelet transform that is designed to overcome the lack of translation-invariance of DWT. Translation-invariance is eliminated by the down-samplers and up-samplers in the DWT and up-sampling the filter coefficients by a particular factor in an algorithm. The SWT is an inherently redundant scheme because the output of each level of SWT offers the same number of samples as the input. Therefore, there is a redundancy of N in the wavelet coefficients for a decomposition of N levels. This algorithm is also known as *"algorithme à trous"* in French. The word *trous* means holes in English that refers to the inserting of zeros in the filters (Charras & Lecroq, 2004).

Figure 20.9 depicts the approaches on SWT.
Table 20.1 tabulates a comparison between spatial and transform methods.

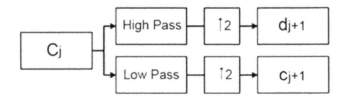

**FIGURE 20.9** Approaches on Stationary Wavelet Transform.

**TABLE 20.1**
**Comparative Methods Between Spatial and Transform Domain**

| S. No. | Image Fusion Method | Domain | Advantages | Disadvantages |
| --- | --- | --- | --- | --- |
| 1. | Average Method | Spatial | It uses straightforward image fusion methodology. | Constituent level image fusion algorithmic program doesn't forever manufacture a transparent output image from the given set of supply pictures. |
| 2. | PCA (Principal Component Analysis) | Spatial | PCA transforms variables into un-correlated variables. | This type of fusion produces spectral degradation. |
| 3. | Intensity Hue Saturation | Spatial | Hue saturation produces sensible visual effects. | It produces color distortion within the output pictures. |
| 4. | High Pass Filtering | Spatial | An N-dimensional signal can benefit from high pass filtering to remove a small amount of high frequency noise. | Uses simply a first-order filter. Does not offer an effective method to determine the cut-off frequency. |
| 5. | Brovey transform | Spatial | Has capability to yield next spatial picture outcome. | It preserves less spectral fidelity. |
| 6. | Distinct ripple transform | Transform | The DWT fusion process outperforms the standard image fusion method. It has a lower spectral distortion and a better signal-to-noise magnitude ratio. | Final amalgamate output image has less spatial resolution. |
| 7. | Pyramid method | Transform | It shows higher spatial still as spectral quality of the amalgamate image. | It causes the block effects and unsought edges throughout fusion. |
| 8. | Stationary ripple transform | Transform | Has the flexibility to repeat the mistake repetitively. | It doesn't collapse the constant at every level of amendment. |

## 20.4 PROPOSED WORK

Figure 20.10 depicts the proposed method.

Step 1. Initially consider a couple of images as Img I1 and Img I2.

Step 2. SWT can be applied directly on a couple of given images and debilitate them into low and high sub categories.

Step 3. Finish disintegration of the pictures.

Step 4. Apply average algorithm on the set of the two pictures, for example Img1 and Img2.

Step 5. Blend the detail part of the blurred pictures according to their weights.

Step 6. Integrate these two distinct pictures, for example, blend and weighted blend utilizing Inverse SWT (ISWT) and obtain latest coefficients (Low-Low$_{new}$, Low-High$_{new}$, High-Low$_{new}$, High-High$_{new}$) likewise of the combined picture.

## 20.5 RESULT AND DISCUSSION

This latest experimented method is tested on unequivocal instructive document 1 of pilot pictures. Apart from medical images, we also implemented the proposed technique on the general-purpose images, for instance on the clock images (as shown in Figure 20.11). Each picture is about 512 × 512 pixels based length. In addition, blending of some of the plane images is also demonstrated in Figure 20.12.

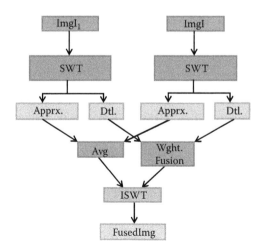

**FIGURE 20.10** Proposed Method.

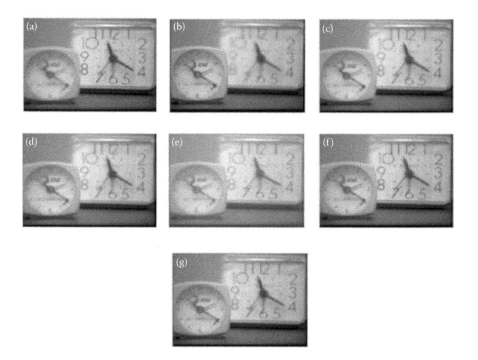

**FIGURE 20.11** Picture Combination Results for Clock Picture Informational Collection: (a) Left Piece Obfuscated; (b) Right Part Obscured; (c) Solidified Picture Utilizing DWT Technique; (d) Weaved Picture by PCA Technique; (e) Blended Picture Utilizing SWT Procedure; (f) Joined Picture using CBF Framework; (g) Final Picture is Developed Through Proposed System.

Figures 20.13 and 20.14

Figure 20.15 presents result utilizing spot commotion, wherein median filter and wiener filters are applied simultaneously to remove the speckle noise. From the results, we can observe that the median filter yielded better results than wiener filter. The accompanying measurements are utilized to assess the outcomes:

### 20.5.1 Average Pixel Intensity (API)

It is utilized to gauge a file of difference, which is addressed as

$$API = \frac{\sum_{i=1}^{m} \sum_{j=1}^{n} f(i,j)}{mn} \tag{20.1}$$

### 20.5.2 Standard Deviation (Sd)

It is portrayed as

# Bidirectional Pattern Matching Algorithm

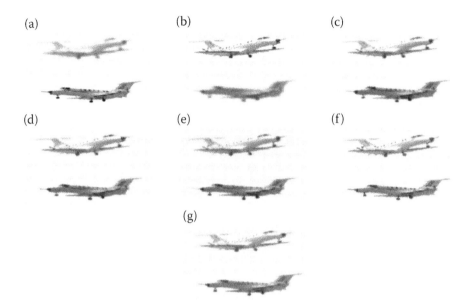

**FIGURE 20.12** Combination Results for Plane Picture Enlightening Rundown: (a) Upper Piece Obscured; (b) Lower Section Obfuscated; (c) Blended Picture Utilizing DWT Framework; (d) Joined Picture Through PCA Strategy; (e) Intertwined Picture by SWT Strategy; (f) Weaved Picture by Cross Bilateral Procedure; (g) Resultant Picture Utilizing Given Updated Proposed Technique.

$$Sd = \sqrt{\frac{\sum_{i=1}^{m} \sum_{j=1}^{n} (f(i,j) - \mu)^2}{mn}} \qquad (20.2)$$

## 20.5.3 Coefficient of Correlation (Cc)

It computes the possibility of combined picture to fundamental unique picture; higher worth shows the significant melded results. This is defined as follows:

$$Cc = \frac{(r_{Xf} + r_{Yf})}{2} \qquad (20.3)$$

Where,

$$r_{Xf} = \frac{\sum_{i=1}^{m} \sum_{j=1}^{n} (X(i,j) - \bar{X})(f(i,j) - \mu)}{\sqrt{(\sum_{i=1}^{m} \sum_{j=1}^{n} (X(i,j) - \bar{X})^2)(\sum_{i=1}^{m} \sum_{j=1}^{n} (f(i,j) - \mu)^2)}}$$

and,

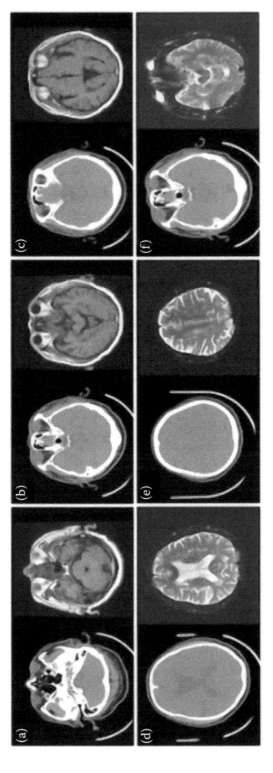

**FIGURE 20.13** The Primary Pictures of Patient no. 1. Now a, b and c these are T1-MRI and CT Pictures and Other Area d, e and f are Now DWI-MRI and CT Pictures.

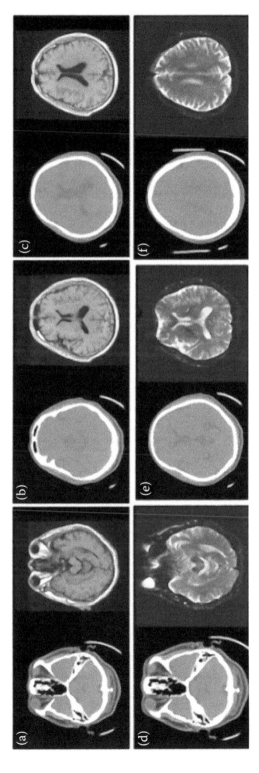

**FIGURE 20.14** The Registered Images of Next Patient no. 2. a, b and c these are Called T1-MRI and Another CT Images and Next Level Area d, e and f are Defined as CT and DWI-MRI Based Pictures.

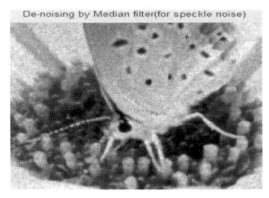 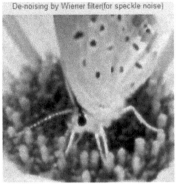

**FIGURE 20.15** Result Utilizing Spot Commotion.

$$r_{yf} = \frac{\sum_{i=1}^{m} \sum_{j=1}^{n} (Y(i,j) - \bar{Y})(f(i,j) - \mu)}{\sqrt{(\sum_{i=1}^{m} \sum_{j=1}^{n} (Y(i,j) - \bar{Y})^2)(\sum_{i=1}^{m} \sum_{j=1}^{n} (f(i,j) - \mu)^2)}}$$

### 20.5.4 Average Gradient (Agr)

It displays the sharpness and smoothness and is characterized as:

$$Ag = \frac{\sum_{i=1}^{m} \sum_{j=1}^{n} ((f(i,j) - f(i+1),j))^2 + (f(i,j) - f(i,j+1))^2)^{1/2}}{mn} \quad (20.4)$$

### 20.5.5 Entropy (En)

Entropy is used to measure a physical property that is most commonly associated with a state of disorder, randomness, or uncertainty. It scales the quantity of data accessible in a melded picture:

$$En = -\sum_{k=0}^{255} p_k \log_2(p_k) \quad (20.5)$$

Where $p_k$ is known as probability of kth state in any picture.

We can see, visually, that the combined images from the current proposed scheme are often much larger than the traditional methods. Only a visual description is not enough to enhance the better quality of the merged image.

In quantitative measuring, the following metrics are used. The findings of previous methods and current methods are in a set of tabular ratio in Table 20.2 after experimenting with all the work tested.

In most cases, in comparison to existing methods, our proposed work gives better results.

**TABLE 20.2**
**Fusion Measured Metrics with Three Image Datasets**

| Info Images | Fusion Methods | Sd | Cc | API | Agr | En |
|---|---|---|---|---|---|---|
| CLOCK | | | | | | |
| | [11] | 46.516 | 0.9711 | 96.038 | 3.279 | 3.7769 |
| | [12] | 46.716 | 0.9744 | 96.037 | 3.278 | 4.4545 |
| | [13] | 47.909 | 0.9962 | 96.035 | 4.572 | 4.4982 |
| | [14] | 48.893 | 0.9349 | 95.548 | 5.7125 | 7.3055 |
| | PROPOSED | 52.317 | 0.9990 | 98.180 | 4.9134 | 7.4861 |
| Airplane | | | | | | |
| | [11] | 52.872 | 0.9846 | 222.44 | 3.2869 | 2.9833 |
| | [12] | 52.874 | 0.9498 | 221.43 | 3.4871 | 3.9173 |
| | [13] | 53.176 | 0.9176 | 220.42 | 4.7307 | 4.3673 |
| | [14] | 51.562 | 0.986 | 224.58 | 4.8419 | 4.2667 |
| | PROPOSED | 56.242 | 0.9804 | 220.55 | 5.9054 | 4.9866 |
| Clinical | | | | | | |
| | [11] | 55.872 | 0.9846 | 212.44 | 3.2269 | 3.9833 |
| | [12] | 54.874 | 0.9498 | 241.43 | 3.4371 | 3.8173 |
| | [13] | 55.176 | 0.9176 | 220.42 | 4.7717 | 3.5673 |
| | [14] | 53.562 | 0.986 | 254.58 | 4.1119 | 4.2897 |
| | PROPOSED | 55.242 | 0.9804 | 290.55 | 5.8054 | 4.9926 |

## 20.6 CONCLUSION

Medical modality has become very popular for the diagnosis of any kind of disease. In this work, Multi modalities have been discussed, and we propose that image fusion techniques are all about the DWT and SWT. Apart from that, we used speckle noise filter method to reduce the unnecessary noise in medical images. Speckle noise filter was used to enhance the quality of the image for better visualization of features. Each method has its own advantages and disadvantages, and some of them are better than others in terms of the outcomes they produced.

In terms of success indicators, we assessed the outcomes in terms of standard deviation, entropy, and average gradient, etc. The proposed schemes are extremely helpful for further diagnosis of patients by revealing the exact and more profound information from a single blended image.

## REFERENCES

Achim, A., Tsakalides, P., & Bezerianos, A. (2003). SAR image denoising via Bayesian wavelet shrinkage based on heavy-tailed modeling. *IEEE Transactions on Geoscience and Remote Sensing*, *41*(8), 1773–1784.

Aguilar, M., & Garrett, A. L. (2001, March). Biologically based sensor fusion for medical imaging. In *Sensor Fusion: Architectures, Algorithms, and Applications V* (Vol. 4385, pp. 149–158). International Society for Optics and Photonics.

Anbarjafari, G., Izadpanahi, S., & Demirel, H. (2015). Video resolution enhancement by using discrete and stationary wavelet transforms with illumination compensation. *Signal, Image and Video Processing*, *9*(1), 87–92.

Balachander, B., & Dhanasekaran, D. (2016). Comparative study of image fusion techniques in spatial and transform domain. *Asian Research Publishing Network (ARPN)*, *11*(9), 5779–5783.

Bhatnagar, G., Wu, Q. J., & Liu, Z. (2013). Directive contrast based multimodal medical image fusion in NSCT domain. *IEEE Transactions on Multimedia*, *15*(5), 1014–1024.

Blum, A. L., & Langley, P. (1997). Selection of relevant features and examples in machine learning. *Artificial Intelligence*, *97*(1–2), 245–271.

Boyer, R. S., & Moore, J. S. (1977). A fast string searching algorithm. *Communications of the ACM*, *20*(10), 762–772.

Calhoun, V. D., & Adali, T. (2008). Feature-based fusion of medical imaging data. *IEEE Transactions on Information Technology in Biomedicine*, *13*(5), 711–720.

Charras, C., & Lecroq, T. (2004). *Handbook of exact string matching algorithms* (pp. 1–17). King's College.

Constantinos, S. P., Pattichis, M. S., & Micheli-Tzanakou, E. (2001, November). Medical imaging fusion applications: An overview. In *Conference Record of Thirty-Fifth Asilomar Conference on Signals, Systems and Computers (Cat. No. 01CH37256)* (vol. 2, pp. 1263–1267). IEEE.

Cover, T., & Hart, P. (1967). Nearest neighbor pattern classification. *IEEE Transactions on Information Theory*, *13*(1), 21–27.

Deshmukh, D. P., & Malviya, A. V. (2015). A review on: Image fusion using wavelet transform. *International Journal of Engineering Trends and Technology*, *21*(8), 376–379.

Gloor, P. A., Lee, I., & Velez-Sosa, A. (1992). Animated algorithms computer science education with algorithm animation. In *Computational Support for Discrete Mathematics* (pp. 41–55).

Horspool, R. N. (1980). Practical fast searching in strings. *Software: Practice and Experience*, *10*(6), 501–506.

Huang, B., Mu, Y., Pan, Z., Bai, L., Yang, H., & Duan, J. (2019). Speckle noise removal convex method using higher-order curvature variation. *IEEE Access*, *7*, 79825–79838.

Jaffery, Z. A., Zaheeruddin, & Singh, L. (2017). Computerised segmentation of suspicious lesions in the digital mammograms. *Computer Methods in Biomechanics and Biomedical Engineering: Imaging & Visualization*, *5*(2), 77–86.

James, A. P., & Dasarathy, B. V. (2014). Medical image fusion: A survey of the state of the art. *Information Fusion*, *19*, 4–19.

Katiyar, A., & Santhi, V. (2017, July). *Region based speckle noise reduction approach using fuzzy techniques* [Conference session]. Intelligent Computing, Instrumentation and Control Technologies (ICICICT), International Conference, IEEE, pp. 673–676.

Kirankumar, Y., & Shenbaga Devi, S. (2007). Transform-based medical image fusion. *International Journal of Biomedical Engineering and Technology*, *1*(1), 101–110.

Knuth, D. E., Morris, Jr., J. H., & Pratt, V. R. (1977). Fast pattern matching in strings. *SIAM Journal on Computing*, *6*(2), 323–350.

Liu, J., Tian, J., & Dai, Y. (2006, August). *Multi-modal medical image registration based on adaptive combination of intensity and gradient field mutual information* [Conference session]. IEEE Engineering in Medicine and Biology Society, International Conference, IEEE, pp. 1429–1432.

Loupas, T., McDicken, W. N., & Allan, P. L. (1989). An adaptive weighted median filter for speckle suppression in medical ultrasonic images. *IEEE Transactions on Circuits and Systems, 36*(1), 129–135.

Malviya, A., & Bhirud, S. G. (2009). Image fusion of digital images. *International Journal of Recent Trends in Engineering, 2*(3), 146.

Mandhare, R. A., Upadhyay, P., & Gupta, S. (2013). Pixel-level image fusion using brovey transforme and wavelet transform. *International Journal of Advanced Research in Electrical, Electronics and Instrumentation Engineering, 2*(6), 2690–2695.

Mishra, H. O. S., & Bhatnagar, S. (2014). MRI and CT image fusion based on wavelet transform. *International Journal of Information and Computation Technology, 4*(1), 47–52.

Qu, G., Zhang, D., & Yan, P. (2001). Medical image fusion by wavelet transform modulus maxima. *Optics Express, 9*(4), 184–190.

Raita, T. (1992). Tuning the boyer-moore-horspool string searching algorithm. *Software: Practice and Experience, 22*(10), 879–884.

Ramandeep, R. K. (2014). Review on different aspects of image fusion for medical imaging. *International Journal of Science and Research, 3*(5), 1887–1889.

Sahu, D. K., & Parsai, M. P. (2012). Different image fusion techniques–A critical review. *International Journal of Modern Engineering Research (IJMER), 2*(5), 4298–4301.

SaiKrishna, V., Rasool, A., & Khare, N. (2012). String matching and its applications in diversified fields. *International Journal of Computer Science Issues (IJCSI), 9*(1), 219.

Shalima, R. V., & Virk, R. (2015). Review of image fusion techniques. *International Research Journal of Engineering and Technology, 2*, 333–339.

Singh, L., & Jaffery, Z. A. (2017). Hybrid technique for the detection of suspicious lesions in digital mammograms. *International Journal of Biomedical Engineering and Technology, 24*(2), 184–195.

Singh, L., & Jaffery, Z. A. (2018). Computer-aided diagnosis of breast cancer in digital mammograms. *International Journal of Biomedical Engineering and Technology, 27*(3), 233–246.

Singh, L., & Jaffery, Z. A. (2018). Computerized detection of breast cancer in digital mammograms. *International Journal of Computers and Applications, 40*(2), 98–109.

Sunday, D. M. (1990). A very fast substring search algorithm. *Communications of the ACM, 33*(8), 132–142.

Wang, Z., & Ma, Y. (2008). Medical image fusion using m-PCNN. *Information Fusion, 9*(2), 176–185.

Yang, B., Luo, J., Guo, L., & Cheng, F. (2016). Simultaneous image fusion and demosaicing via compressive sensing. *Information Processing Letters, 116*(7), 447–454.

Yang, L., Guo, B. L., & Ni, W. (2008). Multimodality medical image fusion based on multiscale geometric analysis of contourlet transform. *Neurocomputing, 72*(1-3), 203–211.

Yu, Y., & Acton, S. T. (2002). Speckle reducing anisotropic diffusion. *IEEE Transactions on Image Processing, 11*(11), 1260–1270.

Zaheeruddin, Z., Jaffery, Z. A., & Singh, L. (2012). *Detection and shape feature extraction of breast tumor in mammograms* [Congress]. World congress on engineering, London, pp. 719–724.

Zhou, S. K., Greenspan, H., Davatzikos, C., Duncan, J. S., van Ginneken, B., Madabhushi, A., ... & Summers, R. M. (2020). A review of deep learning in medical imaging: Image traits, technology trends, case studies with progress highlights, and future promises. *arXiv preprint arXiv:2008.09104*.

# Index

2-D cross correlation, 165
51% Attack, 25

## A

access control, 146, 147, 150, 152, 154–157
accessibility threats, 24
actuator, 123, 125, 127–129, 131, 133
admission control, 184, 188, 196, 197
amyotrophic lateral sclerosis, 164
android, 123–128, 135, 137, 140–143
application programming interface, 32
ARDIC, 224, 226, 228, 233
Arduino, 124, 126–128, 142; IDE, 310
artificial intelligence, 8, 11, 235, 236, 302, 309, 312
artificial neural network, 90, 164
attribute based encryption, 151
AUC, 115, 118
authentication, 255
average methodology, 348

## B

behavioural intention, 320, 321
bilingual, 87, 89, 91, 92, 94, 95, 96, 97, 99, 100, 101, 102, 103, 105, 106
biometric security, 251, 255, 256, 261
blockchain, 4, 5, 6, 7, 8, 9, 10, 11, 12, 13, 14, 15, 16, 17, 18, 19, 20, 21, 22, 23, 24, 25, 26, 27
botherders, 265, 267, 269, 271, 273, 275
botnets, 266, 267, 269, 270, 275
broadcast group-key, 154–155
broker, 53, 54, 55
Brovey Transforms, 349
building blocks, 69, 70

## C

camera mouse, 168

cipher text, 151–152
Class Change Factor (CCF), 109, 115, 116, 118
classification, 90, 91, 95, 100, 106; Report, 334, 335
classifier, 329, 331, 332, 334
cloud, 31, 35, 40–42, 45; computing, 49, 51–53, 55, 57, 59, 61, 63–67, 147, 149, 160, 179–183, 185, 187; deployment models, 147–148; security, 145, 149
CloudSim, 50, 52, 57, 59, 63, 64, 66; simulation, 57, 66
clustering, 201, 202, 203, 207
computer tomography, 339–341, 344, 345; vision, 328–330
Confusion Matrix, 334, 335
control system, 123, 125, 126
controller, 31–33, 35, 37–45
convolutional neural networks, 328, 333
cryptographic technique, 5
cube, 278–281
Customer Insights, 70
cyber forensics, 265, 266, 269

## D

data analytics, 309; mining, 278, 279, 284; modelling, 279, 289; privacy, 243, 244, 245
datacentre, 39–41
decentralized autonomous corporations (DAC), 23
deep learning, 72, 73, 86, 328, 330
detect drowsiness, 328, 332
digital learning, 317–319, 321–324
digitalisation of healthcare, 237, 245
discrete wavelet transform, 341, 349, 350, 352, 353, 357
dynamically changing workload, 52, 53, 55, 57, 59

361

## E

e-commerce, 251, 252, 253, 254, 255, 257, 261, 263, 264, 265
e-health monitoring system, 153
E-LEACH, 202, 207, 208, 210
e-learning, 75, 76, 83, 86, 317–320, 322, 323
electronic health records, 235
embedded sensors, 299, 302
emotional engagement, 71
empowered customers, 240, 242, 242, 243
encryption, 149–158
energy management, 201, 202
e-transactions, 251, 255
evolutionary computational, 87, 92
extended priority based scheduling, 50, 55, 57, 60, 64
eye-tracking, 330

## F

face detection, 329
facial Landmarks Localization, 172
facial recognition, 256, 258, 259, 312
facilitating condition, 319, 321, 322
fingerprints, 256, 257, 260
Flip classroom, 323
forensics, 265, 266, 267, 268, 269
frauds, 251, 252, 253, 256, 261, 262

## G

GMDH, 109, 110, 111, 112, 113, 115, 118
government effort, 243
GPIO, 218
Gurmukhi, 90, 92, 94, 96, 97, 99–103, 105, 107

## H

Haar Cascade, 329, 331, 332
handwriting dataset, 88, 106
handwriting recognition, 87, 88, 89, 90, 91, 92, 96, 97, 99, 100, 101, 103, 105, 106, 107
healthcare, 4, 8, 11, 12, 22, 26
healthcare 4.0, 235, 236, 237
high pass filtering, 346, 350
higher education, 317, 319
hMouse 165
home automation, 299, 303, 309, 313
hybrid learning, 317, 318
hyper-personalization, 69, 70, 77, 78, 79, 80, 81, 82, 84, 85
hypervisors, 31, 35, 45

## I

ICT, 235, 240
image conversion to grayscale, 170
image fusion, 339, 340, 341, 347, 348, 349, 350, 357
Industry 4.0, 235, 237
Infrastructure-as-a- Service (IaaS), 149–150
intensity hue saturation, 348, 350
Internet of Things (IoT), 4, 5, 6, 7, 9, 10, 17, 22, 27, 302, 314
interpretive structural modelling, 235, 237, 241, 242
investigation, 265, 266, 267, 268, 270

## J

jDrumBox, 110
journals, 70, 78, 83
jXLS 110, 114, 117

## K

knowledge base approaches, 89

## L

LCD, 216, 222
LCOM, 114, 116, 117
LEACH, 202, 207–210
learning management system (LMS), 318
load balancing, 184, 186, 190
local binary pattern, 166
locomotor disability, 166
loss function, 213, 222

## M

machine learning, 77, 80, 83, 86, 266, 267, 269–271, 273, 275, 276, 302, 303, 312
magnetic resonance imaging, 340, 341, 344, 345, 354, 355
marketing, 70, 75, 85
MCU, 215, 216, 221, 222
mean memory cost, 61, 62
MICMAC analysis, 235, 243, 244
microcontroller, 127, 128, 130, 131, 134, 135, 299
mining, 6, 15, 17, 25
ML model, 110, 115
mouse movement using pupil detection, 169
multidimensional analysis, 278, 281, 292, 296
multidimensional data, 278, 280
multi-modality, 339
multiple datacentres scheduling, 54

# Index

## N

network simulation, 207
neural network, 114, 115, 118, 331, 333
non-repudiation, 255, 262
northbound interfaces, 31, 37, 45

## O

object detection, 332, 333
offline handwriting recognition, 88, 89, 105
OLAP, 278
online and offline handwriting recognition, 89
online education, 317, 318
online payment, 252, 253
OpenCV, 169, 330, 331, 332
OpenDaylight, 37, 42
OpenFlow, 32, 33, 35, 37–43
optimization, 302
OTP, 254, 255

## P

pattern matching, 339, 341, 342, 343, 345, 347, 349, 351
pattern recognition, 87, 107, 314
performance expectancy, 319, 321, 322, 323
persons with disability (PWD), 123–125, 140, 141
PIC16, 216–218, 220
PIC16F877A, 216–218, 221
Platform-as-a-Service (PaaS), 149–150
PLX-DAQ, 216, 221
PNN, 110, 111, 114, 115, 116, 118
positron emission tomography, 340, 345, 346
Principal Component Analysis, 348, 350, 352, 353
Printed Circuit Board (PCB), 137, 138, 139
privacy issues, 147, 149, 153
privacy preservation, 153
process migration, 53, 65
processing time, 57, 62, 64
Product Identification Device (PID), 215
projection approach, 88, 91

## Q

Quality of Service (QoS), 183, 184, 188

query generation, 278, 280, 281

## R

RBF neural network, 114, 115, 118
Receiver Operating Characteristic (ROC), 115
recommendation system, 69, 71–73, 75, 77, 79, 81, 83, 85, 86
region of interest, 87, 165, 329, 330, 331
Remote sensors, 300
rescheduling, 188, 194, 197
research framework, 265, 267, 268, 270, 271
resource allocation, 180–183, 185–189, 194; Strategies (RAS), 183, 186–189, 195, 196, 199; system, 180–183
resource management, 184, 186, 187, 199
resource provisioning, 181, 184
resource scheduling, 49–52, 55, 63, 65, 67
retina scan, 256, 259
RFID, 215, 216, 220, 224, 228, 233, 302, 306, 307, 309
Roman script, 90, 100, 101, 103, 105

## S

scheduling, 180, 183, 184, 187–189, 194
SDLC, 110, 117, 118
segmentation, 87–95, 99, 100, 102–107
Service Level Agreement (SLA), 52, 182–185, 188–191, 195, 199
single decision tree, 110, 114–116, 118
smart city, 11, 18, 22
smart home, 123–129, 131, 133–135, 137–142
softmax layer, 331, 333
Software as a Service (SaaS), 149–150, 184, 188, 189, 192, 197, 199
Software Defined Networking, 6, 31, 33
southbound interfaces, 31, 35, 45
space shared, 59
spatial domain, 346, 347, 349
speckle noise, 339, 345–348, 352, 357
stationary wavelet transform, 339, 341, 349, 351–353, 357
statistical algorithms, 284, 296
statistical analysis, 278, 284